IFCoLog Journal of Logics and their Applications

Volume 4, Number 3

April 2017

Disclaimer

Statements of fact and opinion in the articles in IfCoLog Journal of Logics and their Applications are those of the respective authors and contributors and not of the IfCoLog Journal of Logics and their Applications or of College Publications. Neither College Publications nor the IfCoLog Journal of Logics and their Applications make any representation, express or implied, in respect of the accuracy of the material in this journal and cannot accept any legal responsibility or liability for any errors or omissions that may be made. The reader should make his/her own evaluation as to the appropriateness or otherwise of any experimental technique described.

ISBN 978-1-84890-239-8
ISSN (E) 2055-3714
ISSN (P) 2055-3706

College Publications
Scientific Director: Dov Gabbay
Managing Director: Jane Spurr

http://www.collegepublications.co.uk

Printed by Lightning Source, Milton Keynes, UK

EDITORIAL BOARD

Scope and Submissions

This journal considers submission in all areas of pure and applied logic, including:

pure logical systems
proof theory
constructive logic
categorical logic
modal and temporal logic
model theory
recursion theory
type theory
nominal theory
nonclassical logics
nonmonotonic logic
numerical and uncertainty reasoning
logic and AI
foundations of logic programming
belief revision
systems of knowledge and belief
logics and semantics of programming
specification and verification
agent theory
databases

dynamic logic
quantum logic
algebraic logic
logic and cognition
probabilistic logic
logic and networks
neuro-logical systems
complexity
argumentation theory
logic and computation
logic and language
logic engineering
knowledge-based systems
automated reasoning
knowledge representation
logic in hardware and VLSI
natural language
concurrent computation
planning

This journal will also consider papers on the application of logic in other subject areas: philosophy, cognitive science, physics etc. provided they have some formal content.

Submissions should be sent to Jane Spurr (jane.spurr@kcl.ac.uk) as a pdf file, preferably compiled in LaTeX using the IFCoLog class file.

CONTENTS

ARTICLES

Proceedings of the Third Workshop: Introduction

Katalin Bimbó

Department of Philosophy, University of Alberta, Edmonton, AB, Canada
`<bimbo@ualberta.ca>`, `www.ualberta.ca/~bimbo`

J. Michael Dunn

School of Informatics and Computing and Department of Philosophy,
Indiana University, Bloomington, IN, U.S.A.
`<dunn@indiana.edu>`

The present collection of papers resulted from the *Third Workshop* that was held at the Department of Philosophy, University of Alberta, on May 16–17, 2016. Are you wondering about when and where the Second and the First workshops took place? Well, we should probably explain why the first workshop is the Third. Since 2014, we have been working on a research project (under the umbrella of an Insight Grant) entitled "The third place is the charm: The emergence, the development and the future of the ternary relational semantics for relevance and some other non-classical logics." To make a long story short, we took the salient word from the long title of the project to create a short catchy brand for the workshop.

The "third place" refers to the third argument place of a three-place relation, specifically, the third argument place of the ternary accessibility relation in the semantics of relevance logics. In the contemporary literature, this semantics is often referred to as "the Routley–Meyer semantics" of relevance logics. The idea of using a ternary relation, instead of the familiar binary relation from the so-called Kripke semantics for modal logic, is fascinating. Our own paper in this collection is an attempt to give an outline of the history while we also aim at describing the variations of the semantics that several logicians produced (quite independently from each other).[1] We give the latter descriptions by "translating" all the semantics into a

---

The support of the *Insight Grant #435–2014–0127*, awarded by the *Social Sciences and Humanities Research Council of Canada*, is gratefully acknowledged.

[1]In addition to what we say in our paper in this proceedings about L. L. Maksimova's work on relational semantics, we wrote [11] concerning her contributions to relevance logics more generally.

uniform language. Our hope is that this will facilitate a greater appreciation toward the potential variations of the "standard" relational semantics.

The development of relational semantics did not stop in the mid 1970s. We cannot give an exhaustive account of what happened later on in a couple of pages, but we can list some pointers. On one hand, the idea of representing a binary operation by a ternary relation can be separated from the aim of giving a semantics for relevance logics. On the other hand, the relational semantics can be informally explained.[2] Some binary operations that have been included into logics are similar to implications. Barwise considered *channels* between situations or information states, sometimes identifying channels with situations that determine them (hence, the ternary relation).[3] Another use of →'s is the arrows (i.e., the maps) of a category. Van Benthem and Venema abstracted the idea of composable arrows into what they labeled "arrow logic," which can be given a ternary relational semantics.[4] Girard introduced a "phase semantics" for his linear logic that used a binary operation, which is of course a special case of a ternary relation. He gave a definition of implication very similar to that of Urquhart, but Girard's operation is not idempotent.[5]

An old idea—compared to situation theory, arrow logic or even the ternary relational semantics—is that grammaticality can be described by exploiting merely one operation, namely, function application. Atomic components of a language (perhaps, words) come with syntactic categories attached to them, and then several words form grammatically a correct expression when their categories can be combined together using function application. Lambek introduced his associative and non-associative calculi in the late 1950s–early 1960s. The two inverses of the composition operation—denoted by \ and /—do share certain properties with →. Namely, \ is monotone decreasing in its first, and monotone increasing in its second argument place. Similarly for /, but with the order of the arguments reversed. Ternary relational semantics for Lambek calculi were given decades after Lambek's work.[6]

At the beginning of the 1990s, Dunn created what he called "gaggle theory" (spelling out "ggl" and abbreviating "generalized Galois logics"). The scope of the theory extends the ternary relational semantics along two dimensions. First of all, a connective in a logic may be *n*-ary (e.g., unary or quaternary) rather than binary.

---

[2]For an article outlining various interpretations of the ternary accessibility relation within the context of relevance logic, see Beall et al. [4]. For an interpretation of the ternary relation in terms of "relevance," see Dunn [20]. Lastly, for a "catalog" of interpretations both inside and outside of relevance logic, see [21].

[3]See [2], [3] and [18]. Tedder's paper in this volume connects channel theory and the ternary relational semantics, building on work by Restall.

[4][28] gives an introduction into arrow logic. For further investigations, see [27] and [19].

[5]See Girard's first paper [24] on linear logic.

[6]Lambek's original papers are [25] and [26]. For semantics of Lambek calculi, see [12] and [15].

Then, the modeling relation will have to have $n + 1$ argument places. Secondly, the algebra of the logic does not need to be (or to have as a reduct) a De Morgan lattice (or even a distributive lattice), but it can be a stronger or weaker structure (e.g., a Boolean algebra or a partial order).[7] An advantage of gaggle theory is that relational semantics no longer appear to be ad hoc contraptions; they have certain components for a reason. Gaggle theory opened the door to approaching the interpretation of a wide range of intensional logics—from modal logics to combinatory logic—in a systematic and uniform way.[8]

We cannot point to all the publications that belong to the *gaggle paradigm*; we limit ourselves to briefly pointing out some papers—and be warned, our list is biased. Allwein and Dunn in [1] gave a relational semantics for a group of logics that do not prove the distributivity of disjunction and conjunction. Dunn and Meyer [23] introduced structurally free logics (in which structural rules are replaced by rules introducing combinators that trace structural changes), and they provided a ternary relational semantics for them. Dunn in [18] gave a ternary relational semantics for relation algebras (which works out smoothly, unlike the representation by binary relations). Bimbó is interested in tweaking and dualizing the ternary relational semantics; some fruits of her efforts are reported in [5, 6, 7, 10, 8] and [9].

This collection of papers is directly related to the *Third Workshop*, as we mentioned at the beginning. However, we should caution against a possible confusion. Although all the authors were speakers at the workshop, not all speakers chose to contribute a paper. Moreover, the papers in this volume are not exactly the papers that their authors presented. On one hand, the papers were finalized months after the workshop, which allowed the authors to absorb the reactions to their papers — indeed, the workshop included lively interactions and plenty of discussion. On the other hand, some authors decided to substitute another paper (with related content).

We feel that we should explain the order of the papers. First of all, we would like to provide an excuse for placing our own paper at the beginning by pointing out that our paper provides the historical setting for the contemporary developments that figure into the other papers.

Alasdair Urquhart utilizes the Routley–Meyer semantics for the logic **KR** directly. Arnon Avron uses the semantics to prove the cut theorem for his hypersequent calculus for **RM**. Guillermo Badia looks at the language of the relational

---

[7]Dunn actually required an underlying distributive lattice as part of a gaggle in [13], his first paper on gaggles. But he quickly came to work with "partial gaggles," "Boolean gaggles," "lattice ordered gaggles," etc.

[8]Dunn's other papers on gaggle theory include [14, 15, 16] and [17]. Gaggles also figure into [22], and they are the sole topic of [10].

semantics from a model-theoretic point of view. Chrysafis Hartonas defines semantics for logics that are based on a lattice. Edwin Mares and Shawn Standefer work out an informal interpretation of the ternary relational semantics and motivate and interpret logics in the neighborhood of **E**. Andrew Tedder considers the channel interpretation of the ternary relation and its logics. Ross Brady summarizes his worries about semantics in general, and the Routley–Meyer semantics in particular. Bryson Brown looks at the ternary relational semantics from the point of view of the preservationist approach. Nicholas Ferenz gives an informal interpretation of the ternary relational semantics using a notion of ambiguity.

We wish to thank these authors for their contributions to this volume, and we also thank the referees.

References

[1] Gerard Allwein and J. Michael Dunn. Kripke models for linear logic. *Journal of Symbolic Logic*, 58:514–545, 1993.

[2] Jon Barwise, editor. *The Situation in Logic*, volume 17 of *CSLI Lecture Notes*. CSLI Publications, Stanford, CA, 1989.

[3] K. Jon Barwise. Constraints, channels and the flow of information. In Peter Aczel, David Israel, Yasuhiro Katagiri, and Stanley Peters, editors, *Situation Theory and its Applications (v. 3)*, volume 37 of *CSLI Lecture Notes*, pages 3–27. CSLI, Stanford, CA, 1993.

[4] Jc Beall, Ross Brady, J. Michael Dunn, A. P. Hazen, Edwin Mares, Robert K. Meyer, Graham Priest, Greg Restall, David Ripley, John Slaney, and Richard Sylvan. On the ternary relation and conditionality. *Journal of Philosophical Logic*, 41:595–612, 2012.

[5] Katalin Bimbó. Semantics for dual and symmetric combinatory calculi. *Journal of Philosophical Logic*, 33:125–153, 2004.

[6] Katalin Bimbó. Relevance logics. In D. Jacquette, editor, *Philosophy of Logic*, volume 5 of *Handbook of the Philosophy of Science* (D. Gabbay, P. Thagard and J. Woods, eds.), pages 723–789. Elsevier (North-Holland), Amsterdam, 2007.

[7] Katalin Bimbó. Functorial duality for ortholattices and De Morgan lattices. *Logica Universalis*, 1:311–333, 2007.

[8] Katalin Bimbó. Dual gaggle semantics for entailment. *Notre Dame Journal of Formal Logic*, 50:23–41, 2009.

[9] Katalin Bimbó. Some relevance logics from the point of view of relational semantics. *Logic Journal of the IGPL*, 24:268–287, 2016. O. Arieli and A. Zamansky (eds.), *Israeli Workshop on Non-classical Logics and their Applications* (IsraLog 2014).

[10] Katalin Bimbó and J. Michael Dunn. *Generalized Galois Logics. Relational Semantics of Nonclassical Logical Calculi*, volume 188 of *CSLI Lecture Notes*. CSLI Publications, Stanford, CA, 2008.

[11] Katalin Bimbó and J. Michael Dunn. Larisa Maksimova's early contributions to relevance logic. In S. Odintsov, editor, *L. Maksimova on Implication, Interpolation and Definability*, Outstanding Contributions to Logic. Springer Nature, Switzerland, 2017. (forthcoming, 28 pages).

[12] Kosta Došen. Sequent-systems and groupoid models. II. *Studia Logica*, 48: 41–65, 1989.

[13] J. Michael Dunn. Gaggle theory: An abstraction of Galois connections and residuation, with applications to negation, implication, and various logical operators. In J. van Eijck, editor, *Logics in AI: European Workshop JELIA '90*, number 478 in Lecture Notes in Computer Science, pages 31–51. Springer, Berlin, 1991.

[14] J. Michael Dunn. Star and perp: Two treatments of negation. *Philosophical Perspectives*, 7:331–357, 1993. (Language and Logic, 1993, J. E. Tomberlin (ed.)).

[15] J. Michael Dunn. Partial gaggles applied to logics with restricted structural rules. In K. Došen and P. Schroeder-Heister, editors, *Substructural Logics*, pages 63–108. Clarendon, Oxford, UK, 1993.

[16] J. Michael Dunn. Gaggle theory applied to intuitionistic, modal and relevance logics. In I. Max and W. Stelzner, editors, *Logik und Mathematik. Frege-Kolloquium Jena 1993*, pages 335–368. W. de Gruyter, Berlin, 1995.

[17] J. Michael Dunn. Generalized ortho negation. In H. Wansing, editor, *Negation: A Notion in Focus*, pages 3–26. W. de Gruyter, New York, NY, 1996.

[18] J. Michael Dunn. A representation of relation algebras using Routley–Meyer frames. In C. A. Anderson and M. Zelëny, editors, *Logic, Meaning and Computation. Essays in Memory of Alonzo Church*, pages 77–108. Kluwer, Dordrecht, 2001.

[19] J. Michael Dunn. Arrows pointing at arrows: Arrow logic, relevance logic and relation algebras. In A. Baltag and S. Smets, editors, *Johan van Benthem on Logic and Information Dynamics*, Outstanding Contributions to Logic, pages 881–894. Springer, New York, NY, 2014.

[20] J. Michael Dunn. The relevance of relevance to relevance logic. In M. Banerjee and S. N. Krishna, editors, *Logic and its Applications: 6th Indian Conference, ICLA 2015, Mumbai, India, January 8–10, 2015*, number 8923 in Lecture Notes in Computer Science, pages 11–29. Springer, Heidelberg, 2015.

[21] J. Michael Dunn. Various interpretations of the ternary accessibility relation, (abstract). In *Devyatye Smirnovskie chteniya po logike: materialy mezhdunarodnoĭ konferencii, 17–19 iyunya 2015g., Moskva*, pages 10–13. Izdatel'stvo Sovremennye Tetradi, Moskva, 2015.

[22] J. Michael Dunn and Gary M. Hardegree. *Algebraic Methods in Philosophical Logic*, volume 41 of *Oxford Logic Guides*. Oxford University Press, Oxford, UK, 2001.

[23] J. Michael Dunn and Robert K. Meyer. Combinators and structurally free logic. *Logic Journal of the IGPL*, 5:505–537, 1997.

[24] Jean-Yves Girard. Linear logic. *Theoretical Computer Science*, 50:1–102, 1987.

[25] Joachim Lambek. The mathematics of sentence structure. *American Mathematical Monthly*, 65:154–169, 1958.

[26] Joachim Lambek. On the calculus of syntactic types. In R. Jacobson, editor, *Structure of Language and its Mathematical Aspects*, pages 166–178. American Mathematical Society, Providence, RI, 1961.

[27] Johan van Benthem. A note on dynamic arrow logic. In J. van Eijck and A. Visser, editors, *Logic and Information Flow*, Foundations of Computing, pages 15–29. MIT Press, Cambridge, MA, 1994.

[28] Yde Venema. *A Crash Course in Arrow Logic*. Studies in Logic, Language and Information. CSLI Publications, Stanford, CA, 1996.

Received 7 April 2017

The Emergence of Set-theoretical Semantics for Relevance Logics around 1970

Katalin Bimbó

Department of Philosophy, University of Alberta, Edmonton, AB, Canada
<bimbo@ualberta.ca>, www.ualberta.ca/~bimbo

J. Michael Dunn

*School of Informatics and Computing, and Department of Philosophy,
Indiana University, Bloomington, IN, U.S.A.*
<dunn@indiana.edu>

Abstract

The early history of the relational ("possible world") semantics for modal logics is well investigated. Successful applications of a relational semantics for relevance logics started to appear about a decade after the first set-theoretical semantics for normal modal logics were designed. This paper gives a brief outline of the results from the late 1960s and the first years of the 1970s. We provide an exposition of three types of attempts (by five people) to provide set-theoretical semantics for relevance logics or some related logics—together with some historical details. The crucial technical features of the semantics can be characterized along the lines of the modeling of the implication connective, in particular, whether the modeling derives from a binary operation or from a ternary relation (which is not assumed to be an operation).

Keywords: entailment, relational semantics, relevant implication, relevance logic, rigorous implication

1 Introduction

The use of a binary "accessibility" relation for what is commonly known as the "Kripke semantics" or "possible world semantics" for modal logic is well-known. There was once a lot of confusion about its origins, but now its history has been written, or to be more precise, there are at least two histories. Copeland and Gold-blatt have each written excellent accounts ([13] and [27], respectively) of the beginnings of the semantics for modal logic, and they agree on the basic history. What is less well-known is the use of a ternary accessibility relation for the semantics of relevance logic and various other non-classical logics. The aim of this paper is to give a brief but relatively fulsome account of the history of this development, with a focus on relevance logic.[1]

Relevance logics emerged in the late 1950s as a new attempt to restrict traditional logic in order to exclude certain undesirable theorems, in particular, implications where there was no "variable sharing" between antecedent and consequent, e.g., $(p \wedge \sim p) \to q$. The logic of *entailment* (**E**) proved to be a coherent logic; so did the logic of *relevant implication* (**R**). Many others followed. Relevance logics originated—exactly like traditional logic, modal logics, etc.—as axiom systems, but natural deduction and consecution calculi for various fragments were soon found too; then, they were algebraized and their metatheory started to be developed.

The 1950s and 1960s had brought about a breakthrough and vast expansion in terms of relational semantics for modal logics. The key idea was to define the modal operators of necessity and possibility using the abstractions of "possible worlds" and a binary relation R on them, which is called "accessibility" or "relative possibility." Valuations are then not simply assignments of truth values to sentences, but rather assignments relativized to possible worlds. The key semantic clause for the necessity operator is that $\Box A$ is true at a world w iff A is true at every world w' such that wRw', and of course, the clause for the possibility operator is that $\Diamond A$ is true at a world w iff A is true at some world w' such that wRw'. Propositions can be viewed as sets of possible worlds, and the proposition assigned to a sentence is the set of possible worlds in which it is true.

The "Kripke semantics" did not stop with modal logic. Kripke in [34], and independently Grzegorczyk in [30], used a binary accessibility relation to define intuitionistic implication and negation. Kripke in his paper did not speak of "possible worlds," but rather of "evidential situations," and Grzegorczyk spoke of "information states." Their core idea was the same. Following Kripke, we define an intuitionistic model structure to be an ordered triple $\langle G, K, R \rangle$, where K is a non-empty set, $G \in K$, and R is a reflexive, transitive relation on K. An intuitionistic model on a model structure $\langle G, K, R \rangle$ is a function φ assigning to each pair of an atomic sentence p and an $H \in K$ a truth value $\mathbf{T}$ or $\mathbf{F}$ ($\varphi(p, H) = \mathbf{T}$ or $\mathbf{F}$). Oops, that would be

---

[1] Dunn has contributed his views on this early history before (see Anderson et al. [3, sec. 48.3] and Dunn [17]), and the scholarly reader might want to go back and look at those. He regrets that he did not say more about Larisa Maksimova's role, but is glad he had the opportunity to correct this here and in another forthcoming publication, Bimbó and Dunn [10].

a modal model for the modal logic **S4**. We must also add that if $\varphi(p, H) = \mathbf{T}$ and HRH', then $\varphi(p, H') = \mathbf{T}$. This "Hereditary Condition" (as it would ultimately be called by Routley and Meyer) is of extreme importance. It captures the idea that once we have enough information to establish ("to prove") that something is true, it continues to hold as we increase our information; the Hereditary Condition was shown using mathematical induction to extend to compound sentences.

Dunn in [16] showed how the Grzegorczyk–Kripke semantics for the intuitionistic logic **J** could be extended to produce a semantics for the semi-relevant logic R-Mingle (**RM**). **RM** is called "semi-relevant" because, as was shown by Meyer, if $\sim\mathcal{A}$ is a theorem and $\mathcal{B}$ is a theorem, then so is $\mathcal{A} \to \mathcal{B}$, even if $\mathcal{A}$ and $\mathcal{B}$ do not share a propositional variable (otherwise, they must). This means, for example, that $\sim(p \to p) \to (q \to q)$ is a theorem of **RM**. This is surprising since **RM** is obtained from **R** by simply adding the seemingly relevant $\mathcal{A} \to (\mathcal{A} \to \mathcal{A})$ as an axiom schema.

Dunn modified the Grzegorczyk–Kripke semantics in five ways. First, the notion of a model was extended to allow for a sentence to be assigned both of the values **T** and **F**. Second, the Hereditary Condition was strengthened so as to require that the relation R not only preserve the assignment of **T** to an atomic sentence but also the assignment of **F**. Third, the satisfaction clause for implication ($\mathcal{A} \to \mathcal{B}$) was restated so as to not just requiring that truth be preserved from $\mathcal{A}$ to $\mathcal{B}$, but also that falsity be preserved from $\mathcal{B}$ to $\mathcal{A}$. Fourth, negation is evaluated so that $\sim\mathcal{A}$ is assigned the value **T** if $\mathcal{A}$ is assigned the value **F**, and $\sim\mathcal{A}$ is assigned the value **F** if $\mathcal{A}$ is assigned the value **T**. So far the changes might be summarized as requiring that evidential situations treat **F** even-handedly with **T**. But the fifth change is different, and made for technical reasons. It requires that R be connected, i.e., that for every H and H', either HRH' or $H'RH$.

Despite [16]'s publication date of 1976, it explains that this "Kripke style" semantics for **RM** was first announced in a talk in the symposium "Natural Language versus Formal Language" at a joint symposium of the Association for Symbolic Logic and the American Philosophical Association in December 1969, and in June 1971, it was presented at the Tarski Symposium at UC–Berkeley. As he confessed in the published paper, Dunn was so "overwhelmed by the ingenuity and the power of ternary semantics" that he considered for a while not publishing his binary semantics for **RM**. But the result and the fact that they could not be extended to obtain semantics for better known relevance logics such as **R** and **E**, implicitly showed the limitations of a binary accessibility relation.

Because of their "Kripke semantics," modal logic (and a bit later intuitionistic logic) enjoyed a reinvigorated interest, but relevance logics were often overlooked. Moreover, some quibbled about their lack of a concrete semantics. The last problem in Alan Ross Anderson's famous list of open problems [1] was the problem of finding a semantics for the Anderson–Belnap system **E** of entailment (p. 16), but despite its being listed at the end, Anderson said that "the writer does *not* regard this question as 'minor'; it is rather the principal 'large' question remaining open." That open problem was solved in the late 1960s and early 1970s when several logicians quite independently invented a "possible world" semantics for relevance logic. The scare quotes are there because these "possible worlds" might be incomplete, inconsistent, or both.[2]

The problem was not always solved for **E** (seemingly, Anderson's favorite logic), but rather the logicians who solved the problem of finding a concrete semantics for relevance logics often put the system **R** of relevant implication on the operating table as they explored solutions. But at least two of these logicians (Fine and Maksimova) did solve the problem for **E** at the roughly the same time. Maybe, Routley and Meyer had the ideas for a solution earlier, but they did not publish it for another decade.

It ironic that despite the series of paper "Semantics of entailment I, II, III, IV," and despite solving the problem for **E+**, the positive fragment of **E** in "I," they did not published a semantics for the whole system **E** until "IV," which had a very delayed publication and appeared as Appendix 1 of [52]. Routley and Meyer say there (p. 430):

> World semantics for system E were found not only by Routley and Meyer, but also, though not independently of the similar earlier semantic for R (of SE I), by Maksimova 73, and subsequently by Fine (see 74) who knew of the first degree theory (of FD) which already included a key idea of the operational semantics, the world-shift in making evaluations.[3]

---

[2]These last are certainly not the possible worlds envisaged in the semantics of modal logic, and hence, researchers have variously called them "situations," "set-ups," "theories," or "information states." And sometimes they have been called just "worlds," or "incomplete or inconsistent worlds." We shall use the term "situations" when we want to be neutral among these.

[3]"Maksimova 73" is [37], and "Fine (see 74)" refers to [19]. "SE I" is of course [50]. Routley is probably thinking of the fact that in the cited paper Maksimova does refer to Routley and Meyer's "II" and "III." (See footnote 11 about the attribution to Routley here.) But he does not mention, and probably did not know about her earlier paper [36], which antedates all of the Routley–Meyer papers. See the section on Maksimova below where we explain why we think this is the first discovery of a ternary semantics for the systems **E** and **R**. Regarding Fine, when Routley says Fine "knew of the first degree theory (of FD) which already included a key idea of the operational semantics, the world-shift in making evaluations," Routley is presumably claiming that Fine knew about the use of the Routleys' * to interpret negation in first-degree entailments. No evidence is given for this, but it is not clear what the point would be anyway. At its heart, * is a binary relation, and Fine must have known about the binary accessibility relation used by Kripke in his—by then very well-known—semantics for modal logic, and where the evaluation of necessity and possibility at

560

There is then an amazing footnote [52, fn. 4 on p. 430], which says:

> As with other intellectual break-throughs, grand or small (e.g. modal
> logic semantics, the infinitesimal calculus, relativity theory, etc.), so with
> semantics for relevant logics, the time was ripe, e.g. requisite techniques
> were sufficiently available, and several people somewhat independently
> reached similar results at around the same time.

The sociologist of science Robert K. Merton in his famous 1961 article [41] wrote
about this general phenomenon of simultaneous discovery, labeling it as "multiples."
He says (p. 473) that "[t]o say that discoveries occur when their time has come is
to say that they occur only under identifiable requisite conditions." He goes on to
develop an explanation for independent simultaneous discoveries of essentially the
same scientific theories by appeal to the accumulation of the requisite conditions in
various places. In a similar spirit, we give a reconstruction of how five different (but
conceptually related) semantics for relevance logics were proposed independently by
a handful of logicians around 1970.

There are other researchers whom we could mention in connection to the inven-
tion of a (ternary) relational semantics for (relevance) logics, first of all, Dana Scott,
but more broadly Bjarni Jónsson and Alfred Tarski ([31] and [32]), and even Charles
S. Peirce (see e.g., [44, p. 148]).[4] However, because of limitations of space, we focus on the work of Robert K. Meyer, Richard Routley, Alasdair Urquhart, Larisa
L. Maksimova, and Kit Fine, as the most "relevant."[5] We shall also discuss some
related and independent work of Dov Gabbay from about the same time, which
though was not focused on relevance logic, still is "almost relevant."

The authors we consider often used different and distinctive terminology and
symbols. This makes it difficult to pinpoint real similarities and real differences. We
will use a kind of "pidgin" of their languages and ours, hoping to convey something
of the essence of their original work with our interpretation of it. This paper contains
no proofs—difficult or easy. The reader who wants to consider proofs in more depth

---

a world involves a similar "world shift" in considering other worlds which are accessible from the
given world. The real issue is the ternary relation, and it seems that Fine came up with his version
of it independently. (See the section on Fine below.)

[4]We found out via communication with both Dov Gabbay and Dana Scott that Scott presented
his idea in a seminar at Stanford University when he was a postdoc there. (Gabbay was in the
audience.) Scott was thinking specifically about Łukasiewicz's many-valued logic, thus the ternary
relation being $x+y=z$. What results is in effect an operational semantics, and it was not published
until 1974 as [53].

[5]Routley changed his name to Sylvan when he remarried in 1983, but we shall refer to him
as Routley given that this is the name he used when he did pioneering work on the semantics of
relevance logic, or "relevant logic," as he and Meyer would call it.

is referred to the papers by the original authors, or to [3], [17] and [9], as well as to some further publications among the references.

Incidentally, while the problem of a semantics for the propositional relevance logics was solved by several researchers about 1970, the problem for quantification relevance logics was left open until some years later. Routley and Meyer had raised the problem of proving completeness with respect to their ternary frames to which a domain of individuals had been added that would be the same for each set-up (so-called constant domain semantics). Fine in [22] showed that this does not work, at least for **EQ** and **RQ** (he does mention that Routley had shown completeness for the system **BQ**). We shall not discuss the details here but do cite a couple of different ways of providing models that do work: first, Fine's in [21], and later, Goldblatt and Mares's in [29] and Goldblatt's in [28].

2   Logic and Algebras

Algebraization is a widely used and highly fruitful approach to increase the efficiency of dealing with knowledge. The Lindenbaum algebra of a logic is the result of the abstraction process that groups together formulas which are provably equivalent into a single element in the algebra. In the case of a propositional logic (meaning a logic with no quantifiers whatsoever), the algebraization typically leads to a structure that is an algebra in a plain sense, that is, a set with finitely many finitary operations (each of which is a total operation on the carrier set). A logic does not reduce to its Lindenbaum algebra, for instance, because there can be different proof systems for a logic with compelling properties. However, the Lindenbaum algebra can be a stepping stone toward a semantics for the logic, exactly, because it focuses on provably equivalent formulas.

The *Lindenbaum algebra* can serve as the blueprint for an algebraic semantics for a logic. An advantage of an algebraic semantics is that it brings with itself results and techniques from (universal) algebra, but it is sometimes perceived as too abstract to constitute the meaning of the logical particles of a logic. Set theory comes to rescue through concrete representations of abstract algebras.

The *concrete set-based semantics* are looked upon favorably, especially, when a more or less informal interpretation is sought. In the case of traditional propositional logic, a proposition A is the set of all those maximally consistent theories that contain A. Equivalently, a proposition A is the set of all those (proper) prime filters that contain $[A]$ (the equivalence class of A).[6]

---

[6]This is the real backbone of Stone's representation of Boolean algebras in his [54].

Traditional logic has a certain simplicity, and this is very clear from facts such as one (well-chosen) connective suffices for this whole logic and the concrete semantics of sets does not need to be enriched with any new operations. We want to emphasize that in the Stone representation of Boolean algebras the algebraic operations turn into *operations* on sets.

It would be preferable to have (if possible) a set-theoretical semantics for relevance logics, in which the connectives, especially, $\rightarrow$ (entailment or relevant implication), would be interpreted by an operation. Philosophically, it would allow us "to read" the operation in the semantics as the interpretation of $\rightarrow$, like $\cup$ interprets $\vee$. From a mathematical point of view, we would be dealing with an algebra of sets over some underlying set U, which would be more simple than having to deal with a relational structure. This can be done for classical logic with the so-called material implication ($A \supset B =_{\mathrm{df}} \neg A \vee B$), where $\vee$ is interpreted as $\cup$ and $\neg$ is interpreted as complement relative to U. But we need some additional structure on the underlying set U to help us define an operation corresponding to the relevant implication $\rightarrow$.

Our running example will be propositional **R**. Although **R** was not necessarily the logic each approach focused on, fixing a logic allows us to provide better comparisons.

Definition 2.1. The *language of* **R** comprises a denumerable set of propositional variables ($\mathbb{P} = \{ p_0, p_1, \dots \}$), a unary connective $\sim$ (negation), binary connectives $\rightarrow$ (implication), $\circ$ (fusion), $\wedge$ (conjunction), $\vee$ (disjunction) and a zero-ary connective **t** (truth).

The set of *well-formed formulas* is generated by the following grammar, with the proviso that $\mathbb{P}$ rewrites to any of its elements.

$$\mathcal{A} := \mathbb{P} \mid t \mid \sim\!\mathcal{A} \mid (\mathcal{A} \rightarrow \mathcal{A}) \mid (\mathcal{A} \circ \mathcal{A}) \mid (\mathcal{A} \wedge \mathcal{A}) \mid (\mathcal{A} \vee \mathcal{A}).$$

We define **R** as an axiom system; its schemas and rules are (1)–(20).

(1) $\mathcal{A} \rightarrow \mathcal{A}$

(2) $(\mathcal{A} \rightarrow \mathcal{B}) \rightarrow ((\mathcal{B} \rightarrow \mathcal{C}) \rightarrow (\mathcal{A} \rightarrow \mathcal{C}))$

(3) $(\mathcal{A} \rightarrow (\mathcal{A} \rightarrow \mathcal{B})) \rightarrow (\mathcal{A} \rightarrow \mathcal{B})$

(4) $(\mathcal{A} \rightarrow (\mathcal{B} \rightarrow \mathcal{C})) \rightarrow (\mathcal{B} \rightarrow (\mathcal{A} \rightarrow \mathcal{C}))$

(5) $(\mathcal{A} \wedge \mathcal{B}) \rightarrow \mathcal{A}$

(6) $(\mathcal{A} \wedge \mathcal{B}) \rightarrow \mathcal{B}$

(7) $((\mathcal{A} \to \mathcal{B}) \wedge (\mathcal{A} \to \mathcal{C})) \to (\mathcal{A} \to (\mathcal{B} \wedge \mathcal{C}))$

(8) $\mathcal{A} \to (\mathcal{A} \vee \mathcal{B})$

(9) $\mathcal{B} \to (\mathcal{A} \vee \mathcal{B})$

(10) $((\mathcal{A} \to \mathcal{B}) \wedge (\mathcal{C} \to \mathcal{B})) \to ((\mathcal{A} \vee \mathcal{C}) \to \mathcal{B})$

(11) $(\mathcal{A} \wedge (\mathcal{B} \vee \mathcal{C})) \to ((\mathcal{A} \wedge \mathcal{B}) \vee (\mathcal{A} \wedge \mathcal{C}))$

(12) $(\mathcal{A} \to (\mathcal{B} \to \mathcal{C})) \to ((\mathcal{A} \circ \mathcal{B}) \to \mathcal{C})$

(13) $((\mathcal{A} \circ \mathcal{B}) \to \mathcal{C}) \to (\mathcal{A} \to (\mathcal{B} \to \mathcal{C}))$

(14) $(\mathcal{A} \to {\sim}\mathcal{B}) \to (\mathcal{B} \to {\sim}\mathcal{A})$

(15) ${\sim}{\sim}\mathcal{A} \to \mathcal{A}$

(16) $(\mathcal{A} \to {\sim}\mathcal{A}) \to {\sim}\mathcal{A}$

(17) $\boldsymbol{t}$

(18) $\mathcal{A} \to (\boldsymbol{t} \to \mathcal{A})$

(19) $\mathcal{A} \to \mathcal{B}, \ \mathcal{A}$ imply $\mathcal{B}$

(20) $\mathcal{A}, \ \mathcal{B}$ imply $\mathcal{A} \wedge \mathcal{B}$

A formula $\mathcal{A}$ is a *theorem* iff it has a *proof*, that is, if there is a finite sequence of formulas ending with $\mathcal{A}$, each element of which is either an instance of an axiom or obtained from earlier elements of the sequence by an application of a rule.

Actually, the original system **R** did not have the propositional constant $\boldsymbol{t}$, but it plays an important role in Dunn's algebraization of **R** in his Ph.D. Thesis [14] (see also §28.2 by him in [2]). $\boldsymbol{t}$ can be added conservatively, that is, in such a way that no new theorems result that do not contain $\boldsymbol{t}$. This constant in **R** corresponds to the identity element of a monoid, and Dunn called the **R** algebras *De Morgan monoids*. (The definition we give below is not the original one, but it is equivalent to that. We use the same symbols for the operations in an algebra as for the connectives; context always determines whether a symbol stands for a connective or an algebraic operation.)

Definition 2.2. A *De Morgan monoid* is an algebra $\mathfrak{D} = \langle A; \boldsymbol{t}, {\sim}, \to, \circ, \wedge, \vee \rangle$ of similarity type $\langle 0, 1, 2, 2, 2, 2 \rangle$ that satisfies (1)–(5). (a, b and c range over A.)

(1) $\langle A; \wedge, \vee \rangle$ is a distributive lattice;

(2) $\sim(a \wedge b) = \sim a \vee \sim b$ and $\sim(a \vee b) = \sim a \wedge \sim b$;

(3) $\sim\sim a = a$;

(4) $a \circ b \leq c$ iff $a \leq b \to c$ (where $a \leq b$ iff $a \wedge b = a$);

(5) $\langle A; t, \circ \rangle$ is an Abelian monoid in which $a \leq a \circ a$.

We are now ready to start discussing the relational semantics of relevance logic. We start with the so-called Routley–Meyer semantics, because it seems to be the most widely known set-theoretical semantics for relevance logics.

3   Routley–Meyer Semantics: Ternary Relation

We have to "preface" our findings, especially in the case of the Routley–Meyer semantics, with some explanations about what we could and what we could not accomplish. The reconstruction of what Routley and Meyer did, respectively, runs into some obstacles. Regrettably, both of them passed away some years ago. Several other people who were privy to the relevant developments in the early 1970s moved to other lines of work. However, we thank Nuel Belnap, Kit Fine, Dov Gabbay, Larisa Maksimova, Sergeĭ Odintsov and Alasdair Urquhart for providing the information (sometimes even documentation) that they did, and we thank Michael McRobbie for providing us with access to his archive that contains many manuscripts and preprints from the time when he was a research assistant to Routley in the mid-1970s.

Historical sketch.   The Routley–Meyer semantics was initiated by Routley, probably in 1970. He penned a 97-page manuscript in longhand, which he entitled as "Semantical analysis of entailment and relevant implication, I." Routley sent a copy of this manuscript to Meyer, likely in January 1971, who was then on the faculty of Indiana University in Bloomington. Both Meyer and Dunn were faculty members in the Philosophy Department, and they discussed Routley's manuscript. Routley's idea was to use an operation in the semantics. Meyer and Dunn discovered that the operational semantics is not complete; then, Meyer set out to develop an idea that was mentioned (but not worked out) by Routley: the use of a three-place relation rather than of a binary operation. By mid February, Meyer has obtained sufficient results to claim completeness for the semantics with a ternary relation. He completed an 11-page typescript on the completeness in the spring of 1971 (see [42]).[7] On page 1 Meyer says:

---

[7]We have a copy of this typescript, thanks to Michael McRobbie.

Fertile new ideas have recently been introduced into the semantical analysis of relevant implication and related concepts by Routley and by Urquhart. The purpose of the present note, to be circulated informally among those who care, is to recast these ideas in a way which yields a simple and straightforward semantical completeness result, at least for R. It is believed by me, anyway, and maybe by Dunn--that this is the first investigation along these lines which actually does yield such a result, though I freely acknowledge the priority of those who have suggested the lines and wish them well in what seems to be their considerably more intricate projects.[8]

Urquhart and Dunn met at the Temple University conference on alternative semantics in December 1970, and kept in touch afterward. As a result, Urquhart contacted Routley, who sent him a copy of his manuscript in early March 1971.[9] By this time, Dunn informed Urquhart that Meyer claimed to have proved a completeness result for the relational semantics.

According to Dunn, Meyer finished his manuscript on the ternary relational semantics sometime in March, and this became the core of the paper [50], "The semantics of entailment." (Later on, an "I" was added to the title in references to distinguish it from the papers with numbers II, III and IV, which are [47], [48] and Appendix 1 in [52].) Also Meyer presented essentially the content of "I" at the annual meeting of the Association for Symbolic Logic in December 1971, and an abstract [49] was published.

These four papers develop their semantics for the **R**, **E**, and some related relevance logics.[10] They were all coauthored by Routley and Meyer, but the first three seems to be more in the style of Meyer's writings, whereas the fourth is more similar to Routley's style.[11] The papers did not appear in the order in which they were

---

[8]Note (by Dunn): At the time I was completely convinced that both Routley's and Urquhart's operational approaches were incomplete for all of **R** (after all Meyer and I together showed this), and I was convinced that Meyer did have a completeness result for all of **R**. But I now believe that Maksimova likely had the first completeness theorem. See more about this in the section on Maksimova below.

[9]We have a scan of Urquhart's copy of Routley's manuscript.

[10]Each of the papers contains some interesting technical results using the semantics developed in that paper. To give just two examples, "I" shows that Ackermann's rule γ is admissible in **R**, and "II" shows that both **R** and **NR** are *reasonable in the sense of Halldén*.

[11]The authorship of "Semantics of entailment, I, II, III" is consistent with what Dunn remembers. Also, the style and content of [50] is very similar to that of [42], which is headed "From the desk of Robert K. Meyer." "Semantics of entailment IV" is Appendix 1 of [52], and on p. xv of that volume it says "The text has been almost entirely written by the first author, contributions from other authors have been, to various degrees, overwritten by him." The first author of that volume

written, with the papers "II" and "III" appearing in 1972, outstripping "I," which appeared in 1973 (in a conference proceedings despite the fact that the paper was not presented at that conference). "I" is devoted to developing a semantics for the system **R** of relevant implication, though it is worth mentioning that it also extends this to a ternary semantics for **RM**. It also examines the positive fragment **R+** of the system **R** and a version of **R** with propositional quantifiers. "II" is focused on developing a semantics for the system **NR** (**R** with an **S4**-type necessity operator N) by adding a binary relation S to interact with the ternary relation R. The system **NR** was developed by Meyer earlier, with the hope that the entailment operator in **E** would turn out to be equivalent to necessary relevant implication $N(A \to B)$. Then, this would provide in effect a semantics for **E**. While the connection holds for the implicational fragments, as Meyer showed, it unfortunately failed for the whole **E** and **NR**, as was proved by Maksimova in [37]. "III" is devoted to developing a semantics for the positive fragment **E+** of **E**, along with semantics for other positive relevance logics (**R+**, **T+** and **B+**). Routley and Meyer say (p. 192):

> The time has come to extend our semantical methods to other systems of relevant logic besides the system R. We shall do so in two stages. The present paper deals only with *positive* systems of entailment, since these may be handled quite simply along previous lines; complications arising from negation are put off until the sequel.

But the "sequel" did not appear until 1982, and not as a paper in the ordinary sense, but rather as Appendix 1 of [52]. There is a long and somewhat personal explanation of the delay given by Routley (cf. [52, p. 430]), but whatever the reason, it seems that Routley and Meyer left the door open for someone else to claim the first publication of a ternary semantics for **E**.

Now we turn to a standard formulation of the Routley–Meyer semantics for **R**.[12]

Definition 3.1. A *structure for* **R** is a quintuple $\mathfrak{F} = \langle U; \leq, I, R, {}^{*} \rangle$, where the elements satisfy (1)–(7). (Lower-case Greek letters range over elements of U.)

(1) $\langle U; \leq \rangle$ is a quasi-ordered set;

---

is Richard Routley.

[12]We do not follow verbatim Routley and Meyer's original papers, though the semantics we present is essentially the same. Terminology and notation in logic has evolved in the last 40 years or so, and we present the semantics in a slightly "modernized" fashion. Also, note that there is a typo in their requirement p4 [50, p. 205] that $R^2abcd \Rightarrow Rabc$. The first a should be 0, and the postulate should look like $R^2 0bcd \Rightarrow Rbcd$. Routley and Meyer abbreviate $\exists e\,(Rabe\ \&\ Recd)$ by R^2abcd.

(2) $\emptyset \neq I \subseteq U$, $\quad R \subseteq U \times U \times U$, $\quad {}^*\colon U \longrightarrow U$;

(3) $\alpha \leq \beta$ iff $\exists \iota \in I\, R\iota\alpha\beta$, $(\delta \leq \alpha\ \&\ \gamma \leq \varepsilon\ \&\ R\alpha\beta\gamma) \Rightarrow R\delta\beta\varepsilon$;

(4) $\alpha^{**} = \alpha$, $\quad R\alpha\beta\gamma \Rightarrow R\alpha\gamma^*\beta^*$;

(5) $\exists\zeta(R\alpha\beta\zeta\ \&\ R\zeta\gamma\delta) \Rightarrow \exists\zeta(R\beta\zeta\delta\ \&\ R\alpha\gamma\zeta)$;

(6) $\exists\zeta(R\alpha\beta\zeta\ \&\ R\zeta\gamma\delta) \Rightarrow \exists\zeta(R\zeta\beta\delta\ \&\ R\alpha\gamma\zeta)$;

(7) $R\alpha\beta\gamma \Rightarrow \exists\zeta(R\alpha\beta\zeta\ \&\ R\zeta\beta\gamma)$.

The last three conditions may be stated concisely as $R^2\alpha\beta\gamma\delta \Rightarrow R^2\beta(\alpha\gamma)\delta$, $R^2\alpha\beta\gamma\delta \Rightarrow R^2\alpha(\beta\gamma)\delta$ and $R\alpha\beta\gamma \Rightarrow R^2\alpha\beta\beta\gamma$.

Definition 3.2. A *model for* **R** is an ordered pair $\mathfrak{M} = \langle \mathfrak{F}, v \rangle$, where $\mathfrak{F}$ is a structure for **R**, and v (of type $v\colon \mathbb{P} \longrightarrow \mathcal{P}(U)$) satisfies (8).

(8) If $\alpha \in v(p)\ \&\ \alpha \leq \beta$, then $\beta \in v(p)$.

The *satisfaction relation* $\vDash$ emerges from v by clauses (9)–(15).

(9) $\mathfrak{M}, \alpha \vDash p$ iff $\alpha \in v(p)$ in $\mathfrak{M}$;

(10) $\mathfrak{M}, \alpha \vDash \boldsymbol{t}$ iff $\exists\iota \in I\, \iota \leq \alpha$ in $\mathfrak{F}$;

(11) $\mathfrak{M}, \alpha \vDash \sim\mathcal{A}$ iff $\mathfrak{M}, \alpha^* \nvDash \mathcal{A}$;

(12) $\mathfrak{M}, \alpha \vDash \mathcal{A} \to \mathcal{B}$ iff for all β, γ, $R\alpha\beta\gamma$ and $\mathfrak{M}, \beta \vDash \mathcal{A}$ imply $\mathfrak{M}, \gamma \vDash \mathcal{B}$;

(13) $\mathfrak{M}, \alpha \vDash \mathcal{A} \circ \mathcal{B}$ iff there are β, γ s.t. $R\beta\gamma\alpha$ and $\mathfrak{M}, \beta \vDash \mathcal{A}$ and $\mathfrak{M}, \gamma \vDash \mathcal{B}$;

(14) $\mathfrak{M}, \alpha \vDash \mathcal{A} \wedge \mathcal{B}$ iff $\mathfrak{M}, \alpha \vDash \mathcal{A}$ and $\mathfrak{M}, \alpha \vDash \mathcal{B}$;

(15) $\mathfrak{M}, \alpha \vDash \mathcal{A} \vee \mathcal{B}$ iff $\mathfrak{M}, \alpha \vDash \mathcal{A}$ or $\mathfrak{M}, \alpha \vDash \mathcal{B}$.

Remark 3.3. There are (hopefully) obvious similarities between the above definitions and Kripke's semantics for, let us say the modal logic **S4** or the intuitionistic logic **J**. U, the set of what Routley and Meyer called "set-ups," is like W, the set of possible worlds.[13] $\leq$ is an *order relation*, which occurs in both the semantics for **S4** (usually denoted as R), and in the semantics of **J** (often denoted as $\leq$). I, the

---

[13]Though it is interesting that in "Semantics of entailment IV" (Appendix 1 of [52]) they are called "situations."

set of *logical set-ups*, is like $\{0\}$, the singleton set of the actual world. v is like the *valuation function* in the semantics of **J**.

Some differences also exist; $\leq$ here is not the relation that figures into the truth condition for $\rightarrow$ or $\sim$. The actual world in the semantics for **S4** can be any possible world and the ι's in I are special set-ups. $\sim$ is not a complementation in a De Morgan monoid, hence, the set-theoretical relative complement is not suitable as its modeling, which leads to the addition of $*$.[14]

Incidentally, there are other ways to treat De Morgan negation, as discussed by Dunn in [18]. First, we may use 4 *truth values*, which can be labeled as *true, false, both* and *neither*. This is a generalization of the two truth values T and F, by not requiring that at least T or F is assigned to a sentence, and not excluding that both are assigned. Then these four truth values can be identified with the subsets of $\{T, F\}$. Negation retains its property of flipping truth with falsity and vice versa. The lattice **4** is characteristic for the implication-free logic of the De Morgan lattices (i.e., for **fde**).

Second, we could use an *orthogonality relation* $\perp$, that originated in Birkhoff and von Neumann's quantum logic. The negation of a proposition holds at a point when that is orthogonal to all points in the proposition. An informal rendering of "orthogonal" is to say that two points are incompatible. In terms of gaggle theory (as in [9]), $\sim$'s distibution type is taken to be $\lor \longrightarrow \land$ when we use the orthogonality relation. However, De Morgan negation has the double negation property (negations count mod 2), and it obeys the De Morgan laws. If we give a preference to the De Morgan law $\sim(\mathcal{A} \land \mathcal{B}) \rightarrow (\sim\mathcal{A} \lor \sim\mathcal{B})$ over the other one, that is, we consider negation with the distribution type $\sim: \land \longrightarrow \lor$, then we get yet another way to model negation in De Morgan lattices, namely, using a *compatibility relation*.[15]

The following theorem has a straightforward proof (which we omit).

Theorem 3.4. *If $\vdash_{\mathbf{R}} \mathcal{A}$, then in any $\mathfrak{M}$ for $\mathbf{R}$, and for any $\iota \in I$, $\mathfrak{M}, \iota \models \mathcal{A}$.*

As we mentioned above, the departure from an operational approach was motivated by the lack of a completeness result for Routley's operational semantics. Thus, it seems imperative that we give a flavor of what the completeness for the

---

[14]The definition of negation using the involution $*$ was used slightly earlier by Richard Routley, together with Val Routley (later Plumwood) in [51] to obtain a semantics for first-degree entailments. It is interesting that both Fine and Maksimova use an involution as well. This can be found much earlier in an algebraic form in Białnycki-Birula and Rasiowa's 1957 [4] representation of what they called "pseudo-Boolean algebras" (now commonly referred to as De Morgan lattices). Dunn discussed these and other representations of De Morgan lattices, in 1966, in his dissertation [14] and in [15].

[15]In addition to Dunn [18], see also Restall [46], and Mares's [39, 40] on these further approaches.

relational semantics hinges on. De Morgan monoids are teeming with patterns of distributivity; beyond the usual $\wedge$–$\vee$ and the De Morgan laws for negation, both $\rightarrow$ and $\circ$ distribute in the underlying lattice. Moreover, $\rightarrow$ is a residual of $\circ$ (what is revealed by axioms (12) and (13) above).

Definition 3.5. The *canonical frame* is $\mathfrak{F}_{\mathfrak{c}} = \langle \mathcal{P}; \subseteq, I_{\mathfrak{c}}, R_{\mathfrak{c}}, {}^{*_{\mathfrak{c}}} \rangle$, where the elements are defined by (1)–(5).

(1) $\mathcal{P}$ is the set of prime filters on the Lindenbaum algebra of $\mathbf{R}$;

(2) $\subseteq$ is the subset relation;

(3) $I_{\mathfrak{c}} = \{\, P \in \mathcal{P} \colon [\boldsymbol{t}]^{\uparrow} \subseteq P \,\}$;

(4) $R_{\mathfrak{c}}\alpha\beta\gamma$ iff for any $[\mathcal{A}], [\mathcal{B}], [\mathcal{A} \circ \mathcal{B}] \in \gamma$, when $[\mathcal{A}] \in \alpha$ and $[\mathcal{B}] \in \beta$;

(5) $\alpha^{*_{\mathfrak{c}}} = \{\, [\mathcal{A}] \colon [\sim\!\mathcal{A}] \notin \alpha \,\}$.

The *canonical valuation* $v_{\mathfrak{c}}$ is defined by (6).

(6) $v_{\mathfrak{c}}([p]) = \{\, P \in \mathcal{P} \colon [p] \in P \,\}$.

The proof of the next theorem is more elaborate and it crucially relies on there being enough many prime filters, moreover, $\mathbf{R}$ being well-defined with respect to the distribution type of $\circ$ (which is $\circ \colon \vee, \vee \longrightarrow \vee$). Details of similar completeness proofs may be found in Routley and Meyer's papers, as well as in [17] and [9]. We only mention a core lemma, which is sometimes labeled as *squeeze lemma*. The multitude of prime filters, certainly, does not mean that all filters are prime. Moreover, taking the fusion of two filters may not result in a prime filter. However, the relation $R_{\mathfrak{c}}$ is in harmony with prime filters in the sense that the following is true.

(7) If $R'_{\mathfrak{c}}xy\gamma$ ($\gamma \in \mathcal{P}$), then there are $\alpha, \beta \in \mathcal{P}$ such that $x \subseteq \alpha$, $y \subseteq \beta$ and $R_{\mathfrak{c}}\alpha\beta\gamma$.

By $R'_{\mathfrak{c}}$ we denoted $R_{\mathfrak{c}}$ on the set of all filters defined like in (4) above, and we assumed that x and y are filters (i.e., $x, y \in \mathcal{F}$). The squeeze lemma is essential in proving that $R_{\mathfrak{c}}$ in the frame in Definition 3.5 has the properties that are special for $\mathbf{R}$, and in proving that $v_{\mathfrak{c}}$ is a valuation. Having set out some pointers, we state the following (without giving a proof here).

Theorem 3.6. *If $\nvdash_{\mathbf{R}} \mathcal{A}$, then there is an $\mathfrak{M}$ for $\mathbf{R}$ in which there is an $\iota \in I$ such that $\mathfrak{M}, \iota \nvDash \mathcal{A}$.*

4 Urquhart's Semi-lattice Semantics: Binary Operation

Historical sketch. Urquhart set out on a path similar to Routley's original path toward a semantics for entailment and relevant implication, namely, using a binary operation. Urquhart told us that he recollects finding the *-semantics for first-degree entailment in the spring of 1970, but was "quite down cast" when Alan Anderson told him it was anticipated by Białynicki-Birula and Rasiowa [4]. Sometime later in the summer he hit upon the semilattice semantics and wrote something out in longhand and showed it to Nuel Belnap. This showed completeness for some fragments of **R** and **E**, and became the basis of a paper he wrote and submitted to the *Journal of Symbolic Logic*. The journal received his paper on November 27, 1970. He told us that "[n]aturally, I thought that it would be a piece of cake to put it [the semilattice semantics] together with the *-semantics to get a semantical analysis for all of **R**. But it was not to be!"

We have already mentioned that Urquhart and Dunn met at the Temple conference in December. (As further "background information" we may mention that Urquhart, Meyer and Dunn had the same Ph.D. advisor when they were students: Nuel Belnap. However, Meyer and Dunn, who overlapped studying at Pitt, left by the time Urquhart started his Ph.D. there.) Urquhart's ideas were heavily influenced by the natural deduction formulation calculi for the implicational fragments of **E** and **R**. One may see a certain similarity between the union operation on the index sets of formulas and the fusion of formulas; it seems that the latter motivated Urquhart to introduce a semi-lattice right away. (Hence, the label by which Urquhart's semantics is often identified: semi-lattice semantics.)

Dunn received a copy of Urquhart's submitted paper soon after the Temple conference, and Meyer and he discovered that while the semantics indeed works for the implicational fragments (even for the implication–conjunction fragments), if ∨ is added with a usual truth condition, then there is a formula that is valid in the semi-lattice semantics, but it is not a theorem of **R** (let alone of **E**). Urquhart's paper was accepted for publication on June 1st, 1971, and the revised version was received by the journal on June 21st, 1971. This is [55]. The semi-lattice semantics also was at the heart of his dissertation [56], written under the supervision of Nuel D. Belnap.

Now we outline the semi-lattice semantics. First, we restrict our language to propositional variables and the arrow. The implicational fragment of **R** is denoted by **R**$_\to$, and it can be axiomatized by the first four axiom schemas above together with detachment as the sole rule.

Definition 4.1. A *structure for* **R**$_\to$ is a three-tuple $\mathfrak{F} = \langle U; \bot, \cup \rangle$ satisfying (1)–(2).

(1) $\langle U; \cup \rangle$ is a semi-lattice;

(2) $\bot \cup \alpha = \alpha \cup \bot = \alpha$.

Urquhart called the elements of U "pieces of information." (It is tempting to draw a parallel between pieces or bits of information and bits (in computers), but instead of bits, tables, records or even databases would be a more appropriate analog from computer science.)

Definition 4.2. A *model for* $\mathbf{R}_\to$ is an ordered pair $\mathfrak{M} = \langle \mathfrak{F}, v \rangle$, where $\mathfrak{F}$ is a structure for $\mathbf{R}_\to$, and $v \colon \mathbb{P} \longrightarrow \mathcal{P}(U)$. The satisfaction relation is defined by (3)–(4), given a valuation v.

(3) $\mathfrak{M}, \alpha \vDash p$ iff $\alpha \in v(p)$;

(4) $\mathfrak{M}, \alpha \vDash \mathcal{A} \to \mathcal{B}$ iff for all β, either $\mathfrak{M}, \beta \nvDash \mathcal{A}$ or $\mathfrak{M}, \alpha \cup \beta \vDash \mathcal{B}$.

Remark 4.3. Set-theoretically speaking, all binary operations are ternary relations, though of a special kind. Thus, it is instructive to glance at (4) written in a form that resembles the matching satisfaction condition in the relational semantics.

(4') $\mathfrak{M}, \alpha \vDash \mathcal{A} \to \mathcal{B}$ iff for all β, γ, $\alpha \cup \beta = \gamma$ and $\mathfrak{M}, \beta \vDash \mathcal{A}$ imply $\mathfrak{M}, \alpha \cup \beta \vDash \mathcal{B}$.

We have not posited specifics about the informal metalanguage, and therefore, we may keep it simple. Then we can rewrite "not–or" in (4) (which is close to Urquhart's original formulation) as "imply" in (4'). The other change we made was the introduction of γ, but it should be obvious that since $\cup$ is an operation, γ can be eliminated in lieu of $\alpha \cup \beta$.

Theorem 4.4. *If* $\vdash_{\mathbf{R}_\to} \mathcal{A}$, *then in any* $\mathfrak{M}$ *for* $\mathbf{R}_\to$, $\mathfrak{M}, \bot \vDash \mathcal{A}$.

For the completeness proof, a canonical structure is defined. Urquhart works with proofs rather than the algebra of $\mathbf{R}_\to$.

Definition 4.5. The *canonical structure* is $\mathfrak{F}_\mathfrak{c} = \langle U_{\text{fin}}; \emptyset, \cup \rangle$, where (1)–(2) specify the elements of the three-tuple.

(1) U_{fin} is the set of finite sets of formulas;

(2) $\emptyset$ and $\cup$ have their usual set-theoretical meaning.

The *canonical valuation* $v_\mathfrak{c}$ is defined by (3).

(3) $v_c(p) = \{\{\mathcal{A}_0, \ldots, \mathcal{A}_n\} \in U_{\text{fin}} : \vdash_{\mathbf{R}_\to} \mathcal{A}_0 \to \cdots \mathcal{A}_n \to p\}$.

Remark 4.6. We commented on the irrelevant proofs in the axiomatic system for **R**. For the implicational fragment, "relevant proofs" can be defined easily using a natural deduction calculus. The first volume of *Entailment* practically starts off with $\mathbf{FR}_\to$, the Fitch-style natural deduction calculus for $\mathbf{R}_\to$ (cf. [2, §3]), which does not allow slipping in an extra (i.e., never used) hypothesis here or there. This should explain the restriction to finite sets of formulas, which are the "premise sets" or the "sets of hypotheses" from which a formula can be proved, for instance, in $\mathbf{FR}_\to$.

Theorem 4.7. *If $\nvdash_{\mathbf{R}_\to} \mathcal{A}$, then there is an $\mathfrak{M}$ for $\mathbf{R}_\to$ in which $\mathfrak{M}, \bot \nvDash \mathcal{A}$.*

The proof is not difficult, because $\mathfrak{M}_c$ is obviously a semi-lattice with a zero. Hence, what needs to be verified is that v_c is a valuation according to (4). (We omit the details which may be found in [55], which uses a slightly different notation.)

Urquhart's idea was to add the usual satisfiability conditions for $\wedge$ and $\vee$, once those connectives are included in the language. His clauses (in our notation) are (5) and (6). (Due to our notational conventions, these two conditions are literally the same as (14) and (15) in the Routley–Meyer semantics, but the $\mathfrak{M}$ here is not the same as the $\mathfrak{M}$ there. Also, earlier, α was a set-up, but now, it is a piece of information.)

(5) $\mathfrak{M}, \alpha \vDash \mathcal{A} \wedge \mathcal{B}$ iff $\mathfrak{M}, \alpha \vDash \mathcal{A}$ and $\mathfrak{M}, \alpha \vDash \mathcal{B}$;

(6) $\mathfrak{M}, \alpha \vDash \mathcal{A} \vee \mathcal{B}$ iff $\mathfrak{M}, \alpha \vDash \mathcal{A}$ or $\mathfrak{M}, \alpha \vDash \mathcal{B}$.

Urquhart noted that although soundness is provable, completeness fails.

Example 4.8. The counterexample developed by Dunn and Meyer that Dunn communicated to Urquhart in his letter of February 17th, 1971 is the following. We added values for all subformulas on the line below the formula; the calculations are carried out according to the matrix M_0, which may be found in [2, pp. 252–253].

$$((\ \mathcal{A}\ \to\ \mathcal{A}\ )\ \wedge\ ((\ \mathcal{A}\ \wedge\ \mathcal{B}\ )\ \to\ \mathcal{C}\ )\ \wedge\ (\ \mathcal{A}\ \to\ (\ \mathcal{B}\ \vee\ \mathcal{C}\ )))\ \to\ \mathcal{A}\ \to\ \mathcal{C}$$
$$+3\ +3\ +3\quad -0\quad +3\ +0\ +0\quad -0\ -0\quad -0\quad +3\ +3\quad +0\ +3\ -0\qquad -3\ +3\ -3\ -0$$

Since -3 is not a distinguished element in M_0, the formula is not a theorem of **R**. On the other hand, the formula is valid on the semi-lattice semantics, when the natural clause for $\vee$ is added.

We quickly go through the validation of this formula in the semi-lattice semantics. Let us assume that $\bot$ does not satisfy the formula. (We omit mentioning $\mathfrak{M}$, since it is the same everywhere.) By (4), there is an α that satisfies the antecedent, but not the consequent. From the former, it is immediate that $\alpha \vDash \mathcal{A} \to \mathcal{A}$, $\alpha \vDash (\mathcal{A} \wedge \mathcal{B}) \to \mathcal{C}$ and $\alpha \vDash \mathcal{A} \to (\mathcal{B} \vee \mathcal{C})$. From the latter, we get that $\beta \vDash \mathcal{A}$ whereas $\alpha \cup \beta \nvDash \mathcal{C}$, for some β. Since $\alpha \vDash \mathcal{A} \to \mathcal{A}$ and $\beta \vDash \mathcal{A}$, $\alpha \cup \beta \vDash \mathcal{A}$. Similarly, we obtain that $\alpha \cup \beta \vDash \mathcal{B} \vee \mathcal{C}$. But $\alpha \cup \beta \nvDash \mathcal{C}$, so it must be that $\alpha \cup \beta \vDash \mathcal{B}$. Combining $\alpha \cup \beta \vDash \mathcal{A}$ and $\alpha \cup \beta \vDash \mathcal{B}$, we have $\alpha \cup \beta \vDash \mathcal{A} \wedge \mathcal{B}$, hence, also $\alpha \cup \beta \vDash \mathcal{C}$, which contradicts to what we have already established. One can see how the subformulas allow us to "move around" with $\alpha \cup \beta$ to arrive at a contradiction.[16]

Although Meyer and Dunn showed that Urquhart's semi-lattice semantics was not complete for the positive fragment of the relevance logic **R**, the question remained as to how the semi-lattice semantics might be axiomatized. This was first addressed by Fine [20] and developed further by Charlwood in his Ph.D. thesis [11] (directed by Urquhart and published in [12]), where he deployed a rule that might charitably be said to be elegant in its complication. Charlwood also developed two natural deduction systems, one with subscripts and the other without. This last is in fact equivalent to the system of Prawitz [45], which Prawitz had wrongly conjectured to be the same as the positive fragment of **R**.

5 Fine: Binary Operation Plus Partial Order

Historical sketch. Fine started to work on a semantics for **R** sometime during 1970 after attending a talk presented by Belnap, who spent a few months at Oxford University. Belnap was visiting at Oxford as a senior research fellow, probably, from January to March, or so, in 1971. He gave a talk at the Mathematical Institute, in which he said the problem of finding a semantics for **R** was open. This stimulated Fine to create his semantics and prove it complete. Fine recalls finishing the work before his junior research fellowship at Oxford ended in June 1971. Fine submitted two manuscripts (one of which was a version of his paper later published as [19]) to an editor of the *Journal of Symbolic Logic*, who received them on April 17th, 1972. About a year later, the editor informed Fine that he still had not received anything from the referee and suggested that instead of waiting longer he send his paper to the recently founded *Journal of Philosophical Logic*. A revised version of his paper—including a brief comparison with the Routley–Meyer semantics—appeared in the *JPL* in 1974.

In what follows, we consider Fine's semantics for **R**; we do not follow his notation or terminology though.

---

[16]There are shorter formulas that do the job. Urquhart gives $((\mathcal{A} \to (\mathcal{B} \vee \mathcal{C})) \wedge (\mathcal{B} \to \mathcal{D})) \to (\mathcal{A} \to (\mathcal{D} \vee \mathcal{C}))$ in [55], whereas Dunn provides $((\mathcal{A} \to (\mathcal{B} \vee \mathcal{C})) \wedge (\mathcal{B} \to \mathcal{C})) \to (\mathcal{A} \to \mathcal{C})$ in [17, §4.6].

Definition 5.1. A *structure for* **R** is a sextuple $\mathfrak{F} = \langle T, U; \leq, i, \cdot, - \rangle$, where the components have properties (1)–(10). (Lower-case Greek letters range over U, whereas $x, y, z, \ldots$ are elements of T.)

(1) $\emptyset \neq U \subseteq T$, $\quad i \in T$, $\quad \cdot : T \times T \longrightarrow T$, $\quad - : U \longrightarrow U$;

(2) $\langle T; \leq \rangle$ is a poset;

(3) $\langle T; \cdot, i \rangle$ is a groupoid in which $\cdot$ is left monotone and i is its left identity;

(4) $x \cdot y \leq \alpha$ implies that there are β, γ such that $x \leq \beta$ & $y \leq \gamma$ and $\beta \cdot y \leq \alpha$ & $x \cdot \gamma \leq \alpha$;

(5) $i \leq \alpha$ implies $-\alpha \leq \alpha$;

(6) $--\alpha = \alpha$, $\quad \alpha \leq \beta$ implies $-\beta \leq -\alpha$;

(7) $x \cdot (y \cdot z) \leq (y \cdot x) \cdot z$;

(8) $x \cdot y \leq y \cdot x$;

(9) $x \cdot x \leq x$;

(10) $x \cdot \alpha \leq \beta$ implies $x \cdot -\beta \leq -\alpha$.

Definition 5.2. A *model for* R is an ordered pair $\mathfrak{M} = \langle \mathfrak{F}, v \rangle$, where $\mathfrak{F}$ is a structure for **R** and v satisfies (11).

(11) If $x \in v(p)$ & $x \leq y$, then $y \in v(p)$.

The *satisfaction relation* is defined from v by (12)–(17).[17]

(12) $\mathfrak{M}, x \vDash p$ iff $x \in v(p)$ in $\mathfrak{M}$;

(13) $\mathfrak{M}, x \vDash \boldsymbol{t}$ iff $i \leq x$ in $\mathfrak{F}$;

(14) $\mathfrak{M}, x \vDash {\sim}\mathcal{A}$ iff for all α, $x \leq \alpha$ implies $\mathfrak{M}, -\alpha \nvDash \mathcal{A}$;

(15) $\mathfrak{M}, x \vDash \mathcal{A} \to \mathcal{B}$ iff for all y, $\mathfrak{M}, y \vDash \mathcal{A}$ implies $\mathfrak{M}, x \cdot y \vDash \mathcal{B}$;

(16) $\mathfrak{M}, x \vDash \mathcal{A} \wedge \mathcal{B}$ iff $\mathfrak{M}, x \vDash \mathcal{A}$ and $\mathfrak{M}, x \vDash \mathcal{B}$.

---

[17]Fine did not consider fusion, hence we do not include a clause for that.

Fine calls the members of T *theories* and the members of U *consolidated* or *saturated* theories (many would call them *prime* theories because they correspond to prime filters as we shall see). Unlike Routley and Meyer, Fine does not take $\vee$ as primitive, but defines $\mathcal{A} \vee \mathcal{B}$ in a customary way as $\sim(\sim\mathcal{A}\wedge\sim\mathcal{B})$. It is straightforward to show that (14) can be replaced with the much simpler clause (17) used by Routley and Meyer.

(17) $\mathfrak{M}, \alpha \vDash \sim\mathcal{A}$ iff $-\alpha \nvDash \mathcal{A}$,

from which the Routley-Meyer clause (18) follows.

(18) $\mathfrak{M}, \alpha \vDash \mathcal{A} \vee \mathcal{B}$ iff $\mathfrak{M}, \alpha \vDash \mathcal{A}$ or $\mathfrak{M}, \alpha \vDash \mathcal{B}$.

Clause (14) might appear at first to have an intuitionistic flavor, but it should be noted that $\mathcal{A}$ is evaluated at $-\alpha$ not at α. It is not difficult to see that $R\alpha\beta\gamma$ in the Routley–Meyer semantics corresponds to $\alpha \cdot \beta \leq \gamma$ here. Condition (4) guarantees that elements of T in a product when it is included in a saturated theory can be replaced in one or the other argument place of $\cdot$ by elements of U, that is, saturated theories. This stipulation is similar to what is established in the so-called "Squeeze Lemma" in the Routley–Meyer semantics for the canonical frame. Fine must have this as a postulate, because his frames contain two sorts of theories, which have to relate to each other appropriately. The operation that Fine denotes as "$-$" is easily seen to be the $*$ in the Routley–Meyer semantics.

Remark 5.3. The missing satisfiability condition for $\circ$ could be given as (19).

(19) $\mathfrak{M}, x \vDash \mathcal{A} \circ \mathcal{B}$ iff there are y, z s.t. $y \cdot z \leq x$ and $\mathfrak{M}, y \vDash \mathcal{A}$ and $\mathfrak{M}, z \vDash \mathcal{B}$.

Theorem 5.4. *If $\vdash_{\mathbf{R}} \mathcal{A}$, then in any model $\mathfrak{M}$ for $\mathbf{R}$, $\mathfrak{M}, i \vDash \mathcal{A}$.*

The presence of prime theories (i.e., the elements of U) in addition to theories (the members of T) makes possible the simpler looking notion of validity by an appeal to i alone.

Definition 5.5. The *canonical frame* is $\mathfrak{F}_{\mathfrak{c}} = \langle \mathcal{F}, \mathcal{P}; \subseteq, i_{\mathfrak{c}}, \cdot_{\mathfrak{c}}, -_{\mathfrak{c}} \rangle$, where the elements are defined by (1)–(6).

(1) $\mathcal{F}$ is the set of (non-empty, proper) filters on the algebra of $\mathbf{R}$;

(2) $\mathcal{P}$ is the set of prime filters on the algebra of $\mathbf{R}$;

(3) $\subseteq$ is the subset relation;

(4) $i_{\mathfrak{c}}$ is $[\boldsymbol{t}]^{\uparrow}$;

(5) $x \cdot y = \{\, [\mathcal{B}] \colon \exists\, [\mathcal{A}] \in y\, [\mathcal{A} \to \mathcal{B}] \in x \,\}$;

(6) $-\alpha = \{\, [\mathcal{A}] \colon [\sim\!\mathcal{A}] \notin \alpha \,\}$.

The *canonical valuation* $v_{\mathfrak{c}}$ is specified by (7).

(7) $v_{\mathfrak{c}}(p) = \{\, F \in \mathcal{F} \colon [p] \in F \,\}$.

Remark 5.6. If $\circ$ is in the language of **R**, then (5) can be replaced by (5′).

(5′) $x \cdot y = \{\, [\mathcal{C}] \colon \exists\, [\mathcal{A}], [\mathcal{B}]([\mathcal{A}] \in x \ \& \ [\mathcal{B}] \in y \ \& \ [\mathcal{A} \circ \mathcal{B}] \leq [\mathcal{C}]) \,\}$.

This may be viewed as a sort of explanation for the two-layered structure of Fine's semantics in which he uses both T and U: the binary operation $\cdot$, which is straightforwardly definable from $\to$ or $\circ$ is not an operation on the set that we want to comprise our situations. $\alpha \cdot_{\mathfrak{c}} \beta$ may not be an element of $\mathcal{P}$, though it is surely an element of $\mathcal{F}$. This obstacle is not new or peculiar to relevance logics. Indeed, in a normal modal logic (whether **K** or **S5**, or some other logic), the operation that is directly definable from $\lozenge$ turns out not to be an operation on the set of prime theories (also known in that context as maximally consistent theories or ultrafilters). Hence, Kripke's relational approach, in a sense, is necessitated for normal modal logics, exactly as the ternary relational approach is entailed for relevance logics.

The proof of the following theorem can be reconstructed from Fine's paper.

Theorem 5.7. *If $\nvdash_{\mathbf{R}} A$, then there is a model $\mathfrak{M}$ for* **R** *in which $i \nvDash A$.*

6 Maksimova: A Variety of Ways

Historical sketch. We described in detail Larisa Maksimova's work on semantics for relevance logics in another paper [10] not long ago. Hence we will limit our exposition here to emphasize the historical aspects. As you will see, Maksimova in a sense did it all, for in various papers she used a ternary relation (much like Routley and Meyer), and a combination of a binary operation and a binary relation (much like Fine), and her early work was completely independent. Despite the "Cold War" and the relative isolation of Novosibirsk, her work came to be known to Anderson and Belnap and other relevance logicians largely through her many publications in the journal *Algebra i logika*, more precisely, through the English translation of that journal *Algebra and Logic*.

Maksimova started to work on Ackermann's $\mathbf{\Pi}'$ calculus and then on relevance logics in 1963, and systematically investigated various features of those logics. First, she dealt with questions about axiomatic formulations and deductions, then she turned to algebraizations and algebraic semantics. She even isolated a sublogic, which she called $\mathbf{SE}$, within $\mathbf{E}$ that she proved decidable.

In 1969, she gave a talk at an algebraic conference in Novosibirsk, which attracted participants from across the USSR. The abstract of her talk was published as [35], and it makes clear that she was the first to hit on the idea of using a *ternary relation* for the modeling of $\rightarrow$. It is impossible to fully reconstruct her models from the half-page abstract, but she stated the satisfiability condition for $\rightarrow$ clearly and claimed soundness and completeness in the form of representation theorems for algebraic models in terms of clopen subsets of a topological space with an involution g, a partial-order $\leq$ and a ternary relation τ. It seems that she never gives colorful names to the points; she does not call them situations, set-ups, pieces of information, theories, information states or whatever. Rather, she immediately proceeds with the mathematics. Maksimova used an involution g on the space to define negation, anticipating the Routleys' *, and most importantly for our present paper, she defined implication using a ternary relation τ on the space. Here is her definition of $\rightarrow$, first verbatim, and then in a notation closer to what we have used above.

(i) $S_1 \rightarrow S_2 = \{\, z \colon \forall x, y\, ((x \in S_1 \,\&\, \tau(x,y,z)) \Rightarrow y \in S_2) \,\}$;

(ii) $\mathfrak{M}, \gamma \vDash \mathcal{A} \rightarrow \mathcal{B}$ iff for all α, β, $R(\alpha, \beta, \gamma)$ and $\mathfrak{M}, \alpha \vDash \mathcal{A}$ imply $\mathfrak{M}, \beta \vDash \mathcal{B}$.

In her paper [36] from (1971), however, she in effect defines a ternary relation using a groupoid operation and a partial order (much in the spirit of Fine, but there is no reason to think she knew of his work, or vice versa). The semantics is developed for use as a tool, to prove interpretation and separation theorems for fragments of $\mathbf{R}$ and $\mathbf{E}$. Despite its title "Interpretation and separation theorems for the logical systems E and R," it turns out that she examines only the implication, implication–conjunction, and implication–conjunction–disjunction ($\mathbf{E+}$, $\mathbf{R+}$) fragments of these systems, and provides semantics only for these. So we find no completeness theorem for whole $\mathbf{E}$ (or $\mathbf{R}$).

In [37], Maksimova provides a semantics for all of $\mathbf{E}$ (and for $\mathbf{E+}$). Again she has an application in mind, this time to provide a counterexample to Dag Prawitz's and Meyer's independent conjecture that entailment in the system $\mathbf{E}$ can be defined as necessary relevant implication. Maksimova uses an explicit ternary accessibility relation to define relevant implication, and defines negation using an involution (just like the * in the Routley–Meyer semantics).

[38] from 1973, focuses on "strict implication lattices." She calls them "strimplas" for *strict implication lattices*.[18] These are the underlying algebraic structures for a wide variety of positive relevance logics, including **E** and **R**. They can also be outfitted with De Morgan complement so as to get algebras ("strimplanas"—the "n" is for negation) corresponding to the full logics **E** and **R**.[19] Maksimova gives representation theorems for both the strimplas (using a ternary relation) and the strimplanas (adding an involution), which amount to completeness theorems for **E+**, **E**, **R+** and **R**. She uses these to show that **E** is a conservative extension of **E+**, and **R** is a conservative extension of **R+**. We suppose it is worth noting that this paper cites Routley and Meyer's "III" [48].[20]

7   Gabbay: Quaternary Relation

Historical sketch.   A very interesting early contribution from 1972 is [23], whose title is "A general theory of the conditional in terms of a ternary operator." The title sounds relevant until one reads it twice and notices that it says "ternary operator," not "ternary relation." Gabbay's goal here is to analyze the subjunctive conditional (not the relevant conditional) $\mathcal{A} > \mathcal{B}$ with the aim of reducing it to the form $\Box_{\mathcal{A},\mathcal{B}}(\mathcal{A} \to \mathcal{B})$. He wants this to be a special case of $\Box_{\mathcal{A},\mathcal{B}}\mathcal{C}$, and he gives a semantics for this in terms of a quaternary relation R. If S is a set of possible worlds, the relation R is $R \subseteq S \times \wp(S) \times \wp(S) \times S$, i.e., R relates a possible world s, a set of possible worlds Q_1, a set of possible worlds Q_2, and a possible world t. Intuitively, s is the possible world at which the sentence $\Box_{\mathcal{A},\mathcal{B}}\mathcal{C}$ is being evaluated, Q_1 is the set of possible worlds in which the sentence $\mathcal{A}$ is true, Q_2 is the same but in relation to the sentence $\mathcal{B}$, and t is a possible world accessible from these via the relation R. The idea is that for $\Box_{\mathcal{A},\mathcal{B}}\mathcal{C}$ to be true at s, $\mathcal{C}$ should be true at all such t.

Spelling this out, for an arbitrary sentence $\mathcal{A}$, let $|\mathcal{A}|$ be the set of worlds in which the sentence $\mathcal{A}$ is true. This is a standard way of representing a proposition expressed by a sentence. Then the pair $\langle |\mathcal{A}|, |\mathcal{B}| \rangle$ determines the binary relation $R_{\langle |\mathcal{A}|, |\mathcal{B}| \rangle}$ which holds between s and t just when $\langle s, |\mathcal{A}|, |\mathcal{B}|, t \rangle \in R$. $\Box_{\mathcal{A},\mathcal{B}}\mathcal{C}$ is then a kind of

---

[18]Do not be confused by the term "strict implication lattice" (or "strimpla") that Maksimova uses in this paper. "Strict implication" is the English translation of the Russian term used to translate Ackermann's "strenge Implikation." Anderson and Belnap translated this into English as "rigorous implication" to avoid a confusion with Lewis's "strict implication" (which is a necessary material implication, requiring no connection between antecedent and consequent).

[19]The **R** strimplanas are closely related to Dunn's De Morgan monoids mentioned above.

[20]She also cites Meyer's [43] as containing a similar proof of the conservative extension result for **R/R+**. It is interesting that Meyer did not prove this for **E/E+** and it would seem to suggest that Routley and Meyer as yet did not have a completeness theorem for **E** with respect to their semantics.

relativized necessity operator, whose binary accessibility relation is determined by the propositions expressed by $\mathcal{A}$ and $\mathcal{B}$.

(1) $\quad s \vDash \Box_{\mathcal{A},\mathcal{B}}\mathcal{C} \quad$ iff $\quad \forall t\,(sR_{\langle|\mathcal{A}|,|\mathcal{B}|\rangle}t \Rightarrow t \in |\mathcal{C}|)$.

Gabbay introduces a ternary logical operator $\Vdash$ so we can write $\mathcal{A},\mathcal{B} \Vdash \mathcal{C}$ instead of $\Box_{\mathcal{A},\mathcal{B}}\mathcal{C}$. The satisfaction clause for this ternary operator (given in (2)) naturally uses the quaternary relation.

(2) $\quad s \vDash (\mathcal{A}_1, \mathcal{A}_2 \Vdash \mathcal{B}) \quad$ iff $\quad$ for all t, such that $\langle s, |\mathcal{A}_1|, |\mathcal{A}_2|, t\rangle \in R$, $t \vDash \mathcal{B}$.

The simplest way to obtain an n-place relation R^- from an $n+1$-place relation R is to ignore one of the arguments; e.g., we can reduce $R^4xyy'z$ to R^-xyz (where R^- is ternary) by omitting y'. Gabbay does something similar—by ignoring $|\mathcal{A}_2|$, he in effect, simplifies the ternary operator $\mathcal{A}_1, \mathcal{A}_2 \Vdash \mathcal{B}$ to a binary operator $\mathcal{A}_1 \to \mathcal{B}$ with the satisfaction clause (3).

(3) $\quad s \vDash (\mathcal{A} \to \mathcal{B}) \quad$ iff $\quad$ for all t, such that $\langle s, |\mathcal{A}|, t\rangle \in R^-$, $t \vDash \mathcal{B}$.

There is still some symbol pushing left to do. Let us define R' so that $R'sat$ iff both $\langle s, |\mathcal{A}|, t\rangle \in R^-$ and $a \in |\mathcal{A}|$, i.e., $a \vDash \mathcal{A}$. It is now straightforward to show that the clause above is equivalent to that in (4).

(4) $\quad s \vDash \mathcal{A} \to \mathcal{B} \quad$ iff $\quad$ for all t, if $a \vDash \mathcal{A}$ and $R'sat$, then $t \vDash \mathcal{B}$.

This at least has the superficial form of the satisfaction clause for implication in the Routley–Meyer semantics. We say "superficial" because it has hidden within it the fact that the relation R' depends upon the proposition $|\mathcal{A}_2|$.

Communications with Gabbay have established that he worked out this semantics while teaching a course called "63B Modal Logic" at Stanford University during the Winter Term, January 4–March 25, 1971. In an email (6/10/2016) he wrote:

> 1. The whole course was devoted to the conditional and the results were obtained while teaching it.
> 2. So after the course, time was needed to write the paper and more time to get it typed it with symbols included. This was done in those days by one secretary in the math dept using an IBM golf ball typewriter. She did the job for everybody, and there was a queue and one had to join the queue. So the timeline is correct. Note that the paper 19 was received by *Theoria* in Dec 1971.

Gabbay ends up in the photo finish in a way because it seems he worked out the idea in the first couple of months of 1971, just like Meyer, and maybe like Fine. Thus Gabbay's contribution falls within that magical first several months of 1971, just like Meyer's bringing together the contribution by Routley and Meyer, and Fine's coming up with his version of the ternary semantics. But he does not really end up in the photo finish because his contribution is not "relevant" (pun intended). He was running in another race since he was not focused on relevant implication but rather on the subjunctive conditional. Also his *Theoria* paper uses a 4-place accessibility relation—RxQ_1, Q_2y, where x and y are possible worlds, and Q_1 and Q_2 are sets of worlds—not a ternary one. It does contain a ternary relation as a special case, say when you ignore Q_2, but as we saw above this seems to give only superficially the Routley–Meyer method of evaluating conditionals.

Incidentally, this semantics is reproduced in Gabbay's 1976 book [26], which also contains a ternary semantics for relevance logic. Footnote 8 on page 301 credits the idea of using a ternary relation to Scott, saying that it "first occurred to Scott in 1964."[21] It goes on to say that "It was found independently and widely applied by R. Routley, R. Meyer, A. Urquhart," and then says to see Anderson and Belnap's *Entailment*, which is listed in the references as "to appear." In fact, the published first volume of the *Entailment* contains no mention of the ternary semantics. It took vol. 2 to do that though vol. 1 does list the early papers on the semantics by those whom we mentioned so far.

Some other papers of Gabbay that might be "relevant" (but we think are not) include [24] and [25]. Section 1 of [24] is titled "General Entailment Type Logics." Also, that section *does* contain Routley–Meyer structures with a ternary relation (and *) and gives truth conditions for → and ∼ in just the way that Routley–Meyer do. But the Bibliography contains [48], and the Introduction cites this paper (along with a paper by Dana Scott, and another by Gabbay himself [25]), saying that the general methods presented are abstracted from these. Because of this citation, [25] is the other paper that attracted our attention as "perhaps relevant." At the first glance, we thought it might well be, because Definition 10 contains the idea of defining a semantics for an n-ary "modality" □. But a closer reading reveals that there is no $n + 1$-place relation hiding behind the scenes.

So we conclude that despite the fact that Gabbay had some original ideas about using relational semantics to interpret conditionals, he did not have relevant conditionals as his primary target, rather he was aiming his modeling at subjunctive conditionals. When he shortly thereafter began to discuss the ternary relational semantics for relevance logic, he did so while citing the work of Routley and Meyer.

---

[21]But it was not published until 1974 in [53]. See footnote 4 above.

Addendum: Gabbay sent us the following email on 10/18/2016 after reading the draft above.

> I remember in 1968 I wrote a Henkin Type completeness proof for implicational relevance logic and gave the proof to Y. Bar-Hillel. Bar-Hillel sent it to Nuel Belnap? (I think and not to Anderson) in a letter saying (he showed me the letter) "I have this young fellow who can write a completeness proof for anything." The proof had to rely on the syntax, i.e., $A \rightarrow B$ was a modality on B dependent on A, in modern notation $\square_A B$. So the accessibility relation depended on the syntactical A. I never got a response from Belnap via Bar-Hillel. I did not think at the time this was "Kosher," and dismissed it.
>
> It did not occur to me that maybe "A" can be identified as a set.
>
> In 1971, I wrote the paper on conditional[s]. There I used dependence on both A and B, but only because of examples like:
>
> If New York were in Georgia then NY were in the South (Georgia were in the North),
>
> you need to see both A and B to evaluate. I did not consider relevance logic, nor Bar Hillel, nor Belnap.

A few days later he warned us that in reading this we should be aware of "the danger of anachronism. Later time way of thinking applied to the past." He went on to say:

> Now let us be very accurate.
>
> 1. The facts I remember is that there was a proof (correct or not) treating essentially "$A \rightarrow B$" as a modality $[A]\,B$.
>
> 2. It was sent to Belnap by Bar-Hillel. (Try Anderson archive as well?)
>
> 3. This introduced the possibility of dependence on A which, if pursued technically could lead to certain semantics.
>
> 4. I considered this "not kosher." Bar-Hillel must have sensed something there, and sent it to Belnap.
>
> 5. However, it may have been a preliminary in my mind for treating the conditional in 1971.

So far neither Gabbay nor we have had any luck locating a copy of Bar-Hillel's letter, however, Gabbay sent us a reconstruction of some of his ideas about relevant implication on 11/02/2016. The following is a sketch that shows the connections to modality as well as syntax.

Let us consider the implicational fragment of the logic $\mathbf{R}$. The models are defined over the set of finite sets of formulas. If Γ, Δ are such sets, then $R_{\mathcal{A}}$ is defined as $\Delta R_{\mathcal{A}} \Gamma$ iff $(\Gamma - \Delta) \to \mathcal{A}$ is a theorem. Of course, $\Gamma - \Delta$ is a finite set of formulas itself, let us say $\{\mathcal{C}_1, \ldots, \mathcal{C}_n\}$, and so we let $(\Gamma - \Delta) \to \mathcal{A}$ be the formula $(\mathcal{C}_1 \to \cdots (\mathcal{C}_n \to \mathcal{A})...)$. The order of the $\mathcal{C}$'s does not matter in the set, nor does in matter in the formula, because of the permutation axiom (axiom (4)) in $\mathbf{R}$.

The "satisfiability" condition for formulas makes explicit both the role of syntax and the necessity-like modality. If p is a proposition letter, then $\Delta \vDash p$ iff $\vdash_{\mathbf{R}_\to} \Delta \to p$. For implicational formulas, we have the next definition with two reformulations.

$$\Delta \vDash \mathcal{A} \to \mathcal{B} \quad \text{iff} \quad \forall \Gamma(\Delta R_{\mathcal{A}} \Gamma \Rightarrow \Gamma \vDash \mathcal{B})$$
$$\text{iff} \quad \forall \Gamma(\vdash (\Gamma - \Delta) \to \mathcal{A} \Rightarrow \Gamma \vDash \mathcal{B})$$
$$\text{iff} \quad \forall \Theta(\vdash \Theta \to \mathcal{A} \Rightarrow \Delta \cup \Theta \vDash \mathcal{B})$$

Then a theorem—which may be viewed as a version of the soundness and completeness theorem—may be proved. It says that $\Delta \vDash \mathcal{A}$ iff $\vdash \Delta \to \mathcal{A}$, and requires that $\vdash \Delta \to (\mathcal{A} \to \mathcal{B})$ iff $\forall \Gamma(\vdash \Gamma \to \mathcal{A} \Rightarrow \vdash \Delta \cup \Gamma \vdash \mathcal{B})$, which is true in $\mathbf{R}_\to$.

From this sketch, we can get soundness and completeness for $\mathbf{R}_\to$, by a straightforward induction. However, it seems that extending it to a semantics for $\mathbf{E}_\to$ would be difficult, despite entailment being the relevant connective that was thought by Anderson and Belnap to incorporate some modality.

8   Conclusions

We gave a short overview of how four set-theoretical semantics were proposed for relevance logics within a few years of each other (Routley–Meyer, Urquhart, Fine, Maksimova), and how a related semantics was proposed by Gabbay about the same time for subjunctive conditionals. There is a clear commonality in the starting point in each case: the aim to provide a set-theoretical semantics for the conditional, and the first four all had the relevant conditional as their target. It is quite remarkable that logicians on four continents came up with comparable results. Our presentation—using similar notation and terminology—intends to help making further comparisons between the semantics. As we mentioned at the beginning of Section 2, in our view the most interesting aspect for a comparison of these semantics is whether they model the arrow from a *binary operation* or a *ternary relation*. The ternary relation worked for Routley and Meyer. The binary operation failed for Urquhart, at least for obtaining a semantics for the whole system $\mathbf{R}$ (and for Routley too in his draft). Fine and Maksimova, in effect, by composing a binary operation and a binary relation ($\leq$) to get a ternary relation achieved the "best of both." Gabbay had a different target and came up with a quaternary relation of mixed type, that includes a ternary accessibility relation, but again of mixed type and so not really like the other four semantics.

Acknowledgments

We have already thanked in the text Nuel Belnap, Kit Fine, Dov Gabbay, Larisa Maksimova, Sergeĭ Odintsov, Alasdair Urquhart and Michael McRobbie, but we want to thank them again, for this paper would not have been possible without their vital information.

We would like to express our gratitude to the audiences at the *2016 Annual Meeting of the Society for Exact Philosophy* in Coral Gables, at the *Logica 2013* conference in Heijnice, at the *UniLog 2013* conference in Rio de Janeiro and at the *Third Workshop* held in Edmonton in May 2016, where talks based on earlier versions of this paper were presented.

The research we report on in this paper is supported by an *Insight Grant* (#435–2014–0127) from the Social Sciences and Humanities Research Council of Canada.

References

[1] Alan R. Anderson. Some open problems concerning the system E of entailment. *Acta Philosophica Fennica*, 16:9–18, 1963.

[2] Alan R. Anderson and Nuel D. Belnap. *Entailment: The Logic of Relevance and Necessity*, volume I. Princeton University Press, Princeton, NJ, 1975.

[3] Alan R. Anderson, Nuel D. Belnap, and J. Michael Dunn. *Entailment: The Logic of Relevance and Necessity*, volume II. Princeton University Press, Princeton, NJ, 1992.

[4] A. Białynicki-Birula and H. Rasiowa. On the representation of quasi-Boolean algebras. *Bulletin de l'Académie Polonaise des Sciences*, 5:259–261, 1957.

[5] Katalin Bimbó. Semantics for dual and symmetric combinatory calculi. *Journal of Philosophical Logic*, 33:125–153, 2004.

[6] Katalin Bimbó. Relevance logics. In D. Jacquette, editor, *Philosophy of Logic*, volume 5 of *Handbook of the Philosophy of Science* (D. Gabbay, P. Thagard and J. Woods, eds.), pages 723–789. Elsevier (North-Holland), Amsterdam, 2007.

[7] Katalin Bimbó. Dual gaggle semantics for entailment. *Notre Dame Journal of Formal Logic*, 50:23–41, 2009.

[8] Katalin Bimbó. Some relevance logics from the point of view of relational semantics. *Logic Journal of the IGPL*, 24:268–287, 2016. O. Arieli and A. Zamansky (eds.), *Israeli Workshop on Non-classical Logics and their Applications* (IsraLog 2014).

[9] Katalin Bimbó and J. Michael Dunn. *Generalized Galois Logics. Relational Semantics of Nonclassical Logical Calculi*, volume 188 of *CSLI Lecture Notes*. CSLI Publications, Stanford, CA, 2008.

[10] Katalin Bimbó and J. Michael Dunn. Larisa Maksimova's early contributions to relevance logic. In S. Odintsov, editor, *L. Maksimova on Implication, Interpolation and Definability*, Outstanding Contributions to Logic. Springer Nature, Switzerland, 2017. (forthcoming, 28 pages).

[11] Gerald W. Charlwood. *Representations of Semi-lattice Relevance Logic*. PhD thesis, University of Toronto, Toronto, Canada, 1978.

[12] Gerald W. Charlwood. An axiomatic version of positive semi-lattice relevance logic. *Journal of Symbolic Logic*, 46:233–239, 1981.

[13] B. Jack Copeland. The genesis of possible worlds semantics. *Journal of Philosophical Logic*, 31:99–137, 2002.

[14] J. Michael Dunn. *The Algebra of Intensional Logics*. PhD thesis, University of Pittsburgh, Ann Arbor (UMI), 1966.

[15] J. Michael Dunn. The effective equivalence of certain propositions about de Morgan lattices, (abstract). *Journal of Symbolic Logic*, 32:433–434, 1967.

[16] J. Michael Dunn. A Kripke-style semantics for R-mingle using a binary accessibility relation. *Studia Logica*, 35:163–172, 1976.

[17] J. Michael Dunn. Relevance logic and entailment. In D. Gabbay and F. Guenthner, editors, *Handbook of Philosophical Logic*, volume 3, pages 117–224. D. Reidel, Dordrecht, 1st edition, 1986.

[18] J. Michael Dunn. A comparative study of various model-theoretic treatments of negation: A history of formal negation. In D. M. Gabbay and H. Wansing, editors, *What is Negation?*, pages 23–51. Kluwer, Dordrecht, 1999.

[19] Kit Fine. Models for entailment. *Journal of Philosophical Logic*, 3:347–372, 1974.

[20] Kit Fine. Completeness for the semi-lattice semantics, (abstract). *Journal of Symbolic Logic*, 41:560, 1976.

[21] Kit Fine. Semantics for quantified relevance logic. *Journal of Philosophical Logic*, 17:27–59, 1988.

[22] Kit Fine. Incompleteness for quantified relevance logics. In J. Norman and R. Sylvan, editors, *Directions in relevant logic*, volume 1 of *Reason and Argument*, pages 205–225. Kluwer Academic Publishers, Dordrecht, 1989.

[23] Dov Gabbay. A general theory of the conditional in terms of a ternary operator. *Theoria*, 38:97–104, 1972.

[24] Dov M. Gabbay. Applications of Scott's notion of consequence to the study of general binary intensional connectives and entailment. *Journal of Philosophical Logic*, 2:340–351, 1973.

[25] Dov M. Gabbay. A generalization of the concept of intensional semantics. *Philosophia*, 4:251–270, 1974.

[26] Dov M. Gabbay. *Investigations in Modal and Tense Logics with Applications to Problems in Philosophy and Linguistics*, volume 92 of *Synthese Library*. D. Reidel, Dordrecht, 1976.

[27] Robert Goldblatt. Mathematical modal logic: A view of its evolution. In Dov M. Gabbay and John Woods, editors, *Modalities in the Twentieth Century*, volume 7 of *Handbook of the History of Logic*, pages 1–98. Elsevier, Amsterdam, 2006.

[28] Robert Goldblatt. *Quantifiers, Propositions and Identity. Admissible Semantics for Quantified Modal and Substructural Logics*, volume 38 of *Lecture Notes in Logic*. Cambridge University Press for the Association for Symbolic Logic, New York, NY, 2011.

[29] Robert Goldblatt and Edwin D. Mares. An alternative semantics for quantified relevant logic. *Journal of Symbolic Logic*, 71:163–187, 2006.

[30] Andrzej Grzegorczyk. A philosophically plausible formal interpretation of intuitionistic logic. *Indagationes Mathematicae*, 26:596–601, 1964.

[31] Bjarni Jónsson and Alfred Tarski. Boolean algebras with operators, I. *American Journal of Mathematics*, 73:891–939, 1951.

[32] Bjarni Jónsson and Alfred Tarski. Boolean algebras with operators, II. *American Journal of Mathematics*, 74:127–162, 1952.

[33] Saul A. Kripke. A completeness theorem in modal logic. *Journal of Symbolic Logic*, 24:1–14, 1959.

[34] Saul A. Kripke. Semantical analysis of intuitionistic logic I. In J. N. Crossley and M. A. E. Dummett, editors, *Formal Systems and Recursive Functions. Proceedings of the Eighth Logic Colloquium*, pages 92–130, Amsterdam, 1965. North-Holland.

[35] Larisa L. Maksimova. Interpretatsiya sistem so strogoĭ implikatsieĭ. In *10th All-Union Algebraic Colloquium (Abstracts)*, page 113, Novosibirsk, 1969. [An interpretation of systems with rigorous implication].

[36] Larisa L. Maksimova. Interpretatsiya i teoremy otdeleniya dlya ischisleniĭ E i R. *Algebra i logika*, 10(4):376–392, 1971. [Interpretations and separation theorems for the calculi E and R].

[37] Larisa L. Maksimova. A semantics for the system E of entailment. *Bulletin of the Section of Logic of the Polish Academy of Sciences*, 2:18–21, 1973.

[38] Larisa L. Maksimova. Struktury s implikatsieĭ. *Algebra i logika*, 12(4):445–467, 1973. [Structures with implication].

[39] Edwin D. Mares. A star-free semantics for R. *Journal of Symbolic Logic*, 60: 579–590, 1995.

[40] Edwin D. Mares. "Four-valued" semantics for the relevant logic **R**. *Journal of Philosophical Logic*, 33:327–341, 2004.

[41] Robert K. Merton. Singletons and multiples in scientific discovery: A chapter in the sociology of science. *Proceedings of the American Philosophical Society*, 105:470–486, 1961. Reprinted in Robert K. Merton, *The Sociology of Science: Theoretical and Empirical Investigations*, University of Chicago Press, Chicago, 1973, pp. 343–370.

[42] Robert K. Meyer. From the desk of Robert K. Meyer: Semantical completeness of relevant implication. (11 pages, typescript), 1971.

[43] Robert K. Meyer. Conservative extension in relevant implication. *Studia Logica*, 31:39–46, 1973.

[44] Charles S. Peirce. *Reasoning and the logic of things. The Cambridge Conferences Lectures of 1898,* (ed. Kenneth Laine Ketner). Harvard University Press, Cambridge, MA, 1992.

[45] Dag Prawitz. *Natural Deduction: A Proof-theoretical Study.* Almqvist & Wiksell, Stockholm, 1965.

[46] Greg Restall. Negation in relevant logics. (How I stopped worrying and learned to love the Routley star). In D. M. Gabbay and H. Wansing, editors, *What is Negation?*, volume 13 of *Applied Logic Series*, pages 53–76. Kluwer Academic Publishers, Dordrecht, 1999.

[47] Richard Routley and Robert K. Meyer. The semantics of entailment – II. *Journal of Philosophical Logic*, 1:53–73, 1972.

[48] Richard Routley and Robert K. Meyer. The semantics of entailment – III. *Journal of Philosophical Logic*, 1:192–208, 1972.

[49] Richard Routley and Robert K. Meyer. A Kripke semantics for entailment, (abstract). *Journal of Symbolic Logic*, 37:42–43, 1972.

[50] Richard Routley and Robert K. Meyer. The semantics of entailment. In H. Leblanc, editor, *Truth, Syntax and Modality. Proceedings of the Temple University Conference on Alternative Semantics*, pages 199–243, Amsterdam, 1973. North-Holland.

[51] Richard Routley and Val Routley. The semantics of first degree entailment. *Noûs*, 6:335–359, 1972.

[52] Richard Routley, Robert K. Meyer, Val Plumwood, and Ross T. Brady. *Relevant Logics and Their Rivals*, volume 1. Ridgeview Publishing Company, Atascadero, CA, 1982.

[53] Dana Scott. Completeness and axiomatizability in many-valued logic. In Leon Henkin, John Addison, C. C. Chang, William Craig, Dana Scott, and Robert Vaugh, editors, *Proceedings of the Tarski Symposium*, volume 25 of *Proceedings of Symposia in Pure Mathematics*, pages 411–435, Providence, RI, 1974. American Mathematical Society.

[54] Marshall H. Stone. The theory of representations for Boolean algebras. *Transactions of the American Mathematical Society*, 40:37–111, 1936.

[55] Alasdair Urquhart. Semantics for relevant logic. *Journal of Symbolic Logic*, 37: 159–169, 1972.

[56] Alasdair Urquhart. *The Semantics of Entailment*. PhD thesis, University of Pittsburgh, 1973.

Received 16 November 2016

The Geometry of Relevant Implication

ALASDAIR URQUHART

Departments of Philosophy and Computer Science, University of Toronto,
Toronto, ON, Canada
<urquhart@cs.toronto.edu>, www.utoronto.academia.edu/AlasdairUrquhart

Abstract

This paper is a continuation of earlier work by the author on the connection between the logic **KR** and projective geometry. It contains a simplified construction of **KR** model structures; as a consequence, it extends the previous results to a much more extensive class of projective spaces and the corresponding modular lattices.

Keywords: KR, modular lattices, projective spaces, relation algebras, relevance logic

1   The Logic KR

The logic **KR** occupies a rather unusual place in the family of relevant logics. In fact, it is questionable whether it should even be classified as a relevant logic, since it is the result of adding to **R** the axiom *ex falso quodlibet*, that is to say, $(A \land \neg A) \to B$. This is of course one of the paradoxes of material implication that relevant logics were devised specifically to avoid, a paradox of consistency. The other type of paradox is a paradox of relevance, of which the paradigm case is the weakening axiom $A \to (B \to A)$. The surprising thing about **KR** is that although it contains the first type of paradox, it avoids the second, contrary to what we might at first suspect. In fact, it is a complex and highly non-trivial system. The credit for its initial investigation belongs to Adrian Abraham, Robert K. Meyer and Richard Routley [13].

The model theory for **KR** is elegantly simple. The usual ternary relational semantics for **R** includes an operation $*$ designed to deal with the truth condition for negation

$$x \vDash \neg A \ \Leftrightarrow \ x^* \nvDash A.$$

The effect of adding *ex falso quodlibet* to **R** is to identify x and x^*; this in turn has a notable effect on the ternary accessibility relation. The postulates for an **R** model structure include the following implication:

$$Rxyz \;\Rightarrow\; (Ryxz \;\&\; Rxz^*y^*).$$

The result of the identification of x and x^* is that the ternary relation in a **KR** model structure (KRms) is *totally symmetric*. In detail, a KRms $\mathcal{K} = \langle S, R, 0 \rangle$ is a 3-place relation R on a set containing a distinguished element 0, and satisfying the postulates:

1. $R0ab \;\Leftrightarrow\; a = b$;

2. $Raaa$;

3. $Rabc \;\Rightarrow\; (Rbac \;\&\; Racb)$ (total symmetry);

4. $(Rabc \;\&\; Rcde) \;\Rightarrow\; \exists f(Radf \;\&\; Rfbe)$ (Pasch's postulate).

The result of adding the weakening axiom $A \to (B \to A)$ to **R** is a collapse into classical logic. The addition of $(A \wedge \neg A) \to B$ does not result in such a collapse — but is the result a trivial or uninteresting system? This is very far from the case, as we shall see in the next section.

2   KR and Modular Lattices

Given a **KR** model structure $\mathcal{K} = \langle S, R, 0 \rangle$, we can define an algebra $\mathfrak{A}(\mathcal{K})$ as follows:

Definition 2.1. The algebra $\mathfrak{A}(\mathcal{K}) = \langle \mathcal{P}(S), \cap, \cup, \neg, \top, \bot, t, \circ \rangle$ is defined on the Boolean algebra $\langle \mathcal{P}(S), \cap, \cup, \neg, \top, \bot \rangle$ of all subsets of S, where $\top = S$, $\bot = \emptyset$, $t = \{\, 0 \,\}$, and the operator $A \circ B$ is defined by

$$A \circ B = \{\, c \colon \exists a \in A \, \exists b \in B (Rabc) \,\}.$$

The algebra $\mathfrak{A}(\mathcal{K})$ is a De Morgan monoid [1], [4] in which $a \wedge \bar{a} = \bot$, where $\bot$ is the least element of the monoid; we shall call any such algebra a **KR**-*algebra*. Hence the fusion operator $A \circ B$ is associative, commutative, and monotone. In addition, it satisfies the square-increasing property, and t is the monoid identity:

$$A \circ (B \circ C) = (A \circ B) \circ C,$$
$$A \circ B = B \circ A,$$

$$(A \subseteq B \wedge C \subseteq D) \Rightarrow A \circ C \subseteq B \circ D,$$
$$A \subseteq A \circ A,$$
$$A \circ t = A.$$

In what follows, we shall assume basic results from the theory of De Morgan monoids, referring the reader to the expositions in Anderson and Belnap [1] and Dunn and Restall [4] for more background. We have defined **KR**-algebras above as arising from De Morgan monoids by the addition of the axiom $a \wedge \bar{a} = \bot$. However, we could also have defined them using the construction of Definition 2.1, since any **KR**-algebra can be represented as a subalgebra of an algebra produced by that construction. This is not hard to prove by using the known representation theorems for De Morgan monoids — see for example [19]. **KR**-algebras are closely related to relation algebras. In fact, they can be defined as square-increasing symmetric relation algebras — for basic definitions the reader can consult the monograph [12] by Roger Maddux.

In a **KR**-algebra, we can single out a subset of the elements that form a lattice; this lattice plays a key role in the analysis of the logic **KR**.

Definition 2.2. Let $\mathfrak{A}$ be a **KR**-algebra. The family $\mathcal{L}(\mathfrak{A})$ is defined to be the elements of $\mathfrak{A}$ that are $\geq t$ and idempotent, that is to say, $a \in \mathcal{L}(\mathfrak{A})$ if and only if $a \circ a = a$ and $t \leq a$. If $\mathcal{K}$ is a **KR** model structure, then we define $\mathcal{L}(\mathcal{K})$ to be $\mathcal{L}(\mathfrak{A}(\mathcal{K}))$.

The following lemma provides a useful characterization of the elements of $\mathcal{L}(\mathfrak{A})$; it is based on some old observations of Bob Meyer.

Lemma 2.3. *Let $\mathfrak{A}$ be a **KR**-algebra. Then the following conditions are equivalent:*

1. $a \in \mathcal{L}(\mathfrak{A})$;

2. $a = (a \rightarrow a)$;

3. $\exists b \, [a = (b \rightarrow b)]$.

Proof. ($\mathbf{1} \Rightarrow \mathbf{2} \Rightarrow \mathbf{3}$): Since $t \leq a$, we have $t \leq (a \rightarrow a) \rightarrow a$, $t \circ (a \rightarrow a) \leq a$, hence $(a \rightarrow a) \leq a$. Since $a \circ a \leq a$, $a \leq (a \rightarrow a)$, so $a = (a \rightarrow a)$, proving the second and hence the third condition.

($\mathbf{3} \Rightarrow \mathbf{1}$): First, we have $t \leq (b \rightarrow b) = a$. Second, $(b \rightarrow b) \leq (b \rightarrow b) \rightarrow (b \rightarrow b)$, so $(b \rightarrow b) \circ (b \rightarrow b) \leq (b \rightarrow b)$, that is to say, $a \circ a \leq a$, so $a \circ a = a$. $\square$

If $\mathcal{K} = \langle S, R, 0 \rangle$ is a **KR** model structure, then a subset A of S is a *linear subspace* if it satisfies the condition

$$(a, b \in A \wedge Rabc) \Rightarrow c \in A.$$

A lattice is *modular* if it satisfies the implication

$$x \geq z \Rightarrow x \wedge (y \vee z) = (x \wedge y) \vee z.$$

For background on modular lattice theory, the reader can consult the texts of Birkhoff [2] or Grätzer [7].

We require a few basic lattice-theoretic definitions here. A *chain* in a lattice L is a totally ordered subset of L; the length of a finite chain C is $|C| - 1$. A chain C in a lattice L is *maximal* if for any chain D in L, if $C \subseteq D$ then $C = D$. If L is a lattice, $a, b \in L$ and $a \leq b$, then the *interval* $[a, b]$ is defined to be the sublattice $\{c : a \leq c \leq b\}$.

Let L be a lattice with least element 0. We define the *height* function: for $a \in L$, let $h(a)$ denote the length of a longest maximal chain in $[0, a]$ if there is a finite longest maximal chain; otherwise put $h(a) = \infty$. If L has a largest element 1, and $h(1) < \infty$, then L has *finite height*.

Let L be a modular lattice with 0 of finite height. Then for $a \in L$, $h(a)$ is the length of *any* maximal chain in $[0, a]$. In addition, the height function in L satisfies the condition

$$h(a) + h(b) = h(a \wedge b) + h(a \vee b),$$

for all $a, b \in L$. For a lattice of finite height, this last condition is equivalent to modularity. These results are proved in the text of Grätzer [7, Chapter IV, §2].

Lemma 2.4. *If $\mathcal{K}$ is a **KR** model structure, then the elements of $\mathcal{L}(\mathcal{K})$ are exactly the non-empty linear subspaces of $\mathcal{K}$.*

Proof. The lemma follows from the definition of $A \circ B$ and the fact that $Raa0$ and $Raaa$ hold in any **KR** model structure. $\square$

Theorem 2.5. *If $\mathfrak{A}$ is a **KR**-algebra, then $\mathcal{L}(\mathfrak{A})$, ordered by containment, forms a modular lattice, with least element t, and the lattice operations of join and meet defined by $a \wedge b$ and $a \circ b$.*

Proof. The fact that $\mathcal{L}(\mathfrak{A})$ forms a lattice, with $\wedge$ as the lattice meet and $\circ$ as the lattice join, can be proved from the basic properties of De Morgan lattices.

We now prove modularity; in the following computation, we use juxtaposition ab for meet $a \wedge b$, and $\bar{a}$ for the Boolean complement. Note that $a \vee b$ is the extensional (Boolean) join, not the lattice join in $\mathcal{L}(\mathfrak{A})$. If $a \geq c$, then

$$
\begin{aligned}
a(b \circ c) &= a[(ba \vee b\bar{a}) \circ c] \\
&= a[(ba \circ c) \vee (b\bar{a} \circ c)] \\
&\leq a[(ba \circ c) \vee (\bar{a} \circ a)] \\
&= a(ba \circ c) \vee a\bar{a} \\
&\leq ab \circ c.
\end{aligned}
$$

The opposite inequality $ab \circ c \leq a(b \circ c)$ follows from the lattice properties of $\mathcal{L}(\mathfrak{A})$, so $a(b \circ c) = ab \circ c$. In the fourth line above, the equation $\bar{a} \circ a = \bar{a}$ follows from Lemma 2.3, since for $a \in \mathcal{L}(\mathfrak{A})$, $a = a \rightarrow a$, so $\bar{a} = \overline{a \rightarrow a} = a \circ \bar{a}$. □

Theorem 2.5 is closely related to Theorem 2.18 of Chin and Tarski [3], [14, p. 268]. The result also appears in a paper of Jónsson [9] from 1959. The result of Chin and Tarski appears somewhat more general than Theorem 2.5 since it does not rely on the square-increasing postulate. However, an examination of the proof above shows that this postulate is not used in the proof, and in fact, the calculation goes through in a more general class of algebras — see, for example, Lemma 5.5 of [18].

The preceding theorem shows that there is a modular lattice canonically associated with any **KR**-algebra. It is natural to ask the question: how general is this construction? That is to say, which modular lattices arise in this way? In earlier papers [15, 16, 17], I provided a partial answer to this question by showing that a very large family of modular lattices, closely connected with classical projective geometries, can be represented as the lattices $\mathcal{L}(\mathcal{K})$ associated with **KR** model structures. This construction made possible the solution of some long-standing problems in the area of relevance logic, particularly those of decidability and interpolability.

The lattices arising from projective spaces, however, are of a rather special type, and the construction given in my earlier work does not make clear whether more general modular lattices can be represented. In this section, I give a very simple construction for **KR** model structures showing that *any* modular lattice can be represented as a sublattice of a lattice $\mathcal{L}(\mathcal{K})$. The earlier representation of geometric lattices can be obtained as a direct corollary of this construction, as is shown in Section 4.

Definition 2.6. Let L be a lattice with least element 0. Define a ternary relation R on the elements of L by:

$$
Rabc \iff a \vee b = b \vee c = a \vee c,
$$

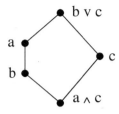

Figure 1: $\mathfrak{N}_5$: The five-element non-modular lattice

and let $\mathcal{K}(L)$ be $\langle L, R, 0 \rangle$.

Theorem 2.7. *$\mathcal{K}(L)$ is a **KR** model structure if and only if L is modular.*

Proof. The first three postulates for a **KR** model structure follow immediately from the definition of R, using only the fact that L is a lattice. Now assume that L is modular; we need to verify the last postulate (the Pasch postulate). Assume that $Rabc$ and $Rcde$, that is to say, $a \vee b = b \vee c = a \vee c$ and $c \vee d = c \vee e = d \vee e$. Define $f = (a \vee d) \wedge (b \vee e)$. We need to show that $Radf$ and $Rfbe$, that is to say, $a \vee f = a \vee d = d \vee f$ and $b \vee f = b \vee e = e \vee f$. We compute

$$
\begin{align}
a \vee f &= a \vee [(a \vee d) \wedge (b \vee e)] \tag{1}\\
&= (a \vee d) \wedge (a \vee b \vee e) \tag{2}\\
&= (a \vee d) \wedge (a \vee c \vee e) \tag{3}\\
&= (a \vee d) \wedge (a \vee c \vee d) \tag{4}\\
&= a \vee d, \tag{5}
\end{align}
$$

where the equality (2) follows by modularity. The remaining three equalities follow by an exactly symmetrical argument.

For the converse implication, assume that the Pasch postulate holds, but L is not modular. Then L has a sublattice isomorphic to $\mathfrak{N}_5$, the five-element nonmodular lattice (see Figure 1). In $\mathfrak{N}_5$, we have $R(a, c, a \vee c)$ and $(b \vee c, b, c)$. Since $a \vee c = b \vee c$, it follows by the Pasch postulate that there is an f so that $R(a, b, f)$ and $R(f, c, c)$. Then $f \leq a \vee f = a \vee b = a$, and $f \leq f \vee c = c \vee c = c$, so $f \leq a \wedge c$. Thus $b \vee f \leq b \vee (a \wedge c)$; hence, $a \leq a \vee b = b \vee f \leq b \vee (a \wedge c) = b$, contradicting $a > b$. $\quad\square$

In Definition 2.6, if a, b and c are distinct points in a projective space, then $Rabc$ holds if and only if the three points are collinear. Hence, the defined ternary relation can be considered as a generalized notion of collinearity that applies to any elements in a modular lattice.

Definition 2.8. If L is a lattice, then an *ideal* of L is a non-empty subset I of L such that

1. If $a, b \in I$ then $a \vee b \in I$;

2. If $b \in I$ and $a \leq b$, then $a \in I$.

The family of ideals of a lattice L, ordered by containment, forms a complete lattice $I(L)$. The original lattice L is embedded in $I(L)$ by mapping an element $a \in L$ into the *principal ideal* containing a, $(a] = \{\, b \colon b \leq a \,\}$. It is easy to verify that the mapping $a \longmapsto (a]$ is a lattice isomorphism between L and a sublattice of $I(L)$.

Theorem 2.9. *Let L be a modular lattice with least element 0, and $\mathcal{K}(L) = \langle L, R, 0 \rangle$ the **KR** model structure constructed from L. Then $\mathcal{L}(\mathcal{K}(L))$ is identical with the lattice of ideals of L.*

Proof. We need to show that the non-empty linear subspaces of $\mathcal{K}(L)$ are exactly the ideals of L. Let $S \subseteq L$ be a non-empty linear subspace of L. If $a, b \in S$, and $a \vee b = c$, then $Rabc$, so $c \in S$. If $a \leq b$ and $b \in S$, then $Rbba$, so that $a \in S$, showing that S is an ideal. Conversely, assume that S is an ideal of L. By definition, S is non-empty. If $a, b \in S$ and $Rabc$, then $a \vee b = a \vee c \in S$, so $c \in S$, since $c \leq a \vee c$. $\square$

Corollary 2.10. *Any modular lattice of finite height (hence any finite modular lattice) is representable as $\mathcal{L}(\mathcal{K})$, for some **KR** model structure $\mathcal{K}$. In addition, any modular lattice is representable as a sublattice of $\mathcal{L}(\mathcal{K})$, for some **KR** model structure $\mathcal{K}$.*

The preceding theorem and corollary constitute a general representation theory for modular lattices. Faigle and Herrmann [5] provided a related representation theorem for modular lattices of finite height. They define a set of axioms for a projective geometry as an incidence structure on partially ordered sets of "points" and "lines," and show that every modular lattice of finite length is isomorphic to the lattice of linear subsets of some finite-dimensional projective geometry.

3   Anticipations of the Main Construction

The construction of Definition 2.6 is very simple and natural, and it is not surprising that it has occurred earlier in the mathematical literature. In a paper of 1959 [9, p. 463], Bjarni Jónsson asked whether every modular lattice is isomorphic to a lattice of commuting equivalence elements of some relation algebra. His question was answered affirmatively by Roger Maddux in a paper published in 1981 [11]. The construction that he used to answer Jónsson's question is the same as that of Definition 2.6; his paper also contains a version of Theorem 2.9. Maddux's monograph on relation algebras also describes the construction [12, pp. 501–502].

A surprising anticipation of Maddux's construction can be found in a paper by D. K. Harrison [8] published in 1979. Harrison defines a *Pasch geometry* (also known as a *multigroup*) to be a set A with a distinguished element e and a ternary relation Δ defined on A satisfying four postulates. His first postulate is:

> For each $a \in A$, there exists a unique $b \in A$ with $(a, b, e) \in \Delta$; denote b by $a^{\#}$.

In Harrison's terminology, a **KR** model structure is a Pasch geometry in which $a^{\#} = a$, for all $a \in A$. Proposition 8 of his paper shows that if the construction of Definition 2.6 is applied to a lattice L with least element e, then the resulting structure (L, Δ, e) is a Pasch geometry if and only if L is modular. The second part of Theorem 2.7 above is adapted from Harrison's proof of his Proposition 8.

4   KR and Projective Spaces

In an earlier paper [15], I showed that there is a close connection between **KR** and projective geometry. More precisely, I proved that every lattice arising from a broad class of projective spaces can be represented as $\mathcal{L}(\mathcal{K})$ for some **KR** model structure $\mathcal{K}$. The proof proceeded by a direct construction of a model structure from a projective space; the construction is essentially the same as that given earlier by Roger C. Lyndon [10] to produce examples of non-representable relation algebras. In the 1983 paper [15], the construction is only sketched; my paper on interpolation from 1993 [17] contains a full exposition.

The present section gives a new proof of the earlier results, based on Theorems 2.7 and 2.9. Before giving the proof, we need some definitions and results relating to projective spaces and the lattices that arise from them; they are adapted from the text of Grätzer [7, Chapter IV, §5].

Definition 4.1. Let A be a set and L a collection of subsets of A. The pair $\langle A, L \rangle$ is a projective space iff the following properties hold:

1. Every $l \in L$ has at least two elements;

2. For any two distinct $p, q \in A$, there is exactly one $l \in L$ so that $p, q \in l$;

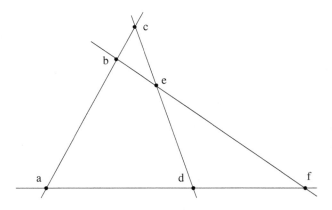

Figure 2: The Pasch Postulate

3. Pasch Postulate: For $a, b, c, d, e \in A$ and $l_1, l_2 \in L$ satisfying $a, b, c \in l_1$ and $c, d, e \in l_2$, there exist $f \in A$ and $l_3, l_4 \in L$ satisfying $a, d, f \in l_3$ and $b, e, f \in l_4$.

We call the members of A *points* and those of L *lines*. For $p, q \in A$, $p \neq q$, let $p + q$ denote the unique line containing p and q; if $p = q$, set $p + q = \{p\}$. Apart from degenerate cases, the Pasch Postulate states that if a line $b + e$ intersects two sides, $a + c$ and $c + d$ of the triangle $\{a, c, d\}$, then it intersects the third side, $a + d$; see Figure 2.

If L is a lattice with least element 0, then $a \in L$ is an *atom* if $h(a) = 1$. An element a of a complete lattice L is *compact* if and only if $a \leq \bigvee X$ for some $X \subseteq L$ implies that $a \leq \bigvee Y$ for some finite $Y \subseteq X$.

Definition 4.2. A lattice L is a modular geometric lattice iff L is complete, every element of L is a join of atoms, all atoms are compact, and L is modular.

A subset X of the set of atoms of a projective space is a *linear subspace* iff $p + q \subseteq X$ whenever $p, q \in X$.

Theorem 4.3. *The linear subspaces of a projective space form a modular geometric lattice, where $A \wedge B = A \cap B$ and*

$$A \vee B = \bigcup \{a + b \colon a \in A, b \in B\}.$$

Proof. See Grätzer [7, Chapter IV, §5, Theorem 5]. □

The construction of a modular geometric lattice from a projective space can be reversed. Given such a lattice L, define a geometry $G(L)$ by defining the points to be the set of atoms of L, while the lines are the elements of L with height 2.

Theorem 4.4. *If L is a modular geometric lattice, then G(L) is a projective space, and L is isomorphic to the lattice of linear subspaces of G(L).*

Proof. See Grätzer [7, Chapter IV, §5]. □

The two preceding theorems show that there is an exact correspondence between projective spaces and modular geometric lattices.

Lemma 4.5. *Let L be a modular geometric lattice. Then the set F of elements of L of finite height is an ideal of L, and every element of F is a finite join of atoms. L is isomorphic to I(F), the lattice of all ideals of F.*

Proof. See Grätzer [7, Corollary 2, p. 179]. □

Theorem 4.6. *Let L be a modular geometric lattice. Then L is isomorphic to $\mathcal{L}(\mathcal{K})$, for some **KR** model structure $\mathcal{K}$.*

Proof. Let F be the family of elements of L of finite height. Then F forms a modular lattice, so we can construct a **KR** model structure $\mathcal{K}(F)$ by Definition 2.6. By Theorem 2.9 and Lemma 4.5, L is isomorphic to $\mathcal{L}(\mathcal{K}(F))$. □

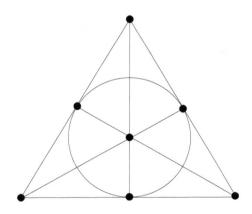

Figure 3: The Fano Plane

The preceding theorem includes the results of [15], but in fact goes further, because the earlier results omitted certain projective spaces and the corresponding geometries. In particular, the construction of **KR** model structures in my 1984 paper required that the underlying projective spaces have at least four points on each line (the construction of Lyndon [10] has the same restriction). This restriction means that the important special case of geometries constructed from the two-element field are not represented. In particular, the best known example of a finite geometry, the Fano plane (Figure 3), is not included in the family of lattices represented in the construction of [15]. The present construction is not only much simpler, but includes these geometries in its scope.

Roger Maddux has reminded me of the fact that Lyndon does treat the case of geometries over the two-element field, though only as an aside [10, p. 24]. The difficulty in the case of the two-element field arises from contraction. If we assume the second postulate in the definition of a **KR** model structure, *Raaa*, then there are not enough points on a line to validate the Pasch postulate. If we omit this postulate, though, we can construct models for contraction-free logics, following Lyndon's method.

5 An Application, a Problem and Acknowledgments

The simple construction of this paper indicates that further results about **KR** and other relevant logics can very likely be obtained by adapting ideas from the well developed and deep theory of modular lattices. As a minor application illustrating these possibilities, we show that if $\mathfrak{A}(G)$ is a **KR**-algebra freely generated by G, then there is a set $G^* \subseteq \mathcal{L}(\mathfrak{A}(G))$ so that G^* freely generates a sublattice of $\mathcal{L}(\mathfrak{A}(G))$. No doubt other such applications can be found, and we include as an open problem another possible use for the construction.

Theorem 5.1. *Let $\mathfrak{A}$ be a **KR**-algebra, and G a subset that freely generates $\mathfrak{A}$. If $G^* = \{\, a \to a \colon a \in G \,\}$, then G^* freely generates a sublattice of $\mathcal{L}(\mathfrak{A})$.*

Proof. Let L be the sublattice of $\mathcal{L}(\mathfrak{A})$ generated by G^*. If M is a modular lattice with least element 0, and $f \colon G^* \longmapsto M$ a function from G^* to M, then we need to show that f can be extended to a lattice homomorphism from L to M.

Using Definition 2.6, we can define the **KR** model structure $\mathcal{K}(M)$, and hence by Definition 2.1, the **KR**-algebra $\mathfrak{B} = \mathfrak{A}(\mathcal{K}(M))$. For $a \in G$, define $g(a) = f(a \to a)$. Since G freely generates $\mathfrak{A}$, g can be extended to a homomorphism h from $\mathfrak{A}$ to $\mathfrak{B}$. By Theorem 2.9, $\mathcal{L}(\mathfrak{B})$ is identical with the lattice of ideals of M, so that we can identify M with a sublattice of $\mathcal{L}(\mathfrak{B})$ by the embedding $a \rightleftharpoons (a]$ that maps an element $a \in M$ into the principal ideal generated by a.

For $a \in G^*$, let $a = b \to b$, for $b \in G$. Then

$$h(a) = h(b \to b) = h(b) \to h(b) = g(b) \to g(b)$$
$$= f(b \to b) \to f(b \to b) = f(a) \to f(a) = f(a).$$

Thus, h restricted to L is a lattice homomorphism from L to M extending f, showing that G^* freely generates L. $\square$

Corollary 5.2. *In the logic **KR**, there are infinitely many distinct formulas built from the formulas $p \to p, q \to q, r \to r$ and $s \to s$ using only the connectives $\wedge$ and $\circ$.*

Proof. Theorem 5.1 shows that the formulas $p \to p, q \to q, r \to r$ and $s \to s$ generate an algebra of formulas isomorphic to the free modular lattice on four generators. This algebra is infinite [2, p. 64]. $\square$

Beth's theorem equating implicit and explicit definability is known to fail in many of the well known relevant logics such as **R**. However, the proof of this result [20] depends on the fact that classical Boolean negation is missing from these logics, and so does not apply to **KR**.

Problem 5.3. Does Beth's definability theorem hold in the logic **KR**?

The construction of this paper suggests a way to attack this problem. The algebraic counterpart of the Beth definability theorem in a variety of algebras is the property that epimorphisms are surjective. Ralph Freese [6, Theorem 3.3] has shown that this property fails in the category of modular lattices and lattice homomorphisms. Consequently, a possible strategy to attack this problem would be to adapt Freese's proof to the algebra of **KR**.

This paper was presented at a special session on algebraic logic (organized by Nick Galatos and Peter Jipsen) at the regional meeting of the AMS in Denver, October 2016. At my talk, Roger Maddux told me of his earlier work on the construction of §2; I am indebted to him for his comments on this paper, and for providing the list of references in §3.

References

[1] Alan R. Anderson and Nuel D. Belnap. *Entailment: The Logic of Relevance and Necessity*, volume I. Princeton University Press, Princeton, NJ, 1975.

[2] Garrett Birkhoff. *Lattice Theory*, volume 25 of *AMS Colloquium Publications*. American Mathematical Society, Providence, RI, 3rd edition, 1967.

[3] L. H. Chin and A. Tarski. Distributive and modular laws in the arithmetic of relation algebras. *Univ. Calif. Publ. Math.*, 1:341–384, 1951.

[4] J. Michael Dunn and Greg Restall. Relevance logic. In D. Gabbay and F. Guenthner, editors, *Handbook of Philosophical Logic*, volume 6, pages 1–128. Kluwer, Amsterdam, 2nd edition, 2002.

[5] Ulrich Faigle and Christian Herrmann. Projective geometry on partially ordered sets. *Transactions of the American Mathematical Society*, 266:319–332, 1981.

[6] Ralph Freese. The variety of modular lattices is not generated by its finite members. *Transactions of the American Mathematical Society*, 255:277–300, 1979.

[7] George Grätzer. *General Lattice Theory*. Academic Press, 1978.

[8] David Kent Harrison. Double coset and orbit spaces. *Pacific Journal of Mathematics*, 60:451–491, 1979.

[9] Bjarni Jónsson. Representation of lattices and relation algebras. *Transactions of the American Mathematical Society*, 92:449–464, 1959.

[10] Roger C. Lyndon. Relation algebras and projective geometries. *Michigan Mathematical Journal*, 8:21–28, 1961.

[11] Roger D. Maddux. Embedding modular lattices into relation algebras. *Algebra Universalis*, 12:242–246, 1981.

[12] Roger D. Maddux. *Relation Algebras*, volume 150 of *Studies in Logic and the Foundations of Mathematics*. Elsevier, Amsterdam, 2006.

[13] Richard Routley, Robert K. Meyer, Val Plumwood, and Ross T. Brady. *Relevant Logics and Their Rivals*, volume 1. Ridgeview Publishing Company, Atascadero, CA, 1982.

[14] Alfred Tarski. *Collected Papers*, volume 3. Birkhäuser, Basel, 1986.

[15] Alasdair Urquhart. Relevant implication and projective geometry. *Logique et Analyse*, 103–104:345–357, 1983.

[16] Alasdair Urquhart. The undecidability of entailment and relevant implication. *Journal of Symbolic Logic*, 49:1059–1073, 1984.

[17] Alasdair Urquhart. Failure of interpolation in relevant logics. *Journal of Philosophical Logic*, 22:449–479, 1993.

[18] Alasdair Urquhart. Decision problems for distributive lattice-ordered semigroups. *Algebra Universalis*, 33:399–418, 1995.

[19] Alasdair Urquhart. Duality for algebras of relevant logics. *Studia Logica*, 56:263–276, 1996.

[20] Alasdair Urquhart. Beth's definability theorem in relevant logics. In E. Orlowska, editor, *Logic at Work: Essays Dedicated to the Memory of Helena Rasiova*, volume 24 of *Studies in Fuzziness and Soft Computing*, pages 229–234. Physica Verlag, Heidelberg, 1999.

Received 30 October 2016

Cut-elimination in RM Proved Semantically

Arnon Avron*

School of Computer Science, Tel Aviv University, Tel Aviv, Israel

<aa@cs.tau.ac.il>

Abstract

A purely semantic proof is presented for the admissibility of the Cut rule in the hypersequential Gentzen-type system for **RM**.

Keywords: cut admissibility, hypersequents, relevant logics, semi-relevance, Sugihara matrices

1 Introduction

There are essentially two main approaches to proving cut-elimination for a given Gentzen-type system G. One is Gentzen's original proof-theoretic method (from [12]). The other is semantic; one shows that the system without the Cut rule is complete for some semantics for which the full system (that is, the system with the Cut rule) is sound. The first method has the advantage that it usually provides an algorithm for converting a proof with cuts to a cut-free proof of the same sequent. In contrast, the semantic method only guarantees the admissibility of Cut, but usually does not provide a clue how to extract a cut-free proof from a given proof that contains cuts. On the other hand, the proof-theoretic method is notoriously difficult to verify, and very error-prone. Due to his very bad experience with mistakes in syntactic proofs of cut-elimination (of himself and of others), the author of this paper tends in recent years to trust semantic proofs of cut-elimination much more than he trusts syntactic ones, to the point in which he simply does not believe a cut-elimination theorem for some system if only syntactic proofs are known for it.

---

*Supported by The Israel Science Foundation under grant agreement no. 817/15

What was said above is particularly true for hypersequential Gentzen-type systems (that is, Gentzen-type systems which employ hypersequents rather than ordinary sequents).[1] A good example is provided by the original proof of cut-elimination for GRM (the hypersequential system for **RM**) in [3]. That proof is extremely complicated (and its author himself, who is thirty years older now, cannot read it anymore), practically impossible to be fully checked, and at best *looks* correct (as the referee of that paper has said in his/her review). To be confident of the validity of this theorem, this paper presents a new, semantic proof of it. Though one cannot claim it is very simple, its main advantage is that it leaves no gaps, and can be (and has been) fully checked to the last detail.

2   The Logic RM and its Characteristic Matrix

The semi-relevant logic **RM** was introduced by Dunn and McCall (see e.g., [1, 11] for more information about **RM**, its history and motivation). It was later extensively investigated by Meyer and Dunn. As noted in [11], it is "by far the best understood of the Anderson–Belnap style systems." A recent detailed description of **RM** and its properties can be found in [9]. We assume that the reader is acquainted with **RM**, and so we review here only material which is needed later in this paper.

In the sequel, $\mathcal{L}$ denotes a propositional language. The set of well-formed formulas of $\mathcal{L}$ is denoted by $\mathcal{W}(\mathcal{L})$, and φ, ψ, σ vary over its elements. $\mathcal{T}$ varies over theories of $\mathcal{L}$ (where by a 'theory' we mean simply a subset of $\mathcal{W}(\mathcal{L})$). We denote by Atoms the set of atomic formulas of $\mathcal{L}$, and by Atoms(φ) (Atoms($\mathcal{T}$)) the set of atomic formulas that appear in φ (in formulas of $\mathcal{T}$).

Definition 2.1. Let $\mathcal{L}$ be a propositional language.

- A *matrix* for $\mathcal{L}$ is a triple $\mathcal{M} = \langle \mathcal{V}, \mathcal{D}, \mathcal{O} \rangle$, where

  - $\mathcal{V}$ is a non-empty set of truth values;
  - $\mathcal{D}$ is a non-empty proper subset of $\mathcal{V}$ (the *designated* elements of $\mathcal{V}$);
  - $\mathcal{O}$ is a function that associates an n-ary function $\tilde{\diamond}_{\mathcal{M}} \colon \mathcal{V}^n \to \mathcal{V}$ with every n-ary connective $\diamond$ of $\mathcal{L}$.

  We say that $\mathcal{M}$ is *(in)finite*, if so is $\mathcal{V}$.

---

[1]This type of systems was first introduced in [15], and independently in [3]. Since then the framework of hypersequential calculi has been applied to many logics of different sorts ([7]). In particular, it is the main framework for the proof theory of fuzzy logics ([14]).

- Let $\mathcal{M} = \langle \mathcal{V}, \mathcal{D}, \mathcal{O} \rangle$ be a matrix for $\mathcal{L}$. An $\mathcal{M}$-*valuation* for $\mathcal{L}$ is a function $\nu : \mathcal{W}(\mathcal{L}) \to \mathcal{V}$ such that for every n-ary connective $\diamond$ of $\mathcal{L}$ and every $\psi_1, \ldots, \psi_n$ in $\mathcal{W}(\mathcal{L})$, $\nu(\diamond(\psi_1, \ldots, \psi_n)) = \tilde{\diamond}_{\mathcal{M}}(\nu(\psi_1), \ldots, \nu(\psi_n))$.

- An $\mathcal{M}$-valuation ν is an $\mathcal{M}$-*model* of a formula ψ, or ν $\mathcal{M}$-*satisfies* ψ (notation: $\nu \vDash_{\mathcal{M}} \psi$), if $\nu(\psi) \in \mathcal{D}$. We say that ν is an $\mathcal{M}$-*model* of a theory $\mathcal{T}$ (notation: $\nu \vDash_{\mathcal{M}} \mathcal{T}$), if it is an $\mathcal{M}$-model of every element of $\mathcal{T}$.

- Let $\mathcal{M}$ be a matrix for $\mathcal{L}$. $\vdash_{\mathcal{M}}$, the consequence relation that is induced by $\mathcal{M}$, is defined by $\mathcal{T} \vdash_{\mathcal{M}} \psi$ if every $\mathcal{M}$-model of $\mathcal{T}$ is an $\mathcal{M}$-model of ψ. We shall denote by $\mathbf{L}_{\mathcal{M}}$ the logic $\langle \mathcal{L}, \vdash_{\mathcal{M}} \rangle$ which is induced by $\mathcal{M}$.

Definition 2.2.

- A *Sugihara chain* is a triple $S = \langle \mathcal{V}, \leq, - \rangle$ such that $\mathcal{V}$ has at least two elements, $\leq$ is a linear order on $\mathcal{V}$, and $-$ is an involution for $\leq$ on $\mathcal{V}$ (i.e., for every $a, b \in \mathcal{V}$, $--a = a$ and $-b \leq -a$ whenever $a \leq b$).

- Let $S = \langle \mathcal{V}, \leq, - \rangle$ be a Sugihara chain, and let $a, b \in \mathcal{V}$.

  - $a < b$ if $a \leq b$ and $a \neq b$.

  - $|a| = \max(-a, a)$.

  - $a \preceq_{+} b$ iff either $|a| < |b|$, or $|a| = |b|$ and $a < b$.

- Let $S = \langle \mathcal{V}, \leq, - \rangle$ be a Sugihara chain. The *Sugihara matrix* $\mathcal{M}(S)$ is the matrix $\langle \mathcal{V}, \mathcal{D}, \mathcal{O} \rangle$ for the language $\mathcal{L}_{RM} = \{\neg, \wedge, \vee, \to\}$ in which:

  - $\mathcal{D} = \{a \in \mathcal{V} : a \geq -a\}$;

  - The functions that $\mathcal{O}$ associates with the connectives of $\mathcal{L}_{RM}$ are the following:

    * $\tilde{\neg}(a) = -a$;
    * $a \, \tilde{\wedge} \, b = \min(a, b)$ and $a \, \tilde{\vee} \, b = \max(a, b)$;
    * $a \, \tilde{\to} \, b = \max_{\preceq_{+}}(-a, b)$.

Definition 2.3. $\mathcal{M}(\mathbb{Z})$ is the Sugihara matrix $\mathcal{M}(\langle \mathbb{Z}, \leq, - \rangle)$, where $\mathbb{Z}$ is the set of the integers, $\leq$ is the standard order relation on $\mathbb{Z}$, and $-$ is the standard involution on $\mathbb{Z}$. (Note that in $\mathcal{M}(\mathbb{Z})$, $\mathcal{D} = \{n \in \mathbb{Z} : n \geq 0\}$.)

Note 2.4. The intensional disjunction $+$ is frequently defined in relevance logics by $\varphi + \psi =_{\mathrm{df}} \neg \varphi \to \psi$. Obviously, in $\mathcal{M}(\mathbb{Z})$ $a \, \tilde{+} \, b = \max_{\preceq_{+}}(a, b)$. It is also easy to see that the above definition of $\tilde{\rightarrow}$ in Sugihara matrices is equivalent to the following original definition of Sugihara ([16]):

$$a \tilde{\rightarrow} b = \begin{cases} \max(-a, b) & \text{if } a \leq b, \\ \min(-a, b) & \text{if } a > b. \end{cases}$$

It easily follows that $a \tilde{\rightarrow} b \in \mathcal{D}$ iff $a \leq b$.

The following key theorem has been proved by Meyer. (See [9] for a proof.)

Theorem 2.5. *If $\mathcal{T}$ is a finite theory then $\mathcal{T} \vdash_{\mathbf{RM}} \varphi$ iff $\mathcal{T} \vdash_{\mathcal{M}(\mathbb{Z})} \varphi$. In particular, $\mathcal{M}(\mathbb{Z})$ is weakly characteristic for $\mathbf{RM}$.*

The following observation proved in [9] will also be useful in the sequel.

Proposition 2.6. *Let $S = \langle \mathcal{V}, \leq, - \rangle$ be a Sugihara chain, and let $a \in \mathcal{V}$. Suppose that ν_1 and ν_2 are valuations in $\mathcal{M}(S)$ such that the following holds for every atom p: if $\max\{ |\nu_1(p)|, |\nu_2(p)| \} \geq a$, then $\nu_1(p) = \nu_2(p)$. Then for every formula φ, if $\max\{ |\nu_1(\varphi)|, |\nu_2(\varphi)| \} \geq a$, then $\nu_1(\varphi) = \nu_2(\varphi)$.*

Another important fact from [9] that we need (originally proved in [2]) is that **RM** has an implication respecting the classical–intuitionistic deduction theorem.

Definition 2.7. $\varphi \supset \psi =_{\mathrm{df}} (\varphi \rightarrow \psi) \vee \psi$.

Proposition 2.8. $\mathcal{T} \vdash_{\mathbf{RM}} \varphi \supset \psi$ iff $\mathcal{T}, \varphi \vdash_{\mathbf{RM}} \psi$.

3 The System GRM

Definition 3.1. Let $\mathcal{L}$ be a propositional language. By a *sequent* for $\mathcal{L}$, we mean in this paper a construct of the form $\Gamma \Rightarrow \Delta$, where Γ, Δ denote finite *sets* of formulas. A *hypersequent* for $\mathcal{L}$ is a non-empty, finite *multiset* of sequents.

Notation 3.2. In the sequel, Γ, Δ, and Σ vary over finite sets of formulas, s varies over sequents, and G varies over hypersequents. We denote by $s_1 \mid s_2 \mid \cdots \mid s_n$ the hypersequent (that is, a multiset) whose elements are $s_1, s_2, \ldots, s_n$.

Figure 1 presents a hypersequential Gentzen-type proof system GRM for **RM**. Γ, Δ denote finite *sets* of formulas. The short names [EC], [EW], [Mi], and [Sp] stand for External Contraction, External Weakening, Mingle, and Splitting, respectively.

The next proposition provides several useful properties of GRM.

Axioms: $G \mid \psi \Rightarrow \psi$

Logical rules:

$$[\neg \Rightarrow] \quad \frac{G \mid \Gamma \Rightarrow \Delta, \varphi}{G \mid \neg\varphi, \Gamma \Rightarrow \Delta} \qquad\qquad [\Rightarrow \neg] \quad \frac{G \mid \varphi, \Gamma \Rightarrow \Delta}{G \mid \Gamma \Rightarrow \Delta, \neg\varphi}$$

$$[\rightarrow \Rightarrow] \quad \frac{G \mid \Gamma_1 \Rightarrow \Delta_1, \varphi \qquad G \mid \psi, \Gamma_2 \Rightarrow \Delta_2}{G \mid \Gamma_1, \Gamma_2, \varphi \rightarrow \psi \Rightarrow \Delta_1, \Delta_2} \qquad [\Rightarrow \rightarrow] \quad \frac{G \mid \Gamma, \varphi \Rightarrow \Delta, \psi}{G \mid \Gamma \Rightarrow \Delta, \varphi \rightarrow \psi}$$

$$[\wedge \Rightarrow] \quad \frac{G \mid \Gamma, \varphi \Rightarrow \Delta \qquad G \mid \Gamma, \psi \Rightarrow \Delta}{G \mid \Gamma, \varphi \wedge \psi \Rightarrow \Delta \qquad G \mid \Gamma, \varphi \wedge \psi \Rightarrow \Delta}$$

$$[\Rightarrow \wedge] \quad \frac{G \mid \Gamma \Rightarrow \Delta, \varphi \qquad G \mid \Gamma \Rightarrow \Delta, \psi}{G \mid \Gamma \Rightarrow \Delta, \varphi \wedge \psi}$$

$$[\vee \Rightarrow] \quad \frac{G \mid \Gamma, \varphi \Rightarrow \Delta \qquad G \mid \Gamma, \psi \Rightarrow \Delta}{G \mid \Gamma, \varphi \vee \psi \Rightarrow \Delta}$$

$$[\Rightarrow \vee] \quad \frac{G \mid \Gamma \Rightarrow \Delta, \varphi \qquad G \mid \Gamma \Rightarrow \Delta, \psi}{G \mid \Gamma \Rightarrow \Delta, \varphi \vee \psi \qquad G \mid \Gamma \Rightarrow \Delta, \varphi \vee \psi}$$

Structural rules:

$$[\text{EC}] \quad \frac{G \mid s \mid s}{G \mid s} \qquad\qquad\qquad [\text{EW}] \quad \frac{G}{G \mid s}$$

$$[\text{Sp}] \quad \frac{G \mid \Gamma_1, \Gamma_2 \Rightarrow \Delta_1, \Delta_2}{G \mid \Gamma_1 \Rightarrow \Delta_1 \mid \Gamma_2 \Rightarrow \Delta_2} \qquad [\text{Mi}] \quad \frac{G \mid \Gamma_1 \Rightarrow \Delta_1 \qquad G \mid \Gamma_2 \Rightarrow \Delta_2}{G \mid \Gamma_1, \Gamma_2 \Rightarrow \Delta_1, \Delta_2}$$

$$[\text{Cut}] \quad \frac{G \mid \Gamma_1 \Rightarrow \Delta_1, \varphi \qquad G \mid \varphi, \Gamma_2 \Rightarrow \Delta_2}{G \mid \Gamma_1, \Gamma_2 \Rightarrow \Delta_1, \Delta_2}$$

Figure 1: The proof system GRM

Proposition 3.3.

1. $\vdash_{GRM} G \mid \neg\varphi, \Gamma \Rightarrow \Delta$ *iff* $\vdash_{GRM} G \mid \Gamma \Rightarrow \Delta, \varphi$.

2. $\vdash_{GRM} G \mid \Gamma \Rightarrow \Delta, \neg\varphi$ *iff* $\vdash_{GRM} G \mid \varphi, \Gamma \Rightarrow \Delta$.

3. $\vdash_{GRM} G \mid \Gamma \Rightarrow \Delta, \varphi \rightarrow \psi$ *iff* $\vdash_{GRM} G \mid \varphi, \Gamma \Rightarrow \Delta, \psi$.

4. $\vdash_{GRM} G \mid \varphi \wedge \psi, \Gamma \Rightarrow \Delta$ *iff* $\vdash_{GRM} G \mid \varphi, \Gamma \Rightarrow \Delta \mid \psi, \Gamma \Rightarrow \Delta$.

5. $\vdash_{GRM} G \mid \Gamma \Rightarrow \Delta, \varphi \wedge \psi$ *iff* $\vdash_{GRM} G \mid \Gamma \Rightarrow \Delta, \varphi$ *and* $\vdash_{GRM} G \mid \Gamma \Rightarrow \Delta, \psi$.

6. $\vdash_{GRM} G \mid \varphi \lor \psi, \Gamma \Rightarrow \Delta$ *iff* $\vdash_{GRM} G \mid \varphi, \Gamma \Rightarrow \Delta$ *and* $\vdash_{GRM} G \mid \psi, \Gamma \Rightarrow \Delta$.

7. $\vdash_{GRM} G \mid \Gamma \Rightarrow \Delta, \varphi \lor \psi$ *iff* $\vdash_{GRM} G \mid \Gamma \Rightarrow \Delta, \varphi \mid \Gamma \Rightarrow \Delta, \psi$.

Proof. We show the last item as an example, leaving the others to the reader. One direction is easy; $G \mid \Gamma \Rightarrow \Delta, \varphi \lor \psi$ is derived in GRM from $G \mid \Gamma \Rightarrow \Delta, \varphi \mid \Gamma \Rightarrow \Delta, \psi$ by using two applications of $[\Rightarrow \lor]$, followed by an application of [EC]. For the converse, note that $\varphi \Rightarrow \psi \mid \psi \Rightarrow \varphi$ can be derived from $\varphi \Rightarrow \varphi$ and $\psi \Rightarrow \psi$ using [Mi] and [Sp], and then $\varphi \lor \psi \Rightarrow \psi \mid \varphi \lor \psi \Rightarrow \varphi$ can be derived from these three sequents by using two applications of $[\lor \Rightarrow]$. A cut on $\varphi \lor \psi$ of this sequent and $G \mid \Gamma \Rightarrow \Delta, \varphi \lor \psi$ yields $G \mid \varphi \lor \psi \Rightarrow \varphi \mid \Gamma \Rightarrow \Delta, \psi$. Another cut on $\varphi \lor \psi$ of this last sequent and $G \mid \Gamma \Rightarrow \Delta, \varphi \lor \psi$ yields $G \mid G \mid \Gamma \Rightarrow \Delta, \varphi \mid \Gamma \Rightarrow \Delta, \psi$. From this $G \mid \Gamma \Rightarrow \Delta, \varphi \mid \Gamma \Rightarrow \Delta, \psi$ can be derived using applications of [EC]. $\square$

Note 3.4. Proposition 3.3 can be used for reducing the provability of a hypersequent G to the provability of a finite set of hypersequents, the components of each have only atomic formulas on their right-hand side, and only atomic formulas or implications on their left-hand side.

Next we turn our attention to the semantics of the hypersequents of GRM.

Definition 3.5. Let ν be a valuation in $\mathcal{M}(\mathbb{Z})$.

- Let $\Gamma \Rightarrow \Delta$ be a non-empty sequent. Define

  - $d_\nu(\Gamma \Rightarrow \Delta) =_{\mathrm{df}} \max\{ |\nu(\varphi)| : \varphi \in \Gamma \cup \Delta \}$

  - $\nu(\Gamma \Rightarrow \Delta) =_{\mathrm{df}} \begin{cases} d_\nu(\Gamma \Rightarrow \Delta) & \text{if } \exists \varphi \in \Gamma (\nu(\varphi) = - d_\nu(\Gamma \Rightarrow \Delta)), \\ d_\nu(\Gamma \Rightarrow \Delta) & \text{if } \exists \psi \in \Delta (\nu(\psi) = d_\nu(\Gamma \Rightarrow \Delta)), \\ - d_\nu(\Gamma \Rightarrow \Delta) & \text{otherwise.} \end{cases}$

- We say that ν is a *model* of a sequent $\Gamma \Rightarrow \Delta$ (in symbols, $\nu \vDash \Gamma \Rightarrow \Delta$), if either $\Gamma \Rightarrow \Delta$ is empty and there is an atom p such that $\nu(p) = 0$, or $\Gamma \Rightarrow \Delta$ is not empty and $\nu(\Gamma \Rightarrow \Delta) \geq 0$ (i.e., there is $\varphi \in \Gamma$ such that $\nu(\varphi) = - d_\nu(\Gamma \Rightarrow \Delta)$, or $\psi \in \Delta$ such that $\nu(\psi) = d_\nu(\Gamma \Rightarrow \Delta)$).

- We say that ν is a *model* of a hypersequent G (in symbols, $\nu \vDash G$), if ν is a model of at least one component of G.

Definition 3.6 (**RM**-validity of hypersequents). A hypersequent G is **RM**-*valid* if every valuation in $\mathcal{M}(\mathbb{Z})$ is a model of G.

610

Note 3.7. Let the translation σ_s of $s = \varphi_1, \ldots, \varphi_n \Rightarrow \psi_1, \ldots, \psi_k$ (where $n + k > 0$) be the sentence $\neg\varphi_1 + \cdots + \neg\varphi_n + \psi_1 + \cdots + \psi_k$, and let the translation σ_G of a hypersequent G be the $\vee$-disjunction of the translations of its components. It is easy to check that $\nu \vDash G$ iff $\nu \vDash \sigma_G$. In particular, if φ is a sentence then $\nu \vDash \varphi$ iff $\nu \vDash \Rightarrow \varphi$. Hence, $\Rightarrow \varphi$ is **RM**-valid iff φ is valid in $\mathcal{M}(\mathbb{Z})$. By Theorem 2.5, this implies that $\Rightarrow \varphi$ is **RM**-valid iff $\vdash_{\mathbf{RM}} \varphi$.

Proposition 3.8 (Soundness of GRM). *Let G be a hypersequent in $\mathcal{L}_R$. If $\vdash_{GRM} G$, then G is **RM**-valid.*

Proof. Using an induction on the length of proofs, we can prove something stronger, namely, that if G follows in GRM from a set S of hypersequents, then every model ν (in $\mathcal{M}(\mathbb{Z})$) of all the elements of S is also a model of G. For this, we need to show that all axioms of GRM are **RM**-valid, and that if a valuation ν in $\mathcal{M}(\mathbb{Z})$ is a model of all premises of some application of a rule of GRM, then it is also a model of the conclusion of that application. This is straightforward but tedious, so we omit the details. We only note one case that requires special attention. An application of the cut rule in the unusual case in which both of the contexts $\Gamma_1 \Rightarrow \Delta_1$ and $\Gamma_2 \Rightarrow \Delta_2$ are empty. In that case, $G \mid \Rightarrow$ is derived from $G \mid \Rightarrow \varphi$ and $G \mid \varphi \Rightarrow$. Let ν be a model of the premises. Then either it is a model of some component of G, in which case it is a model of $G \mid \Rightarrow$ too, or it is a model of both $\Rightarrow \varphi$ and $\varphi \Rightarrow$. The latter is possible only if $\nu(\varphi) = 0$. A necessary condition for this is that $\nu(p) = 0$ for some atom $p \in \mathsf{Atoms}(\varphi)$. Hence, in that case ν is model of $\Rightarrow$, and so also of $G \mid \Rightarrow$. $\square$

4 Completeness and the Admissibility of Cut

In this section, we simultaneously show that GRM is also complete, and that the cut-elimination theorem holds for it. We do this by constructing a refuting $\mathcal{M}(\mathbb{Z})$-valuation for every hypersequent G that does not have a cut-free proof in GRM. This is done by first extending G to a hypersequent G^*, for which a refuting valuation ν can be constructed in stages from its components and formulas. For this, we define a strictly decreasing (and so finite) sequence of natural numbers $k_1, k_2, \ldots, k_m$ and a corresponding sequence $\nu_1, \nu_2, \ldots, \nu_m$ of partial valuations. The desired ν is ν_m, the last element of the latter sequence. The construction of k_i and ν_i is done in a way that ensures that ν_i has the following properties for each i ($1 \leq i \leq m$):

1. If $i < m$, then ν_{i+1} is a proper extension of ν_i;

2. if $\nu(\varphi)$ is defined, then $k_i \leq |\nu(\varphi)| \leq k_1$;

3. every component of G^* which contains a formula φ such that $\nu_i(\varphi)$ is defined is necessarily refuted by any $\mathcal{M}(\mathbb{Z})$-valuation which extends ν_i.

Let us turn to the details.

Notation 4.1. $\vdash^{cf}_{GRM} G$ means that G has a cut-free proof in GRM.

Lemma 4.2. *If* $\Sigma \neq \emptyset$ *and* $\mathsf{Atoms}(\Gamma \cup \Delta) \subseteq \Sigma$, *then* $\vdash^{cf}_{GRM} \Sigma, \Gamma \Rightarrow \Delta, \Sigma$.

Proof. Induction on the complexity of $\Gamma \cup \Delta$. The base case, where $\Gamma \cup \Delta$ consists of Atoms, is proved using the *Mingle* rule. The induction steps are straightforward. $\square$

Definition 4.3. A hypersequent G is *maximal*, if $\nvdash^{cf}_{GRM} G$ but $\vdash^{cf}_{GRM} G \mid s$ when

- s consists only of subformulas of formulas in G;

- s is not a component of G.

Lemma 4.4. *If* $\nvdash^{cf}_{GRM} G$, *then* G *can be extended to a maximal hypersequent* G^*.

Proof. This follows from the fact that the number of hypersequents which consist only of subformulas of formulas in G is finite. $\square$

Definition 4.5. Let ν be a partial function from Atoms to $\mathbb{Z}$. We call ν a *k-semivaluation*, if the following two conditions are satisfied:

1. $|\nu(p)| \geq k$, for every $p \in \mathrm{Dom}(\nu)$.

2. If $\mathrm{Dom}(\nu) \neq \mathsf{Atoms}$, then $\sharp(\mathsf{Atoms} - \mathrm{Dom}(\nu)) < k$.

Note 4.6. If ν is a 0-semivaluation or a 1-semivaluation, then ν is a full valuation, i.e., $\nu \colon \mathsf{Atoms} \to \mathbb{Z}$.

Definition 4.7. A valuation $\nu^* \colon \mathsf{Atoms} \to \mathbb{Z}$ is a *k-completion of a k-semivaluation* ν, if $\nu^*(p) = \nu(p)$, for $p \in \mathrm{Dom}(\nu)$, while $|\nu^*(p)| < k$, otherwise.

Definition 4.8. Let ν be an *l*-semivaluation, and let $k \geq l$. A formula φ is a $\nu^l_{<k}$-*formula* (in symbols, $\varphi \in \nu^l_{<k}$), if $|\nu^*(\varphi)| < k$ for every *l*-completion ν^* of ν. φ is a $\nu^l_{\geq k}$-*formula* (in symbols, $\varphi \in \nu^l_{\geq k}$), if $|\nu^*(\varphi)| \geq k$ for every *l*-completion ν^* of ν. φ is a $\nu_{<l}$-*formula* (in symbols, $\varphi \in \nu_{<l}$) or a $\nu_{\geq l}$-*formula* (in symbols, $\varphi \in \nu_{\geq l}$), if it is a $\nu^l_{<l}$ or $\nu^l_{\geq l}$-formula, respectively.

Lemma 4.9. *Let* ν *be an l-semivaluation, and let* $k \geq l$.

1. *Every formula is either a* $\nu^l_{<k}$-*formula or a* $\nu^l_{\geq k}$-*formula.*

2. φ is a $\nu^l_{\geq k}$-formula iff $|\nu^*(\varphi)| \geq k$ for some l-completion ν^* of ν.
φ is a $\nu^l_{<k}$-formula, if $|\nu^*(\varphi)| < k$ for some l-completion ν^* of ν.

3. Let ν_1 and ν_2 be l-completions of ν. If φ is an $\nu^l_{\geq k}$-formula, then $\nu_1(\varphi) = \nu_2(\varphi)$.

Proof. Immediate from Proposition 2.6. $\qquad\qquad\qquad\qquad\qquad\qquad\qquad$ $\square$

Definition 4.10. Let G be a hypersequent, and ν be a k-semivaluation. We say that ν is *k-adequate* for G, if the following conditions are satisfied:

1. If $\Gamma \Rightarrow \Delta$ is a component of G, and $\Gamma \cup \Delta$ contains some $\nu_{\geq k}$-formula, then every k-completion of ν refutes $\Gamma \Rightarrow \Delta$.

2. If $k = 0$ and the empty sequent $\Rightarrow$ is a component of G, then $\nu(p) \neq 0$ for every $p \in$ Atoms.

3. Let Σ be the set of atoms for which ν is not defined. Suppose $\Sigma \neq \emptyset$, and that $\Gamma \Rightarrow \Delta$ is a component of G which consists only of $\nu_{<k}$-formulas. Then $\vdash^{cf}_{GRM} G \mid \varphi, \Sigma \Rightarrow \Sigma$ for $\varphi \in \Gamma$, and $\vdash^{cf}_{GRM} G \mid \Sigma \Rightarrow \Sigma, \psi$ for $\psi \in \Delta$. (By using repeated applications of *Mingle*, this implies that $\vdash^{cf}_{GRM} G \mid \Sigma, \Gamma' \Rightarrow \Delta', \Sigma$ whenever $\Gamma' \subseteq \Gamma$ and $\Delta' \subseteq \Delta$.)

Lemma 4.11. *If G is a maximal hypersequent, and ν is a k-semivaluation which is k-adequate for G, then ν has a k-completion which is not a model of G.*

Proof. By induction on k.

The case $k = 0$ is trivial, since if ν is a 0-semivaluation, then ν is a full valuation, and every formula is a $\nu_{\geq 0}$-formula. Hence, if ν is 0-adequate for G, then (by definition) ν is a 0-completion of itself which is not a model of G.

Similarly, the case where $k > 0$ and $\text{Dom}(\nu) =$ Atoms is trivial, since in this case ν is a k-completion of itself, every formula is a $\nu_{\geq k}$-formula, and $\nu(p) \neq 0$ for every p. Hence, the k-adequacy of ν implies that ν itself is a k-completion as required. It follows that the lemma is true in the particular case where $k = 1$.

Now assume that $k > 1$, and the claim is true for every $l < k$. Let ν be a k-semivaluation such that $\text{Dom}(\nu) \neq$ Atoms, and ν is k-adequate for G. Let $I = \{ \Gamma_j \Rightarrow \Delta_j : 1 \leq j \leq m \}$ be the set of all components of G which consist only of $\nu_{<k}$-formulas.

Suppose first that I is empty (i.e., $m = 0$). Then every component of G contains some $\nu_{\geq k}$-formula, and so every k-completion of ν refutes G, by definition of k-adequacy.

Now assume that I is not empty, and let $\Gamma_0 = \bigcup_{1 \leq j \leq m} \Gamma_j$, $\Delta_0 = \bigcup_{1 \leq j \leq m} \Delta_j$. Suppose that $\Gamma_0 \Rightarrow \Delta_0$ is not a component of G. Then the maximality of G entails that there exists a cut-free proof in GRM of $\Gamma_0 \Rightarrow \Delta_0 \mid G$. By using the *Splitting* rule [Sp], we can get from such a proof a cut-free proof of $I \mid G$, and so of G (since $I \subseteq G$). A contradiction. It follows that $\Gamma_0 \Rightarrow \Delta_0 \in G$. Since this sequent consists only of $\nu_{<k}$-formulas, this entails that $\Gamma_0 \Rightarrow \Delta_0 \in I$, and so it is actually the maximal element of I. Let Σ be the set of atoms for which ν is not defined. Since $\Sigma \neq \emptyset$ by our assumption, and ν is k-adequate for G, the third item of Definition 4.10 implies that $\vdash^{cf}_{GRM} \Sigma, \Gamma_0 \Rightarrow \Delta_0, \Sigma \mid G$. It follows that $\Sigma^{(=)} = \Sigma - \Gamma_0 \cap \Delta_0$ is not empty, because otherwise $\Sigma, \Gamma_0 \Rightarrow \Delta_0, \Sigma \mid G$ is identical to G. Let $\Sigma^{(<)} = \Sigma - \Sigma^{(=)} = \Sigma \cap \Gamma_0 \cap \Delta_0$, and let $l = 1 + \sharp(\Sigma^{(<)})$. Now $0 < l < k$, since $\Sigma^{(<)}$ is a *proper* subset of Σ, while $\sharp(\Sigma) < k$, because ν is a k-semivaluation. Extend ν to a semivaluation $\tilde{\nu}$ on $\mathrm{Dom}(\nu) \cup \Sigma^{(=)}$ as follows:

$$\tilde{\nu}(p) = \begin{cases} \nu(p) & \text{if } p \in \mathrm{Dom}(\nu), \\ l & \text{if } p \in \Sigma^{(=)} \cap \Gamma_0, \\ -l & \text{if } p \in \Sigma^{(=)} - \Gamma_0. \end{cases}$$

Note that $\Sigma^{(<)}$ is the set of atoms for which $\tilde{\nu}$ is not defined. It follows that $\tilde{\nu}$ is an l-semivaluation, and its definition implies that every l-completion of $\tilde{\nu}$ is a k-completion of ν. Moreover, for every l-completion ν^* of $\tilde{\nu}$ and every φ, either $|\nu^*(\varphi)| \geq k$, or $|\nu^*(\varphi)| \leq l$. Now we show that the following two claims are true for every $\varphi \in \Gamma_0 \cup \Delta_0$ and every l-completion ν^* of $\tilde{\nu}$:

(a) If $\varphi \in \Gamma_0$, then $\nu^*(\varphi) = l$, or $\varphi \in \tilde{\nu}_{<l}$ and $\vdash^{cf}_{GRM} \varphi, \Sigma^{(<)} \Rightarrow \Sigma^{(<)} \mid G$.

(b) If $\varphi \in \Delta_0$, then $\nu^*(\varphi) = -l$, or $\varphi \in \tilde{\nu}_{<l}$ and $\vdash^{cf}_{GRM} \Sigma^{(<)} \Rightarrow \Sigma^{(<)}, \varphi \mid G$.

What follows is a proof of these claims by an induction on the complexity of φ. We assume that ν^* is some l-completion of $\tilde{\nu}$, and so also a k-completion of ν. By Lemma 4.9, this implies that $|\nu^*(\varphi)| < l$ iff $\varphi \in \tilde{\nu}_{<l}$, and $|\nu^*(\varphi)| < k$ iff $\varphi \in \tilde{\nu}^l_{<k}$.

$\varphi \in \Gamma_0$, and φ is atomic.

Then $\varphi \in \Sigma$, because ν is a k-semivaluation, and so an atom is a $\nu_{<k}$-formula iff ν is not defined for it. It follows that either $\varphi \in \Sigma^{(=)} \cap \Gamma_0$, or $\varphi \in \Sigma^{(<)}$. In the first case $\nu^*(\varphi) = l$, by definition of $\tilde{\nu}$. In the second $\varphi \in \tilde{\nu}_{<l}$ (since $\tilde{\nu}$ is an l-semivaluation, and $\Sigma^{(<)}$ is the set of atoms for which $\tilde{\nu}$ is not defined), and $\varphi, \Sigma^{(<)} = \Sigma^{(<)}$. Hence, $\vdash^{cf}_{GRM} \varphi, \Sigma^{(<)} \Rightarrow \Sigma^{(<)} \mid G$, by using [Mi] and [EW].

$\varphi \in \Delta_0$, and φ is atomic.

Then $\varphi \in \Sigma$, and so either $\varphi \in \Sigma^{(=)} - \Gamma_0$, or $\varphi \in \Sigma^{(<)}$. In the first case, $\nu^*(\varphi) = -l$, by definition of $\tilde{\nu}$. In the second case, $\varphi \in \tilde{\nu}_{<l}$, and again we have $\vdash^{cf}_{GRM} \Sigma^{(<)} \Rightarrow \Sigma^{(<)}, \varphi \mid G$, by using [Mi] and [EW].

$\varphi \in \Gamma_0$, **and** $\varphi = \neg\psi$.

Then φ is a $\nu_{<k}$-formula, and so ψ is a $\nu_{<k}$-formula, since $|\nu^*(\neg\psi)| = |\nu^*(\psi)|$. It is impossible that $\vdash^{cf}_{GRM} \Gamma_0 \Rightarrow \Delta_0, \psi \mid G$, since otherwise, we would have got that $\vdash^{cf}_{GRM} G$, by using $(\neg \Rightarrow)$. It follows that $\Gamma_0 \Rightarrow \Delta_0, \psi$ is in G (because G is maximal), and since it consists only of $\nu_{<k}$-formulas, necessarily $\psi \in \Delta_0$. Therefore, we get by the induction hypothesis on ψ that either $\nu^*(\psi) = -l$, or $\vdash^{cf}_{GRM} \Sigma^{(<)} \Rightarrow \Sigma^{(<)}, \psi \mid G$ and $\psi \in \tilde{\nu}_{<l}$. Hence, either $\nu^*(\varphi) = l$, or $\varphi \in \tilde{\nu}_{<l}$ and (using $[\neg \Rightarrow]$) $\vdash^{cf}_{GRM} \varphi, \Sigma^{(<)} \Rightarrow \Sigma^{(<)} \mid G$.

$\varphi \in \Delta_0$, **and** $\varphi = \neg\psi$.

This case is similar to the previous one.

$\varphi \in \Gamma_0$, **and** $\varphi = \sigma \rightarrow \psi$.

Then φ is a $\nu_{<k}$-formula, and so ψ and σ are $\nu_{<k}$-formulas, because $|\nu^*(\varphi)| = \max\{|\nu^*(\psi)|, |\nu^*(\sigma)|\}$ in this case. It is impossible that both $\vdash^{cf}_{GRM} \psi, \Gamma_0 \Rightarrow \Delta_0 \mid G$ and $\vdash^{cf}_{GRM} \Gamma_0 \Rightarrow \Delta_0, \sigma \mid G$, since otherwise, we would have gotten that $\vdash^{cf}_{GRM} G$, by using $[\rightarrow \Rightarrow]$. Assume e.g., that $\nvdash^{cf}_{GRM} \psi, \Gamma_0 \Rightarrow \Delta_0 \mid G$. Then $\psi, \Gamma_0 \Rightarrow \Delta_0$ is in G (because G is maximal), and since it consists only of $\nu_{<k}$-formulas, necessarily $\psi \in \Gamma_0$. It follows by the induction hypothesis on ψ that either $\nu^*(\psi) = l$, or $\psi \in \tilde{\nu}_{<l}$ and $\vdash^{cf}_{GRM} \psi, \Sigma^{(<)} \Rightarrow \Sigma^{(<)} \mid G$. In the first case, $\nu^*(\varphi) = l$ (since σ is a $\nu_{<k}$-formula, and so $|\nu^*(\sigma)| \leq l$). In the second case, we consider the following two subcases:

- Suppose that $\sigma \notin \Delta_0$. Then $\Gamma_0 \Rightarrow \Delta_0, \sigma \notin I$, and since this sequent consists only of $\nu_{<k}$-formulas, $\Gamma_0 \Rightarrow \Delta_0, \sigma \notin G$. Hence, $\vdash^{cf}_{GRM} \Gamma_0 \Rightarrow \Delta_0, \sigma \mid G$. Using $[\rightarrow \Rightarrow]$, this and our assumption $\vdash^{cf}_{GRM} \psi, \Sigma^{(<)} \Rightarrow \Sigma^{(<)} \mid G$ imply $\vdash^{cf}_{GRM} \varphi, \Sigma^{(<)}, \Gamma_0 \Rightarrow \Delta_0, \Sigma^{(<)} \mid G$. Since, $\Sigma^{(<)} \subseteq \Gamma_0 \cap \Delta_0$, and $\varphi \in \Gamma_0$, we get that $\vdash^{cf}_{GRM} G$. A contradiction. Hence, this subcase is actually impossible.

- If $\sigma \in \Delta_0$, then by the induction hypothesis, either $\nu^*(\sigma) = -l$, or $\sigma \in \tilde{\nu}_{<l}$ and there is a cut-free proof of $\Sigma^{(<)} \Rightarrow \Sigma^{(<)}, \sigma \mid G$. In the first case, again $\nu^*(\varphi) = l$. In the second case, $\varphi \in \tilde{\nu}_{<l}$ (because $\psi \in \tilde{\nu}_{<l}$ and $\sigma \in \tilde{\nu}_{<l}$), and we can get a cut-free proof of $\varphi, \Sigma^{(<)} \Rightarrow \Sigma^{(<)} \mid G$ by applying $[\rightarrow \Rightarrow]$ to $\psi, \Sigma^{(<)} \Rightarrow \Sigma^{(<)} \mid G$ and $\Sigma^{(<)} \Rightarrow \Sigma^{(<)}, \sigma \mid G$.

$\varphi \in \Delta_0$, **and** $\varphi = \sigma \rightarrow \psi$.

Then φ is a $\nu_{<k}$-formula, and so ψ and σ are $\nu_{<k}$-formulas. It is impossible that $\vdash^{cf}_{GRM} \Gamma_0, \sigma \Rightarrow \Delta_0, \psi \mid G$, since otherwise, we would have got that $\vdash^{cf}_{GRM} G$ by using $[\Rightarrow \rightarrow]$. Hence, $\Gamma_0, \sigma \Rightarrow \Delta_0, \psi$ is in G, and since it consists only of $\nu_{<k}$-formulas, necessarily $\psi \in \Delta_0$ and $\sigma \in \Gamma_0$. Hence, by the induction hypothesis

(b) applies to ψ and (a) to σ. If $\nu^*(\psi) = -l$ and $\nu^*(\sigma) = l$, or $\nu^*(\psi) = -l$ and $|\nu^*(\sigma)| < l$, or $\nu^*(\sigma) = l$ and $|\nu^*(\psi)| < l$, then $\nu^*(\varphi) = -l$. Otherwise, $\psi \in \tilde{\nu}_{<l}$, $\sigma \in \tilde{\nu}_{<l}$, $\vdash_{GRM}^{cf} \Sigma^{(<)} \Rightarrow \Sigma^{(<)}, \psi \mid G$, and $\vdash_{GRM}^{cf} \sigma, \Sigma^{(<)} \Rightarrow \Sigma^{(<)} \mid G$. It follows that $\varphi \in \tilde{\nu}_{<l}$, and $\vdash_{GRM}^{cf} \sigma, \Sigma^{(<)} \Rightarrow \Sigma^{(<)}, \psi \mid G$ (using *Mingle*). By using $[\Rightarrow \rightarrow]$, we get $\vdash_{GRM}^{cf} \Sigma^{(<)} \Rightarrow \Sigma^{(<)}, \varphi \mid G$.

$\varphi \in \Gamma_0$, **and** $\varphi = \psi \vee \sigma$.

It is impossible that $\vdash_{GRM}^{cf} \psi, \Gamma_0 \Rightarrow \Delta_0 \mid G$ and $\vdash_{GRM}^{cf} \sigma, \Gamma_0 \Rightarrow \Delta_0 \mid G$ are both true, since this would have implied $\vdash_{GRM}^{cf} G$ by using $[\vee \Rightarrow]$. Assume e.g., that $\nvdash_{GRM}^{cf} \psi, \Gamma_0 \Rightarrow \Delta_0 \mid G$. Then $\psi, \Gamma_0 \Rightarrow \Delta_0$ is in G. Assume for contradiction that $\psi \in \nu_{\geq k}$. Then ν^* refutes $\psi, \Gamma_0 \Rightarrow \Delta_0$ (since ν^* is a k-completion of ν, and ν is k-adequate). This is possible only if $\nu^*(\psi) \geq k$, because $\Gamma_0 \cap \Delta_0 \subseteq \nu_{<k}$, and so $\nu^*(\psi, \Gamma_0 \Rightarrow \Delta_0) = -\nu^*(\psi)$ under our assumptions. But in such a case, also $\nu^*(\varphi) \geq k$, contradicting the fact that $\varphi \in \Gamma_0$, and so $\varphi \in \nu_{<k}$. It follows that $\psi \in \nu_{<k}$, hence, $\psi, \Gamma_0 \Rightarrow \Delta_0 \in I$, and so $\psi \in \Gamma_0$. By applying the induction hypothesis to ψ, we get that either $\nu^*(\psi) = l$, or $\psi \in \tilde{\nu}_{<l}$ (and so $|\nu^*(\psi)| < l$) and $\vdash_{GRM}^{cf} \psi, \Sigma^{(<)} \Rightarrow \Sigma^{(<)} \mid G$. In the first case, $\nu^*(\varphi) \geq l$, and since $\varphi \in \nu_{<k}$ (because $\varphi \in \Gamma_0$), $\nu^*(\varphi) = l$. In the second case, we consider the following three subcases:

- Assume that $\sigma \in \nu_{\geq k}$. Then either $\nu^*(\sigma) \geq k$ or $\nu^*(\sigma) \leq -k$. In the first case, also $\nu^*(\varphi) \geq k$ as well, contradicting the fact that $\varphi \in \nu_{<k}$. So $\nu^*(\sigma) \leq -k$. Since $|\nu^*(p)| < l$ for every $p \in \Sigma^{(<)}$, $\nu^*(\sigma, \Sigma^{(<)} \Rightarrow \Sigma^{(<)}) = -\nu^*(\sigma) \geq k$ in this case, and so ν^* is a model of the sequent $\sigma, \Sigma^{(<)} \Rightarrow \Sigma^{(<)}$. Hence, $\sigma, \Sigma^{(<)} \Rightarrow \Sigma^{(<)}$ is not in G (since ν is k-adequate for G, ν^* is a k-completion of ν, and $\sigma, \Sigma^{(<)} \Rightarrow \Sigma^{(<)}$ contains the $\nu_{\geq k}$-formula σ). It follows that $\vdash_{GRM}^{cf} \sigma, \Sigma^{(<)} \Rightarrow \Sigma^{(<)} \mid G$. From this and our assumption that $\vdash_{GRM}^{cf} \psi, \Sigma^{(<)} \Rightarrow \Sigma^{(<)} \mid G$, it follows that $\vdash_{GRM}^{cf} \varphi, \Sigma^{(<)} \Rightarrow \Sigma^{(<)} \mid G$ (using $[\vee \Rightarrow]$). Moreover, $|\nu^*(\varphi)| < l$ because $\nu^*(\sigma) \leq -k < -l$ while $|\nu^*(\psi)| < l$. Hence, $\varphi \in \tilde{\nu}_{<l}$.

- Assume that $\nu^*(\sigma) = l$. Then $\nu^*(\varphi) = l$ (since $|\nu^*(\psi)| < l$).

- Assume that $\sigma \in \nu_{<k}$ and $\nu^*(\sigma) \neq l$. Then $\nu^*(\sigma) = -l$ or $|\nu^*(\sigma)| < l$ (recall that $|\nu^*(\alpha)| \geq k$ or $|\nu^*(\alpha)| \leq l$, for every α). In the first case, $\nu^*(\varphi) = \nu^*(\psi)$ and so $|\nu^*(\varphi)| < l$. In the second case, our assumption on $\nu^*(\psi)$ again implies that $|\nu^*(\varphi)| < l$. It remains to prove that $\vdash_{GRM}^{cf} \varphi, \Sigma^{(<)} \Rightarrow \Sigma^{(<)} \mid G$. Since $\vdash_{GRM}^{cf} \psi, \Sigma^{(<)} \Rightarrow \Sigma^{(<)} \mid G$, it suffices to prove that $\vdash_{GRM}^{cf} \sigma, \Sigma^{(<)} \Rightarrow \Sigma^{(<)} \mid G$. If $\sigma \in \Gamma_0$, this would follow from our induction hypothesis applied to σ, since $\nu^*(\sigma) \neq l$. If $\sigma \notin \Gamma_0$ then

$\sigma, \Sigma^{(<)} \Rightarrow \Sigma^{(<)}$ is not in I, and since we are assuming that $\sigma \in \nu_{<k}$, it follows that $\sigma, \Sigma^{(<)} \Rightarrow \Sigma^{(<)}$ is not in G. Hence, the maximality of G implies that $\vdash_{GRM}^{cf} \sigma, \Sigma^{(<)} \Rightarrow \Sigma^{(<)} \mid G$ in this case too.

$\varphi = \psi \vee \sigma$ and $\varphi \in \Delta_0$.

Without a loss of generality, we may assume that $\nu^*(\sigma) \leq \nu^*(\psi)$, and so $\nu^*(\varphi) = \nu^*(\psi)$. Now $\varphi \in \Delta_0 \subseteq \nu_{<k}$, and so $|\nu^*(\varphi)| < k$. It follows that $|\nu^*(\psi)| < k$, and so $\psi \in \nu_{<k}$. Suppose that $\psi \notin \Delta_0$. Then $\Gamma_0 \Rightarrow \Delta_0, \psi$ is not in I, and since this sequent consists only of formulas in $\nu_{<k}$, $\Gamma_0 \Rightarrow \Delta_0, \psi$ is not in G either. Hence, $\vdash_{GRM}^{cf} \Gamma_0 \Rightarrow \Delta_0, \psi \mid G$, and so $\vdash_{GRM}^{cf} G$ (by using $[\Rightarrow \vee]$). A contradiction. It follows that $\psi \in \Delta_0$, and so by the induction hypothesis either $\nu^*(\psi) = -l$, or $\psi \in \tilde{\nu}_{<l}$ and $\vdash_{GRM}^{cf} \Sigma^{(<)} \Rightarrow \Sigma^{(<)}, \psi \mid G$. In the first case, $\nu^*(\varphi) = -l$. In the second case, $|\nu^*(\varphi)| = |\nu^*(\psi)| < l$, and so $\varphi \in \tilde{\nu}_{<l}$. Using $[\Rightarrow \vee]$, that $\vdash_{GRM}^{cf} \Sigma^{(<)} \Rightarrow \Sigma^{(<)}, \varphi \mid G$ in this case follows from the assumption that $\vdash_{GRM}^{cf} \Sigma^{(<)} \Rightarrow \Sigma^{(<)}, \psi \mid G$.

$\varphi = \psi \wedge \sigma$ and $\varphi \in \Gamma_0$.

The proof in this case is similar to that in the previous one.

$\varphi = \psi \wedge \sigma$ and $\varphi \in \Delta_0$.

The proof is similar to the case where $\varphi = \psi \vee \sigma$ and $\varphi \in \Gamma_0$.

This concludes the proof of (a) and (b). Next we show that $\tilde{\nu}$ is l-adequate for G. Since $l < k$, this will end the proof of the lemma by the induction hypothesis for l (and the fact that any l-completion of $\tilde{\nu}$ is a k-completion of ν).

Now the second condition in Definition 4.10 is vacuously satisfied (since $l > 0$), while the third is immediate from (a) and (b), since if $\Gamma \Rightarrow \Delta$ is a component of G which consists only of $\tilde{\nu}_{<l}$-formulas, then $\Gamma \subseteq \Gamma_0$ and $\Delta \subseteq \Delta_0$. This follows from the definitions of I, Γ_0, and Δ_0, and the fact that $\tilde{\nu}_{<l} \subset \tilde{\nu}_{<k} = \nu_{<k}$.

To show the first condition, assume that ν^* is an l-completion of $\tilde{\nu}$, $\Gamma \Rightarrow \Delta$ is a component of G, and $\Gamma \cup \Delta$ contains some $\tilde{\nu}_{\geq l}$-formula. If $\Gamma \Rightarrow \Delta \notin I$, then ν^* refutes $\Gamma \Rightarrow \Delta$, because ν is k-adequate, and ν^* is a k-completion of ν. If $\Gamma \Rightarrow \Delta \in I$, then $\Gamma \subseteq \Gamma_0$ and $\Delta \subseteq \Delta_0$, and so (a) and (b) imply that $\nu^*(\varphi) = l$ for every $\tilde{\nu}_{\geq l}$-formula $\varphi \in \Gamma$, and $\nu^*(\varphi) = -l$ for every $\tilde{\nu}_{\geq l}$-formula $\varphi \in \Delta$. Now since $\Gamma \cup \Delta$ contains some $\tilde{\nu}_{\geq l}$-formula, only such formulas determine whether ν^* is a model of $\Gamma \Rightarrow \Delta$ or not. It follows that ν^* refutes $\Gamma \Rightarrow \Delta$. □

Theorem 4.12. *Let G be hypersequent in $\mathcal{L}_R$. Then either $\vdash_{GRM}^{cf} G$, or G is not **RM**-valid.*

Proof. Suppose that $\nvdash_{GRM}^{cf} G$. Let Σ be the set of atomic formulas which occur in G, and assume that the number of elements in Σ is n. Let G^* be a maximal hypersequent which extends G (Lemma 4.4). Define $\nu(p) = n + 1$ for every $p \in$ Atoms $- \Sigma$. Obviously, ν is an $n+1$-semivaluation, and if φ occurs in G^*, then φ is a $\nu_{<n+1}$-formula such that Atoms$(\varphi) \subseteq \Sigma$. These facts and Lemma 4.2 easily imply (using *External Weakenings*) that ν is $n + 1$-adequate for G^*. Therefore, it follows from Lemma 4.11 that ν has an $n + 1$-completion ν^* which is not a model of G^*, and so not a model of G either. Hence, G is not **RM**-valid. $\square$

Theorem 4.13 (Adequacy of GRM). $\vdash_{GRM} G$ *iff* G *is* **RM**-*valid.*

Proof. Immediate from Proposition 3.8 and Theorem 4.12. $\square$

Corollary 4.14. $\vdash_{GRM} \Rightarrow \varphi$ *iff* $\vdash_{\mathbf{RM}} \varphi$.

Proof. Immediate from Theorem 4.13 and Note 3.7. $\square$

Corollary 4.14 shows how GRM can be used to characterize the validity of formulas in **RM**. The next theorem generalizes it by showing how GRM can be used to characterize the consequence relation of **RM**. Since the latter is finitary, it suffices to treat the case of inferences from finite theories.

Theorem 4.15. *Let* $\mathcal{T} = \{\varphi_1, \ldots, \varphi_n\}$. *Then* $\mathcal{T} \vdash_{\mathbf{RM}} \psi$ *iff the hypersequent* $\varphi_1 \Rightarrow \psi \mid \cdots \mid \varphi_n \Rightarrow \psi \mid \Rightarrow \psi$ *is provable in* GRM.

Proof. Since $\supset$ (Definition 2.7) is an implication for **RM** (Proposition 2.8), and $\wedge$ is a conjunction for it, $\mathcal{T} \vdash_{\mathbf{RM}} \psi$ iff $\vdash_{\mathbf{RM}} \varphi_1 \wedge \cdots \wedge \varphi_n \supset \psi$. By Corollary 4.14, this is equivalent to $\vdash_{GRM} \Rightarrow \varphi_1 \wedge \cdots \wedge \varphi_n \supset \psi$, which in turn is equivalent to $\vdash_{GRM} \varphi_1 \Rightarrow \psi \mid \cdots \mid \varphi_n \Rightarrow \psi \mid \Rightarrow \psi$, by Proposition 3.3 and the Definition of $\supset$. $\square$

Theorem 4.16 (Cut-elimination for GRM). *If* $\vdash_{GRM} G$, *then* G *has in* GRM *a cut-free proof.*

Proof. Immediate from Proposition 3.8 and Theorem 4.12. $\square$

Note 4.17. Theorems 4.12, 4.13 and 4.16 and their proofs can easily be generalized to derivations from a set of hypersequents. Thus a hypersequent G follows in GRM from a set $\mathcal{S}$ of hypersequents iff every model of $\mathcal{S}$ is a model of G. Moreover, the strong cut-elimination theorem applies to GRM. That is, if $\mathcal{S} \vdash_{GRM} G$, then there is a proof in GRM of G from $\mathcal{S}$ in which all cuts are on formulas from $\bigcup_{H \in \mathcal{S}} \mathcal{F}_H$, where $\mathcal{F}_H$ is the set of *formulas* (not subformulas!) which appear in some component of H.

Note 4.18. We have proved that $\vdash_{GRM} \varphi$ iff $\vdash_{\mathbf{RM}} \varphi$ by using the semantics of the systems. However, it is not difficult to prove it syntactically, by directly showing that $\vdash_{GRM} G$ iff $\vdash_{HRM} \sigma_G$ (see Note 3.7), where HRM is the standard Hilbert-type system for **RM** (using which **RM** was actually originally defined. See [1, 11, 9]).

Here are two examples of the use of the cut-elimination theorem for GRM in order to prove a property of **RM**.

Example 4.19. Let $\mathbb{Z}^* = \mathbb{Z} - \{\,0\,\}$. The matrix $\mathcal{M}(\mathbb{Z}^*)$ for $\mathcal{L}_{RM}$ is defined similarly to $\mathcal{M}(\mathbb{Z})$, but with $\mathbb{Z}^*$ as its set of truth-values. Definition 3.5 can be adapted to $\mathcal{M}(\mathbb{Z}^*)$ in a straightforward way. It then follows from the definitions that $G \mid \Rightarrow$ is valid in $\mathcal{M}(\mathbb{Z})$ iff G is valid in $\mathcal{M}(\mathbb{Z}^*)$. Now, it is easy to prove by induction on the length of a cut-free proof of a sequent of the form $G \mid \Rightarrow \mid \cdots \mid \Rightarrow$ that such a sequent is provable in GRM iff G is provable in GRM. It follows that a hypersequent G is valid in $\mathcal{M}(\mathbb{Z})$ iff it is valid in $\mathcal{M}(\mathbb{Z}^*)$. Hence, the weak completeness of **RM** for $\mathcal{M}(\mathbb{Z})$ (see Theorem 2.5) implies the weak completeness of **RM** for $\mathcal{M}(\mathbb{Z}^*)$.

Example 4.20. As already noted in [3], Dunn and Meyer's well-known result that the disjunctive syllogism is admissible in **RM** can easily be proved using GRM as follows. Suppose that $\vdash_{\mathbf{RM}} \neg\varphi$, and $\vdash_{\mathbf{RM}} \varphi \vee \psi$. Then $\vdash_{GRM} \Rightarrow \neg\varphi$, and $\vdash_{GRM} \Rightarrow \varphi \vee \psi$. By Proposition 3.3, this implies that $\vdash_{GRM} \varphi \Rightarrow$ and $\vdash_{GRM} \Rightarrow \varphi \mid \Rightarrow \psi$. Using a cut, we infer that $\vdash_{GRM} \Rightarrow \mid \Rightarrow \psi$. It follows, by the observation made in the previous example, that $\vdash_{GRM} \Rightarrow \psi$, and so $\vdash_{\mathbf{RM}} \psi$.

5 Conclusion and Further Research

The work presented in this paper is a continuation of an ongoing project, whose goal is to provide semantic proofs of the admissibility of the cut rule in calculi of hypersequents, especially those that are due to the author.[2] As said in the introduction, we strongly believe that these proofs are more reliable and easier to be followed and verified than the original syntactic proofs of cut-elimination for those systems. The project started in [8], which treats the case of GLC, the hypersequential system for Gödel–Dummett logic (first introduced in [5]).[3] We intend next to attack the case of $GRMI$ (the hypersequential system for the purely relevant logic RMI), which is described in [4]. We believe that a proof which is similar to the one given here, but more complicated, should be possible. (Note that our proof here for GRM is itself

---

[2]In the case of our hypersequential system for the modal logic S5, such a semantic proof has already been presented in [6].

[3]M. Baaz and A. Ciabattoni have discovered an annoying gap in the treatment of $\vee$ in the syntactic proof of cut-elimination given in [5]. Their discovery has been the trigger for this project.

far more complicated than that given for *GLC* in [8].) Then it will be natural to see to what extent can the method used in this paper be extended to hypersequent calculi for fuzzy substructural logics (cf. [13]) other than **RM**, like **UML**.[4] (See [14] for several such systems.)

Another interesting line of research is to compare the approach taken here to other semantic completeness proofs for hypersequent calculi, like the general method of Ciabattoni, Galatos, and Terui in [10]. Note that the latter so far has been developed only for single-conclusion hypersequent calculi.

References

[1] A. R. Anderson and N. D. Belnap. *Entailment: The Logic of Relevance and Necessity, Vol.I.* Princeton University Press, 1975.

[2] A. Avron. On an implication connective of RM. *Notre Dame Journal of Formal Logic*, 27:201–209, 1986.

[3] A. Avron. A constructive analysis of RM. *Journal of Symbolic Logic*, 52:939–951, 1987.

[4] A. Avron. Relevance and paraconsistency – A new approach. Part III: Cut-free Gentzen-type systems. *Notre Dame Journal of Formal Logic*, 32:147–160, 1991.

[5] A. Avron. Hypersequents, logical consequence and intermediate logics for concurrency. *Annals of Mathematics and Artificial Intelligence*, 4:225–248, 1991.

[6] A. Avron. The structure of interlaced bilattices. *Journal of Mathematical Structures in Computer Science*, 6:287–299, 1996.

[7] A. Avron. The method of hypersequents in proof theory of propositional non-classical logics. In W. Hodges, M. Hyland, C. Steinhorn, and Truss J., editors, *Logic: Foundations to Applications*, pages 1–32. Oxford Science Publications, 1996.

[8] A. Avron. Simple proof of completeness and cut-elimination for propositional Gödel logic. *Journal of Logic and Computation*, 21:813–821, 2011.

[9] A. Avron. RM and its nice properties. In K. Bimbó, editor, *J. Michael Dunn on Information Based Logics*, volume 8 of *Outstanding Contributions to Logic*, pages 15–43. Springer, 2016.

---

[4]**RM** itself is presented as a fuzzy logic in Section 7 of [9].

[10] A. Ciabattoni, N. Galatos, and K. Terui. Algebraic proof theory for substructural logics: cut-elimination and completions. *Annals of Pure and Applied Logic*, 163:266–290, 2012.

[11] J. M. Dunn and G. Restall. Relevance logic. In D. Gabbay and F. Guenther, editors, *Handbook of Philosophical Logic*, volume 6, pages 1–136. Kluwer, 2002. Second edition.

[12] G. Gentzen. Investigations into logical deduction, 1934. In German. An English translation appears in 'The Collected Works of Gerhard Gentzen', edited by M. E. Szabo, North-Holland, 1969.

[13] G. Metcalfe and F. Montagna. Substructural fuzzy logics. *Journal of Symbolic Logic*, 72:834–864, 2007.

[14] G. Metcalfe, N. Olivetti, and D. Gabbay. *Proof Theory for Fuzzy Logics*. Springer, 2009.

[15] G. Pottinger. Uniform, cut-free formulations of T, S4 and S5 (abstract). *Journal of Symbolic Logic*, 48:900, 1983.

[16] T. Sugihara. Strict implication free from implicational paradoxes. *Memoirs of the Faculty of Liberal Arts, Fukui University, Series I*, pages 55–59, 1955.

 Received 13 October 2016

MODEL DEFINABILITY IN RELEVANT LOGIC

GUILLERMO BADIA
Department of Knowledge-Based Mathematical Systems,
Johannes Kepler Universität, Linz, Austria
<guillebadia89@gmail.com>

Abstract

It is shown that the classes of Routley–Meyer models which are axiomatizable by a theory in a propositional relevant language with fusion and the
Ackermann constant can be characterized by their closure under certain model-
theoretic operations involving prime filter extensions, relevant directed bisimulations and disjoint unions.

Keywords: directed bisimulations, model definability, model theory, prime filter
extensions, relevant logic, Routley–Meyer semantics

1 Introduction

In this note we give a non-classical answer to the question of which classes of
Routley–Meyer models (cf. [20, 19, 17, 18]) can be defined (or, in other words,
axiomatized) in the language of relevant logic. In particular, we give a criterion in
terms of closure under certain model-theoretic operations for a class of Routley–
Meyer models to be axiomatizable.

By a "non-classical answer" we mean that we use constructions intrinsically belonging to the Routley–Meyer semantic framework as opposed to more generic ones
like ultraproducts (although we also provide similar answers employing these). A
good general recent reference (with a well-developed duality theory) on the Routley–
Meyer framework and similar relational semantics is [3].

The history of this kind of questions in the setting of non-classical logics seems to
have started with [11], a review by Kaplan of one of Kripke's foundational papers on
possible worlds semantics. Kaplan asked which properties of the binary relation on
Kripke frames were expressible by formulas of a propositional modal language. Some
years later, an answer to this problem was provided in [8] (Theorem 3) using closure
under "non-trivial disjoint unions" and "SA-constructions" (a construction similar

to what is also known as ultrafilter extensions). Moreover, the authors also provided what came to be known as the "Goldblatt–Thomason theorem" (Theorem 8 from [8]), which gave a characterization of those Kripke frames definable by propositional modal formulas which were also definable by a first order theory. It involved closure under subframes, p-morphic images, disjoint unions and reflection under ultrafilter extensions. The original proof of this theorem was obtained by algebraic methods via duality theory, but later J. van Benthem showed that it could be established by direct model-theoretic methods (see the proof of Theorem 3.19 from [4]). Similar results were obtained for intuitionistic logic in §13–§17 of the, sadly hard to find, [16]. In particular, Theorem 15.3 from [16] established a Goldblatt–Thomason theorem for intuitionistic logic generalizing the ultrafilter extension construction to prime filters.

In [16], the author also moved the characterization question to classes of models rather than just frames and provided a result for intuitionistic logic involving ultraproducts and bisimulations among other things (Theorem 13.8). M. de Rijke later did a similar thing for arbitrary pointed Kripke relational structures (Corollary 6.2 from [15]) again involving ultraproducts and bisimulations. In [23], Y. Venema aimed to do away with ultraproducts (an "impure" modal model theoretic construction) in the case of these characterizations for Kripke models and modal logic. The replacement was ultrafilter extensions. More recently, [7] provided an analogue to the main result of [23] for the setting of intuitionistic logic.

Our goal is to generalize the ideas and results from [7] for intuitionistic logic to another class of logics. This comes to complement — for the particular case of relevant logic — the sort of results obtained in [2], where a Goldblatt–Thomason theorem was established for a wide variety of substructural logics (Theorem 7.6) using co-algebraic methods inspired by [12]. This is why the focus of our paper has been on classes of models rather than frames −since the latter question has already been answered.[1]

The layout of the paper is as follows. In Section 2, we provide some necessary background regarding relevant languages and the Routley–Meyer framework as well as how they fit in the more general setting of first order logic. In Section 3, we introduce the key model-theoretic relations and constructions that will be used in the paper, namely, relevant directed bisimulations, bounded morphisms, (definable) prime filter extensions and disjoint unions. We also establish in this section a group of simple results that would be used latter on. In Section 4, we provide the promised characterization of axiomatizability or definability using closure conditions under the model-theoretic *dramatis personae* introduced in the previous section. Finally, in

---

[1]We are thankful to the referee who pressed us to clarify this point.

Section 5, we summarize our work and provide some final observations.

2   Preliminaries

2.1   Routley–Meyer Semantics

We will work with the relevant language L which has a countable set PROP of propositional variables together with the following logical symbols: $\perp$ (an absurdity constant), $\top$ (a truth constant), $\vee$ (disjunction), $\wedge$ (conjunction), $\rightarrow$ (implication), $\circ$ (fusion) and $\mathbf{t}$ (the Ackermann constant). Formulas are constructed in the expected manner:

$$\phi ::= p \mid \perp \mid \top \mid \mathbf{t} \mid \sim\phi \mid \phi \wedge \psi \mid \phi \vee \psi \mid \phi \rightarrow \psi \mid \phi \circ \psi,$$

where $p \in$ PROP. The set of all formulas is sometimes called $\text{Fmla}(L)$ in these pages.

A structure $M = \langle W^M, R^M, *^M, O^M, V^M \rangle$ is called a $\mathbf{B}^{\circ\mathbf{t}}$-*model* for L, if for any $x, y, z, v \in W$, letting $x \leqslant^M y$ abbreviate that there is $z \in O^M$ such that $R^M zxy$, we have that:

(i) $x \leqslant^M x$;

(ii) if $x \leqslant^M y$ and $R^M yzv$, then $R^M xzv$;

(iii) if $x \leqslant^M y$ and $R^M zyv$, then $R^M zxv$;

(iv) if $x \leqslant^M y$ and $R^M zvx$, then $R^M zvy$;

(v) if $x \leqslant^M y$, then $y^{*^M} \leqslant^M x^{*^M}$;

(vi) $x = x^{*^M *^M}$.

while $V^M :$ PROP $\longrightarrow \wp(W^M)$ is a valuation function such that for every $p \in$ PROP, $V^M(p)$ is upwards closed under the $\leqslant^M$ relation, that is,

(viii) $x \in V^M(p)$ and $x \leqslant y$ implies that $y \in V^M(p)$. Moreover, we also require that O^M be upwards closed under the $\leqslant^M$.

The system known as $\mathbf{B}^{\circ\mathbf{t}}$ (essentially Routley and Meyer's $\mathbf{B}$ system with the appropriate axioms for $\mathbf{t}$, $\top$ and $\perp$ plus the residuation rule between $\circ$ and $\rightarrow$) based on L is sound and complete with respect to the class of $\mathbf{B}^{\circ\mathbf{t}}$-models (cf. [5]). In this paper, *all* models will be $\mathbf{B}^{\circ\mathbf{t}}$-models so sometimes we just talk about "Routley–Meyer models" or simply "models."

We define the expression ϕ *is satisfied at w in M* recursively as follows:

$$
\begin{array}{lll}
M, w \Vdash \bot & \text{never,} \\
M, w \Vdash \top & \text{always,} \\
M, w \Vdash \mathbf{t} & \text{iff} \quad w \in O^M, \\
M, w \Vdash p & \text{iff} \quad w \in V(p), \\
M, w \Vdash \sim\!\phi & \text{iff} \quad M, w^* \nVdash \phi, \\
M, w \Vdash \phi \wedge \psi & \text{iff} \quad M, w \Vdash \phi \text{ and } M, w \Vdash \psi, \\
M, w \Vdash \phi \vee \psi & \text{iff} \quad M, w \Vdash \phi \text{ or } M, w \Vdash \psi, \\
M, w \Vdash \phi \rightarrow \psi & \text{iff} \quad \text{for every } a, b \text{ such that } R^M wab, \text{ if } M, a \Vdash \phi \text{ then } M, b \Vdash \psi, \\
M, w \Vdash \phi \circ \psi & \text{iff} \quad \text{there are } a, b \text{ such that } R^M abw, M, a \Vdash \phi \text{ and } M, b \Vdash \psi.
\end{array}
$$

The valuation V^M can be extended to a function from *all* formulas of L to $\wp(W^M)$ using this definition. So, $V^M(\phi)$ will denote the collection of all worlds in M that satisfy ϕ. We will denote by $U(\leqslant^M)$ the collection of all subsets of W^M upward closed under the $\leqslant^M$ relation.

Now, if M is a model, we will say that a formula ϕ *is true in M* if for all $w \in O^M$ we have that $M, w \Vdash \phi$. Given a set Γ of relevant formulas we will denote the class of all models where all the formulas in Γ are true as $\text{Mod}(\Gamma)$. This leads us to the following definition, fundamental for the purposes of this paper.

Definition 2.1. A class K of Routley–Meyer models is said to be *definable* or *axiomatizable* in a given relevant language L, if there is a set Θ of formulas of L such that $\text{Mod}(\Theta) = K$.

2.2 Embedding Relevant Languages into First Order Logic

Take now a first order language (that is, a language with a countable list of individual variables $x, y, z, \ldots$ and the following logical symbols: $\wedge$ (conjunction), $\vee$ (disjunction), $\neg$ (boolean negation), $\supset$ (material implication), $\exists$ (existential quantifier), $\forall$ (universal quantifier) and (if necessary) $=$ (equality or identity)) that comes with one function symbol $*$, a distinguished three place relation symbol R, a unary predicate O, and a unary predicate P for each $p \in \mathsf{PROP}$. We might call this a *correspondence language L^{corr}* for L. Now we can read a model M as a first order model for L^{corr} in a straightforward way: W is taken as the domain of the structure, V specifies the denotation of each of the predicates $P, Q, \ldots$, while $*$ is the denotation of the function symbol $*$ of L^{corr} (in what follows we will write x^* instead of the customary $*(x)$), the subset O of W the denotation of the obvious unary predicate and R the denotation of the relation R of L^{corr}.

Where t is a term in the first order correspondence language, we write $\phi^{t/x}$ for the result of replacing x with t everywhere in the formula ϕ. As expected, it is easy to specify a translation from the formulas of the basic relevant language with absurdity (and fusion) into formulas of first order logic with *one free variable* as follows:

$$ST_x(\bot) = Rxxx \wedge \neg Rxxx$$
$$ST_x(\top) = Rxxx \vee \neg Rxxx$$
$$ST_x(\mathbf{t}) = Ox$$
$$ST_x(p) = Px$$
$$ST_x(\sim\phi) = \neg ST_x(\phi)^{x^*/x}$$
$$ST_x(\phi \wedge \psi) = ST_x(\phi) \wedge ST_x(\psi)$$
$$ST_x(\phi \vee \psi) = ST_x(\phi) \vee ST_x(\psi)$$
$$ST_x(\phi \rightarrow \psi) = \forall y, z(Rxyz \wedge ST_x(\phi)^{y/x} \supset ST_x(\psi)^{z/x})$$
$$ST_x(\phi \circ \psi) = \exists y, z(Ryzx \wedge ST_x(\phi)^{y/x} \wedge ST_x(\psi)^{z/x})$$

In the above, it is worth noting that in $ST_x(\phi \rightarrow \psi)$ and $ST_x(\phi \circ \psi)$, y needs to be free at the free occurrences of the variable x in the formula $ST_x(\phi)$ and, similarly, z has to be free at the free occurrences of the variable x in the formula $ST_x(\psi)$. This can easily be arranged by taking y, z to be new variables every time but one can, of course, also economize by reusing variables.

Next we prove a lemma to the effect that our proposed translation is adequate. Note that while $\Vdash$ stands for satisfaction as defined for basic relevant languages with absurdity, $\vDash$ is the usual Tarskian satisfaction relation from classical logic. In particular, when ϕ is a first order formula and $\overline{a}$ a sequence of elements of a first order model $M = \langle D, P_0^M, P_1^M, \ldots \rangle$, $M \vDash \phi[\overline{a}]$ means that the sequence $\overline{a}$ satisfies the formula ϕ in M according to the usual recursive definition:

$$M, w \vDash P_n[\overline{a}] \quad \text{iff} \quad \overline{a} \in P_n^M$$
$$M, w \vDash \neg\phi[\overline{a}] \quad \text{iff} \quad M, w \nvDash \phi[\overline{a}]$$
$$M, w \vDash (\phi \wedge \psi)[\overline{a}] \quad \text{iff} \quad M, w \vDash \phi[\overline{a}] \text{ and } M, w \vDash \psi[\overline{a}]$$
$$M, w \vDash \forall x\phi[\overline{a}] \quad \text{iff} \quad \text{for every } b \in D, \ M, w \vDash \phi[\overline{a}b]$$

The reader should keep in mind the difference between $\Vdash, \vDash$ and $\vdash$ since we will make use of all three below.

Lemma 2.2 (Switch Lemma). *For any w, $M, w \Vdash \phi$ if and only if $M \vDash ST_x(\phi)[w]$.*

Proof. Simply note that ϕ and $ST_x(\phi)$ express the same thing about w in M. Strictly speaking, this would be established by a routine induction on formula complexity. $\square$

In what follows we will make free use of this lemma since it works as a useful bridge between first order logic and relevant logic. In particular, it lets us bring all the classical machinery of ultraproducts and ultrapowers into our setting.

2.3 Countably Saturated Structures

We start by recalling some useful ideas from the model theory of first order logic. The following notions can be found in any of the standard references such as [1, 6, 10]. They are included here only for the sake of completeness. For the ultraproduct construction the best places are [1, 6]. The symbols $\bar{a}$, $\bar{b}$, and $\bar{x}$, $\bar{y}$ are used to denote sequences of elements of a model and variables, respectively.

We speak of a set of first order formulas $\Phi(\bar{x})$ as being *realizable* in a model M, if there is some sequence $\bar{a}$ of elements of M such that $M \models \Phi[\bar{a}]$. $\Phi(\bar{x})$ is said to be *refutable* in M if $\Phi'(\bar{x}) = \{\,\neg\phi \colon \phi \in \Phi(\bar{x})\,\}$ is realizable in M.

When M is a model of first order logic, by $\mathrm{dom}(M)$ we denote the domain of M. If $X \subseteq \mathrm{dom}(M)$, then $(M, a)_{a \in X}$ is the expansion of M obtained by adding a constant c_a for each $a \in X$.

Definition 2.3. Let λ be a cardinal. A model M of first order logic is said to be λ-*saturated* if whenever $X \subseteq \mathrm{dom}(M)$ and $|X| < \lambda$, then the expansion $(M, a)_{a \in X}$ of M realizes every set of formulas $\Phi(x)$ of the language of $(M, a)_{a \in X}$ which is consistent with the set of all first order sentences true in $(M, a)_{a \in X}$.

Note that if $\kappa < \lambda$, λ-saturation implies κ-saturation.

Definition 2.4. Let M and N be two first order models. N is an *elementary extension* of M if M is (isomorphic to) a submodel N' of N such that where $\bar{a}$ is a sequence of elements of N', for each first order formula ϕ,

$$N' \models \phi[\bar{a}] \quad \text{iff} \quad N \models \phi[\bar{a}].$$

Proposition 2.5. *Each model of the correspondence first order language of a basic propositional relevant language with absurdity has an ω-saturated elementary extension.*

Proof. Let M be one such model. Since the correspondence language is countable, by a result due to Keisler (Theorem 11.2.1 in [1] and Theorem 6.1.1 in [6]), we see that we can form an ω_1-saturated ultrapower of M. But the canonical embedding between M and the ultrapower obtained is elementary (Lemma 5.2.3 in [1]). Finally, recall that ω_1-saturation implies ω-saturation. $\qquad\square$

Proposition 2.6. *Let M be an ω-saturated model for the correspondence first order language of basic relevant logic, and $\Phi(\bar{x}, \bar{y})$ a set of formulas. If $M \models \exists \bar{x} \bigwedge \Psi[\bar{a}]$ for every finite $\Psi \subseteq \Phi$, $M \models \Phi[\bar{a}, \bar{b}]$ for some finite tuple $\bar{b}$ of elements of M.*

Proof. Immediate from Theorem 10.1.7 in [10]. $\qquad\square$

The important thing to keep in mind about elementary extensions of a given model is that we have a transfer principle, which is an immediate consequence of the fact that the extension is elementary. In other words, if we can show that first order statements involving parameters only from the original model are true in the extended model, we know that those statements are also true in the original model.

3 Some Model Theory

In this section, we will discuss some model-theoretic relations and constructions that will play a fundamental role in our characterization result (Theorem 4.5 below). We will start with the relevant analogue of the concept of bisimulation from modal logic. Bisimulations for the setting of relevant logic were originally introduced in [14].

3.1 Relevant Directed Bisimulations

Definition 3.1. Let $M_1 = \langle W_1, R_1, *_1, O_1, V_1 \rangle$ and $M_2 = \langle W_2, R_2, *_2, O_2, V_2 \rangle$ be two Routley–Meyer models for a relevant language L. A *relevant directed bisimulation* for L between M_1 and M_2 is a pair of non-empty relations $\langle Z_1, Z_2 \rangle$ where $Z_1 \subseteq W_1 \times W_2$ and $Z_2 \subseteq W_2 \times W_1$, such that (1)–(5) hold when $i, j \in \{1, 2\}$, $i \neq j$:

(1) $x Z_i y$ only if $y^{*_j} Z_j x^{*_i}$;

(2) if $x Z_i y$ and $R_j y b c$ for some $b, c \in W_j$, there are $b', c' \in W_i$ such that $R_i x b' c'$, $b Z_j b'$ and $c' Z_i c$;

(3) if $x Z_i y$ and $p \in \text{PROP} \subset L$, then $M_i, x \Vdash p$ only if $M_j, y \Vdash p$;

(4) if $x Z_i y$ and $R_i b c x$ for some $b, c \in W_i$, there are $b', c' \in W_j$ such that $R_j b' c' y$, $b Z_i b'$ and $c Z_i c'$;

(5) if $x Z_i y$ and $x \in O_i$, then $y \in O_j$.

The bisimulation is said to be *surjective with respect to O_i* if for each $w \in O_i$ there is $w' \in O_j$ such that $w' Z_j w$.

Proposition 3.2. *Let $M_1 = \langle W_1, R_1, *_1, O_1, V_1 \rangle$ and $M_2 = \langle W_2, R_2, *_2, O_2, V_2 \rangle$ be two Routley–Meyer models for a relevant language L and $\langle Z_1, Z_2 \rangle$ be a relevant directed bisimulation between these models. If $i \neq j, i, j \in \{1, 2\}$ and $w Z_i v$, then $M_i, w \Vdash \phi$ only if $M_j, v \Vdash \phi$ for ϕ a formula of L.*

Proof. We proceed by induction on formula complexity. The case of propositional variables uses (3) from Definition 3.1. The case of $\mathbf{t}$ uses (5). The cases of $\rightarrow$ and $\circ$ use (2) and (4), respectively. $\qquad\square$

Corollary 3.3. *Let $M_1 = \langle W_1, R_1, *_1, O_1, V_1 \rangle$ and $M_2 = \langle W_2, R_2, *_2, O_2, V_2 \rangle$ be two Routley–Meyer models for a relevant language L. Suppose that $\langle Z_1, Z_2 \rangle$ is a relevant directed bisimulation surjective w.r.t. O_2 between these models, then $M_1 \Vdash \phi$ only if $M_2 \Vdash \phi$ for every formula ϕ of L.*

Proof. Suppose that $M_2 \nVdash \phi$. Let $w \in O_2$ be such that $M_2, w \nVdash \phi$. Then contrapose Proposition 3.2. □

Corollary 3.4 (Hereditary Lemma). *Let M be a Routley–Meyer model for a relevant language L. Then for every $w, v \in W^M$, $w \leqslant^M v$ only if for all formulas ϕ of L, $M, w \Vdash \phi$ implies that $M, v \Vdash \phi$.*

Proof. Simply check that $\langle \leqslant^M, \leqslant^M \rangle$ is a relevant directed auto-bisimulation (i.e., a relation between M and itself). □

Let us write $rel\text{-}tp_S(e)$ for the relevant type of a world e in a model S, i.e., the set of all first order translations of relevant formulas that e satisfies. If M_1, M_2 are models, we will write $M_1, w \Rightarrow_L M_2, v$ if for every formula ϕ of L, $M_1, w \Vdash \phi$ only if $M_2, v \Vdash \phi$.

Proposition 3.5. *Let L be a basic relevant language with absurdity and M_1 and M_2 two models for L. Suppose that M_1 and M_2 are ω-saturated as first order models. Then the relation $\Rightarrow_L$ induces a relevant directed bisimulation $\langle Z_1, Z_2 \rangle$ between M_1 and M_2 defined as follows:*

$$xZ_1y \quad \text{iff} \quad rel\text{-}tp_{M_1}(x) \subseteq rel\text{-}tp_{M_2}(y),$$
$$xZ_2y \quad \text{iff} \quad rel\text{-}tp_{M_2}(x) \subseteq rel\text{-}tp_{M_1}(y).$$

Proof. Suppose that $k, m \in \{1, 2\}$, $k \neq m$, and for every formula ϕ of L, $M_k, w \Vdash \phi$ only if $M_m, u \Vdash \phi$.

Let us start by noting that Z_k is non-empty since $rel\text{-}tp_{M_k}(w) \subseteq rel\text{-}tp_{M_m}(u)$. By the argument for clause (1) in the definition of a relevant directed bisimulation in the paragraph below, we also obtain that Z_m is non-empty for $rel\text{-}tp_{M_m}(u^{*m}) \subseteq rel\text{-}tp_{M_k}(w^{*k})$, i.e., $u^{*m} Z_m w^{*k}$. Since $k \neq m$ we must have that both Z_1 and Z_2 are non-empty.

In what follows, let $i, j \in \{1, 2\}$. If xZ_iy, i.e., $rel\text{-}tp_{M_i}(x) \subseteq rel\text{-}tp_{M_j}(y)$, we will see that $M_j, y^{*j} \Vdash \psi$ then $M_i, x^{*i} \Vdash \psi$ for any formula ψ of L, so we will obtain that $rel\text{-}tp_{M_j}(y^{*j}) \subseteq rel\text{-}tp_{M_i}(x^{*i})$, i.e., $y^{*j} Z_j x^{*i}$. Suppose that $M_i, x^{*i} \nVdash \psi$, so $M_i, x \Vdash (\sim\psi)$ and since $rel\text{-}tp_{M_i}(x) \subseteq rel\text{-}tp_{M_j}(y)$, $M_j, y \Vdash (\sim\psi)$. Consequently, $M_j, y^{*j} \nVdash \psi$ as we wanted. This proves (1) in Definition 3.1.

For clause (2) in Definition 3.1, suppose that xZ_iy, i.e., $rel\text{-}tp_{M_i}(x) \subseteq rel\text{-}tp_{M_j}(y)$, and R_jybc for some b, c. Consider

$$nrel\text{-}tp_{M_j}(y) = \{\, \neg ST_x(\psi) \colon M_j, y \not\Vdash \psi, \psi \in \mathrm{Fmla}(L) \,\},$$

where $\mathrm{Fmla}(L)$ denotes the set of formulas of L. We claim that the set of formulas $rel\text{-}tp_{M_j}(b) \cup nrel\text{-}tp_{M_j}(c)$ (we need some care and make sure by renaming that the variable free in $rel\text{-}tp_{M_j}(b)$ is different from the variable free in $nrel\text{-}tp_{M_j}(c)$) is satisfiable in M_i by a pair b', c' of elements such that $R_ixb'c'$. Take any finite subset of $rel\text{-}tp_{M_j}(b) \cup rel\text{-}tp_{M_j}(c)$. Say it is $\{\, ST_z(\delta_1), \ldots, ST_z(\delta_n) \,\} \cup \{\neg ST_v(\sigma_1), \ldots, \neg ST_v(\sigma_m) \}$, where $\{\, ST_z(\delta_1), \ldots, ST_z(\delta_n) \,\} \subseteq rel\text{-}tp_{M_j}(b)$ and $\{\neg ST_v(\sigma_1), \ldots, \neg ST_v(\sigma_m) \} \subseteq nrel\text{-}tp_{M_j}(c)$. It is clear that $M_j, y \not\Vdash \bigwedge\{\delta_1, \ldots, \delta_n\} \to \bigvee\{\sigma_1, \ldots, \sigma_m\}$, i.e., $M_j \not\models ST_z(\bigwedge\{\delta_1, \ldots, \delta_n\} \to \bigvee\{\sigma_1, \ldots, \sigma_m\})[y]$, so given that $rel\text{-}tp_{M_i}(x) \subseteq rel\text{-}tp_{M_j}(y)$, $M_i \not\models ST_z(\bigwedge\{\delta_1, \ldots, \delta_n\} \to \bigvee\{\sigma_1, \ldots, \sigma_m\})[x]$. It follows that $\{ST_z(\delta_1), \ldots, ST_z(\delta_n)\} \cup \{\neg ST_v(\sigma_1), \ldots, \neg ST_v(\sigma_m)\}$ is satisfiable in M_i by a pair of elements b_0, c_0 such that $R_ixb_0c_0$. By the ω-saturation of M_i, there must be a pair b', c' such that $R_ixb'c'$ realizing the whole of $rel\text{-}tp_{M_j}(b) \cup nrel\text{-}tp_{M_j}(c)$. Since $rel\text{-}tp_{M_j}(b)$ is realized by b', we have that bZ_jb' and since c' realizes $nrel\text{-}tp_{M_j}(c)$, i.e., $M_j, c \not\Vdash \psi$ only if $M_i, c' \not\Vdash \psi$, by contraposing it must be the case that $rel\text{-}tp_{M_i}(c') \subseteq rel\text{-}tp_{M_j}(c)$, i.e., $c'Z_ic$.

Condition (3) in Definition 3.1 is obvious; if xZ_iy, i.e., $rel\text{-}tp_{M_i}(x) \subseteq rel\text{-}tp_{M_j}(y)$, then a fortiori, $M_i, x \Vdash p$ only if $M_j, y \Vdash p$ for any propositional variable p of L.

For clause (4) in Definition 3.1, suppose that xZ_iy, i.e., $rel\text{-}tp_{M_i}(x) \subseteq rel\text{-}tp_{M_j}(y)$, and R_ibcx for some b, c. We claim that the set of formulas $rel\text{-}tp_{M_i}(b) \cup rel\text{-}tp_{M_i}(c)$ is satisfiable in M_j by a pair b', c' of elements such that $R_jb'c'y$. Take any finite subset of $rel\text{-}tp_{M_i}(b) \cup rel\text{-}tp_{M_i}(c)$. Say it is $\{ ST_z(\delta_1), \ldots, ST_z(\delta_n) \} \cup \{ ST_v(\sigma_1), \ldots, ST_v(\sigma_m) \}$, where $\{ ST_z(\delta_1), \ldots, ST_z(\delta_n) \} \subseteq rel\text{-}tp_{M_i}(b)$ and $\{ST_v(\sigma_1), \ldots, ST_v(\sigma_m)\} \subseteq rel\text{-}tp_{M_i}(c)$. It is clear that $M_i, x \Vdash \bigwedge\{\delta_1, \ldots, \delta_n\} \circ \bigwedge\{\sigma_1, \ldots, \sigma_m\}$, so given that $rel\text{-}tp_{M_i}(x) \subseteq rel\text{-}tp_{M_j}(y)$, $M_j, y \Vdash \bigwedge\{\delta_1, \ldots, \delta_n\} \circ \bigwedge\{\sigma_1, \ldots, \sigma_m\}$. It follows that $\{ ST_z(\delta_1), \ldots, ST_z(\delta_n) \} \cup \{ ST_v(\sigma_1), \ldots, ST_v(\sigma_m) \}$ is satisfiable in M_j by a pair of elements b_0, c_0 such that $R_jb_0c_0y$. By the ω-saturation of M_i, there must be a pair b', c' such that $R_jb'c'y$ realizing the whole of $rel\text{-}tp_{M_i}(b) \cup rel\text{-}tp_{M_i}(c)$. Since $rel\text{-}tp_{M_i}(b)$ is realized by b', we have that bZ_ib' and since c' realizes $rel\text{-}tp_{M_i}(c)$, cZ_ic'.

Finally, clause (5) in Definition 3.1 is obvious by the semantics of **t**. $\qquad\square$

3.2 Morphisms and Disjoint Unions

The following definition has appeared in places like [21] and [13].

Definition 3.6. Let $M_1 = \langle W_1, R_1, *_1, O_1, V_1 \rangle$ and $M_2 = \langle W_2, R_2, *_2, O_2, V_2 \rangle$ be two Routley–Meyer models for a relevant language L. Let us denote by $\leqslant^1, \leqslant^2$ the partial orders of M_1 and M_2, respectively. A map $f \colon W_1 \longrightarrow W_2$ is a *bounded morphism* if for all $a, b, c \in W_1$ and $a', b', c' \in W_2$:

  (i) $f(a^{*_1}) = f(a)^{*_2}$,

  (ii) $R_2 f(a) b' c'$ only if there are $b, c \in W_1$ such that $R_1 abc$ while $b' \leqslant^2 f(b)$ and $f(c) \leqslant^2 c'$,

  (iii) $a \in V_1(p)$ iff $f(a) \in V_2(p)$ $(p \in \mathsf{PROP})$,

  (iv) $R_2 b' c' f(a)$ only if there are $b, c \in W_1$ such that $R_1 bca$ while $b' \leqslant^2 f(b)$ and $c' \leqslant^2 f(c)$,

  (v) $a \in O_1$ only if $f(a) \in O_2$,

  (vi) $R_1 abc$ only if $R_2 f(a) f(b) f(c)$.

Proposition 3.7. *Let $M_1 = \langle W_1, R_1, *_1, O_1, V_1 \rangle$ and $M_2 = \langle W_2, R_2, *_2, O_2, V_2 \rangle$ be two Routley–Meyer models for a relevant language L. Furthermore, let $f \colon W_1 \longrightarrow W_2$ be a bounded morphism. Then, the pair $\langle Z_1, Z_2 \rangle$ is a relevant directed bisimulation, where:*

$$x Z_1 y \quad \text{iff} \quad f(x) \leqslant^2 y,$$
$$x Z_2 y \quad \text{iff} \quad x \leqslant^2 f(y).$$

Moreover, if for each $x \in O_2$ there is a $y \in O_1$ such that $f(y) \leqslant^2 x$, then the relevant directed bisimulation is surjective w.r.t. both O_1 and O_2.

Proof. To establish (1) from Definition 3.1 simply use (i) from Definition 3.6. For if $x Z_1 y$, i.e., $f(x) \leqslant^2 y$, then, by properties of Routley–Meyer $\mathbf{B}^{\circ \mathbf{t}}$-models, it must be that $y^{*_2} \leqslant^2 f(x)^{*_2} = f(x^{*_1})$, i.e., $y^{*_2} Z_2 x^{*_1}$. On the other hand, if $x Z_2 y$, i.e., $x \leqslant^2 f(y)$ then, by properties of Routley–Meyer $\mathbf{B}^{\circ \mathbf{t}}$-models, it must be that $f(y)^{*_2} = f(y^{*_1}) \leqslant^2 x^{*_2}$, i.e., $y^{*_1} Z_1 x^{*_2}$.

Clauses (2) and (4) in Definition 3.1 follow along similar lines applying clauses (ii) and (iv), respectively (in conjunction with (vi)) from Definition 3.6. Clause (iii) from Definition 3.1 is straightforward from (iii) of Definition 3.6 and the Hereditary Lemma.

Finally, we know by the hypothesis of the proposition and the definition of Z_1 that $\langle Z_1, Z_2 \rangle$ is surjective w.r.t. O_2. On the other hand, if $w \in O_1$, then since $f(w) \leqslant^2 f(w)$ we must have that $f(w) Z_2 w$. But also by (v) in Definition 3.6, we have that indeed $f(w) \in O_2$, so $\langle Z_1, Z_2 \rangle$ is surjective w.r.t. O_1. $\square$

Definition 3.8. Let $M_i = \langle W_i, R_i, *_i, O_i \rangle (i \in I)$ be a collection of frames indexed by the set I such that $W_i \cap W_j = \emptyset$ for distinct $i, j \in I$. The *disjoint union* of $\{ M_i : i \in I \}$, in symbols, $\bigoplus_{i \in I} M_i$ is the structure $\langle W^{\oplus}, R^{\oplus}, *^{\oplus}, O^{\oplus} \rangle$ such that

$$W^{\oplus} = \bigcup_{i \in I} W_i,$$

$$R^{\oplus} = \bigcup_{i \in I} R_i,$$

$$*^{\oplus} = \bigcup_{i \in I} *_i,$$

$$O^{\oplus} = \bigcup_{i \in I} O_i,$$

$$V^{\oplus} = \bigcup_{i \in I} V_i.$$

If our initial family is not disjoint, in order to apply this construction, we can replace it by a disjoint family of isomorphic copies of the models in the original family, that is, we replace each M_i by an isomorphic copy M_i' such that for all $i, j \in I$, $W_i' \cap W_j' = \emptyset$.

From now on, we will make the tacit assumption that *all our classes of model are closed under isomorphic images*. This means that in Proposition 3.9 below the general case where the family $\{ M_i : i \in I \}$ is not disjoint follows taking the appropriate isomorphic copies of the models in the family in conjunction with the fact that the class defined by ϕ is closed under isomorphic images.

Proposition 3.9. *Let ϕ be a formula of L and $\{ M_i : i \in I \}$ a collection of models. If $M_i \Vdash \phi$ (for all $i \in I$), then also $\bigoplus_{i \in I} M_i \Vdash \phi$.*

Proof. This can be seen noting that the inclusion relation induces a relevant directed bisimulation between each M_i and $\bigoplus_{i \in I} M_i$ relating the worlds in O_i. Thus, if $M_i \Vdash \phi$ (for all $i \in I$), when we take an arbitrary $w \in O^{\oplus}$, we will have that, indeed, $\bigoplus_{i \in I} M_i, w \Vdash \phi$, by Proposition 3.2. $\square$

3.3 (Definable) Prime Filter Extensions

Definition 3.10. A structure $\langle A, \rightarrow, \circ, -, \cap, \cup, \top, \bot, 1 \rangle$ is called a *relevant algebra* ([22, p. 264]) if the following holds:

(i) $\langle A, \cup, \cap, \top, \bot \rangle$ is a bounded distributive lattice,

(ii) $x \circ (y \cup z) = (x \circ y) \cup (x \circ z)$,

(iii) $(y \cup z) \circ x = (y \circ x) \cup (z \circ x)$,

(iv) $-(x \cup y) = -x \cap -y$,

(v) $-(x \cap y) = -x \cup -y$,

(vi) $-\top = \bot$ and $-\bot = \top$,

(vii) $x \circ \bot = \bot \circ x = \bot$,

(viii) $1 \circ x = x$,

(ix) $x \circ y \leqslant z$ iff $x \leqslant y \to z$,

where $\leqslant$ in this context is the standard lattice order in terms of either $\cap$ or $\cup$.

Definition 3.11. Given a model $M = \langle W^M, R^M, *^M, O^M, V^M \rangle$, we will use M^+ to denote the *dual algebra* of M, which is the structure $\langle U(\leqslant^M), \to, \circ, -, \cap, \cup, \top, \bot, 1 \rangle$ where $\cap, \cup$ are simply set intersection and union respectively, $1 = O^M$, $\bot = \varnothing$, $\top = W^M$ and the remaining operations of sets are as follows:

(i) $X \circ Y = \{ w \in W^M : \exists v, u \in W^M (R^M vuw \wedge v \in X \wedge u \in Y) \}$,

(ii) $X \to Y = \{ w \in W^M : \forall v, u \in W^M (R^M wvu \wedge v \in X \supset u \in Y) \}$,

(iii) $-X = \{ w \in W^M : w^{*M} \notin X \}$.

Proposition 3.12. *Let M be a Routley–Meyer $\mathbf{B}^{\circ t}$-model for L. Then M^+ is a relevant algebra.*

Proof. Left to the reader. Straightforward from the definition of M^+ and the properties of a Routley–Meyer model. $\square$

Definition 3.13. Given a model $M = \langle W^M, R^M, *^M, O^M, V^M \rangle$, we will use $(M^+)_+$ to denote the *dual model* of M^+, which is the structure $\langle PF(U(\leqslant^M)), R, *, O, V \rangle$ where:

(i) $PF(U(\leqslant^M))$ is the set of all prime filters of $U(\leqslant^M)$ in the algebra M^+,

(ii) $O = \{ w \in PF(U(\leqslant^M)) : O^M \in w \}$,

(iii) $Rwvu$ iff for all $x, y \in U(\leqslant^M)$, if $x \to y \in w$ and $x \in v$ then $y \in u$,

(iv) $w^* = \{\, x \in U(\leqslant^M) \colon -x \notin w \,\}$,

(v) $V(p) = \{\, w \in PF(U(\leqslant^M)) \colon V^M(p) \in w \,\}$.

$(M^+)_+$ will be also called the prime filter extension of M and denoted by $\mathfrak{pe}(M)$. This is because we can define the following map $\pi \colon W^M \longrightarrow PF(U(\leqslant^M))$:

$$w \mapsto \{\, x \in U(\leqslant^M) \colon w \in x \,\},$$

which is an embedding (in the standard model-theoretic sense) of the structure M into $\mathfrak{pe}(M)$.

Proposition 3.14. *Let M be a Routley–Meyer $\mathbf{B}^{\circ t}$-model for L. Then for any prime filter u of $U(\leqslant^M)$ and formula ϕ of L we must have that $V^M(\phi) \in u$ iff $\mathfrak{pe}(M), u \Vdash \phi$. Hence, $M, w \Vdash \phi$ iff $\mathfrak{pe}(M), \pi(w) \Vdash \phi$ for any formula ϕ of L. Moreover, $M \Vdash \phi$ iff $\mathfrak{pe}(M) \Vdash \phi$ for any such ϕ.*

Proof. We establish that $V^M(\phi) \in u$ iff $\mathfrak{pe}(M), u \Vdash \phi$ by induction on the complexity of ϕ. The case when ϕ is a propositional variable is immediate by definition of $V^{\mathfrak{pe}(M)}$. The case when $\phi = \bot$ is trivial since $V^M(\bot) = \varnothing$, which can never belong to a filter u. When $\phi = \mathbf{t}$, we have that $V(\mathbf{t}) = O^M \in u$ iff $u \in O^{\mathfrak{pe}(M)}$ iff $\mathfrak{pe}(M), u \Vdash \mathbf{t}$, as desired. The cases for $\vee$ and $\wedge$ are obvious.

Let now $\phi = \sim\psi$. Then $V^M(\sim\psi) = \{\, w \in W^M \colon w^{*M} \notin V^M(\psi) \,\} = \sim V^M(\psi) \in u$ iff $V^M(\psi) \notin u^{*\mathfrak{pe}(M)}$ iff $\mathfrak{pe}(M), u^{*\mathfrak{pe}(M)} \not\Vdash \psi$ iff $\mathfrak{pe}(M), u \Vdash \sim\psi$, where the second biconditional follows by inductive hypothesis.

Let $\phi = \psi \to \chi$. Suppose that $\mathfrak{pe}(M), u \not\Vdash \psi \to \chi$, i.e., there are u_1, u_2 such that $R^{\mathfrak{pe}(M)} u u_1 u_2$ while $\mathfrak{pe}(M), u_1 \Vdash \psi$ and $\mathfrak{pe}(M), u_2 \not\Vdash \chi$. By inductive hypothesis, $V^M(\psi) \in u_1$ and $V^M(\chi) \notin u_2$. Hence $V^M(\phi) \to V^M(\chi) = V^M(\psi \to \chi) \notin u$ by definition of $R^{\mathfrak{pe}(M)}$. On the other hand, if $V^M(\phi) \to V^M(\chi) = V^M(\psi \to \chi) \notin u$, then by Lemma 4.1 from [21], we see that indeed $\mathfrak{pe}(M), u \not\Vdash \phi$. Finally, the case $\phi = \psi \circ \chi$ is established similarly only this time appealing to Lemma 4.2 from [21].

For the last part of the proposition suppose that $M \Vdash \phi$, so $O^M \subseteq V^M(\phi)$, but O^M is a member of any prime filter u of $U(\leqslant^M)$ which belongs to $O^{\mathfrak{pe}(M)}$. By general properties of filters then $V^M(\phi) \in u$, so indeed $\mathfrak{pe}(M) \Vdash \phi$. Conversely, if $M \not\Vdash \phi$, there is some $w \in O^M$ such that $M, w \not\Vdash \phi$. Thus, $\mathfrak{pe}(M), \pi(w) \not\Vdash \phi$. But $w \in O^M$ iff $M, w \Vdash \mathbf{t}$ iff $\mathfrak{pe}(M), \pi(w) \Vdash \mathbf{t}$ iff $\pi(w) \in O^{\mathfrak{pe}(M)}$. Consequently, $\mathfrak{pe}(M) \not\Vdash \phi$. $\square$

Definition 3.15. Given a model $M = \langle W^M, R^M, *^M, O^M, V^M \rangle$, we will use $M^{+\delta}$ to denote the *definable dual algebra* of M, which is the structure $\langle U(M), \to, \circ, -, \cap, \cup, \top, \bot, 1 \rangle$ where $U(M) = \{\, V^M(\phi) \colon \phi \in \mathrm{Fmla}(L) \,\}$ and the operations are defined as in Definition 3.11. In particular, note that $O^M = V^M(\mathbf{t}), V^M(\top) = W^M$ and $V^M(\bot) = \varnothing$.

Proposition 3.16. *Let M be a Routley–Meyer $\boldsymbol{B}^{\circ t}$-model for L. Then $M^{+\delta}$ is a relevant algebra.*

Definition 3.17. Given a model $M = \langle W^M, R^M, *^M, O^M, V^M \rangle$, we will use $(M^{+\delta})_+$ to denote the *dual model of* of $M^{+\delta}$, which is the structure $\langle PF(U(M)), R, *, O, V \rangle$, where:

(i) $PF(U(M))$ is the set of all prime filters of $U(M)$ in the algebra $M^{+\delta}$;

(ii) $O = \{\, w \in PF(U(M)) : O^M \in w \,\}$;

(iii) $Rwvu$ iff for all $x, y \in U(M)$, if $x \to y \in w$ and $x \in v$ then $y \in u$;

(iv) $w^* = \{\, x \in U(M) : -x \notin w \,\}$;

(v) $V(p) = \{\, w \in PF(U(M)) : V^M(p) \in w \,\}$.

Once more, $(M^{+\delta})_+$ (or simply M^δ) will be also called the *definable prime filter extension* of M since again we can define the following embedding $\pi : W^M \longrightarrow PF(U(M))$:

$$w \mapsto \{\, x \in U(M) : w \in x \,\}.$$

Proposition 3.18. *Let M be a Routley–Meyer $\boldsymbol{B}^{\circ t}$-model for L. The map $x \mapsto x \cap U(M)$ is a bounded morphism from $\mathfrak{pe}(M)$ onto M^δ. Moreover, for each $x \in O^{M^\delta}$ there is $y \in O^{\mathfrak{pe}(M)}$ such that $x = y \cap U(M)$.*

Proof. Consider the inclusion homomorphism i from $M^{+\delta}$ into M^+. Recall from [21, p. 106] that the dual mapping $i_+ : (M^+)_+ \longrightarrow (M^{+\delta})_+$ is defined by the equation:

$$i_+(x) = i^{-1}(x).$$

But $i^{-1}(x) = \{\, y \in U(M) : y \in x \,\}$, so i_+ is indeed the map mentioned in the statement of the proposition. However, i_+ is a bonded morphism, by Lemma 6.1 of [21]. The only thing that we need to notice is that $\mathfrak{pe}(M), x \Vdash p$ iff $V^M(p) \in x$ iff $V^M(p) \in x \cap U(M)$ iff $M^\delta, x \cap U(M) \Vdash p$, for any propositional variable p of L.

Finally, let $x \in O^{M^\delta}$. Since x is a prime filter of $U(M)$ it is not difficult to see that x is separated from $U(M) \backslash x$, that is for any finite $x' \subseteq x$, $Y \subseteq U(M) \backslash x$, we have that $\bigcap x'$ is not a subset of $\bigcup Y$. Hence, using the prime filter theorem ([9, p. 186], or Lemma 6.1 in [7]), there must be a prime filter of $U(\leqslant^M)$, $y \supseteq x$ such that $y \cap U(M) \backslash x = \varnothing$, so indeed $y = x \cap U(M)$, as desired. Furthermore, since $O^M \in x \subseteq y$, we have that $y \in O^{\mathfrak{pe}(M)}$ as well. A similar argument establishes that the mapping under consideration is surjective. $\qquad \square$

We will say that a class K is closed under surjective relevant directed bisimulations w.r.t. O, if whenever $\langle Z_1, Z_2 \rangle$ is a relevant directed bisimulation surjective w.r.t. O_2 between two models M_1, M_2, then $M_1 \in K$ only if $M_2 \in K$.

Corollary 3.19. *Suppose K is a class of Routley–Meyer $\boldsymbol{B}^{\circ t}$-models for L closed under surjective relevant directed bisimulations w.r.t. O. Then $\mathfrak{pe}(M) \in K$ iff $M^\delta \in K$.*

Proof. Consider the bounded morphism $x \mapsto x \cap U(M)$ (call it g) given by Proposition 3.18. By Proposition 3.7, the pair $\langle Z_1, Z_2 \rangle$ is a relevant directed bisimulation which is surjective w.r.t. both O^{M^δ} and $O^{\mathfrak{pe}(M)}$, where:

$$
\begin{aligned}
x Z_1 y & \quad \text{iff} \quad x \leqslant^{M^\delta} g(y), \\
x Z_2 y & \quad \text{iff} \quad g(x) \leqslant^{M^\delta} y.
\end{aligned}
$$

Consequently, $\mathfrak{pe}(M) \in K$ iff $M^\delta \in K$ by the closure assumption on K. □

Corollary 3.20. *Let M be a Routley–Meyer $\boldsymbol{B}^{\circ t}$-model for L. Then for any prime filter u of $U(W)$ and formula ϕ of L we must have that $V^M(\phi) \in u$ iff $M^\delta, u \Vdash \phi$. Moreover, $M \Vdash \phi$ iff $M^\delta \Vdash \phi$ for any such ϕ.*

Proof. Consider the mapping given in Proposition 3.18. We have that $\mathfrak{pe}(M), x \Vdash \phi$ iff $M^\delta, x \cap U(M) \Vdash \phi$ and that indeed all elements of M^δ are of the form $x \cap U(M)$. Therefore, using Proposition 3.14, we see that if $u = x \cap U(M)$ is a prime filter of $U(W)$, $V^M(\phi) \in u$ iff $V^M(\phi) \in x$ iff $\mathfrak{pe}(M), x \Vdash \phi$ iff $M^\delta, u \Vdash \phi$.

The last part of the result follows from Proposition 3.14 again using Corollary 3.19. □

So, according to Corollaries 3.3, 3.20 and Propositions 3.14, 3.9 we can see that a class of models K definable by a theory of the relevant language L is always going to be closed under surjective relevant directed bisimulations, (definable) prime extensions and disjoint unions. Indeed, both K and $\overline{K}$ will be closed under (definable) prime extensions.

4 Characterizing Definable Classes of Models

In this section, we finally tackle the main result of the paper (Theorem 4.5). We start with a simple lemma, which was first established by Goldblatt for modal logic.

We will call a submodel — in the classical sense of first order logic — M' of a given model M an *inner submodel* if the pair $\langle I, I \rangle$, where I is just the identity relation on M', is a relevant directed bisimulation between M and M'. Indeed, then $\langle I, I \rangle$ is a relevant directed bisimulation surjective w.r.t. $O^{M'}$. Note that, using Corollary 3.3, it follows that a definable class of models is always going to be closed under inner submodels, that is, given a class K, if $M \in K$ and M' is an inner submodel of M, then $M' \in K$.

Lemma 4.1. *Let $\{ M_i \colon i \in I \}$ be a family of $\boldsymbol{B}^{\circ t}$-models and $\prod M_i/U$ an ultraproduct. Then $\prod M_i/U$ is isomorphic to an inner submodel of the ultrapower $(\bigoplus_{i \in I} M_i)^I/U$. Hence, a class closed under disjoint unions, inner submodels and ultrapowers is closed under ultraproducts.*

Proof. Simply consider the mapping $f/U \mapsto f'/U$ where $f' \colon I \longrightarrow \bigcup_{i \in I} W_i$ is defined by $f'(i) = f(i)$. $\square$

Lemma 4.2. *Let L^{corr+} be obtained by adding a list of constants to L^{corr}. Suppose K is a class of Routley–Meyer $\boldsymbol{B}^{\circ t}$-models which is closed under ultraproducts. Then for any set Θ of formulas of L^{corr+}, if Θ is finitely satisfiable in K then Θ is indeed satisfiable in K.*

Proof. This is simply the proof of the compactness theorem using ultraproducts which can be found, for example, in [1] (Theorem 4.1). $\square$

Proposition 4.3. *Let M be a Routley–Meyer $\boldsymbol{B}^{\circ t}$-model and N an ω-saturated Routley–Meyer $\boldsymbol{B}^{\circ t}$-model such that $N \Vdash \phi$ iff $M \Vdash \phi$ for any formula ϕ of L. Then there is a relevant directed bisimulation surjective w.r.t. both O^N and O^{M^δ} between N and M^δ.*

Proof. Given $x \in W^N$, let $f(x) = \{ M(\phi) \colon N, x \Vdash \phi, \phi \in \mathrm{Fmla}(L) \}$. It is not difficult to verify that $f(x)$ is indeed in the domain of M^δ. Now define the following pair of relations $Z_1 \subseteq W^{M^\delta} \times W^N, Z_2 \subseteq W^N \times W^{M^\delta}$:

$$
\begin{aligned}
x Z_1 y &\quad\text{iff}\quad x \subseteq f(y), \\
x Z_2 y &\quad\text{iff}\quad f(x) \subseteq y.
\end{aligned}
$$

Next we show that $\langle Z_1, Z_2 \rangle$ is a relevant directed bisimulation surjective w.r.t. both O^N and O^{M^δ}.

To establish (1) from Definition 3.1 we start by noting that

$$
\begin{aligned}
f(x)^{*\delta} &= \{ M(\phi) \colon M(\sim\phi) \notin f(x), \phi \in \mathrm{Fmla}(L) \} \\
&= \{ M(\phi) \colon N, x \nVdash \sim\phi, \phi \in \mathrm{Fmla}(L) \} \\
&= \{ M(\phi) \colon N, x^{*N} \Vdash \phi, \phi \in \mathrm{Fmla}(L) \} \\
&= f(x^{*N}).
\end{aligned}
$$

Suppose next that xZ_1y, i.e., $x \subseteq f(y)$. Then $f(y^{*_N}) = f(y)^{*_\delta} = \{M(\phi) \colon M(\sim\phi)$ $\notin f(y), \phi \in \text{Fmla}(L)\} \subseteq \{M(\phi) \colon M(\sim\phi) \notin x, \phi \in \text{Fmla}(L)\}$, which means that $f(y)^{*_\delta} Z_2 x^{*_N}$. On the other hand, if xZ_2y, i.e., $f(x) \subseteq y$, we have that $y^{*_\delta} = \{M(\phi) \colon M(\sim\phi) \notin y, \phi \in \text{Fmla}(L)\} \subseteq \{M(\phi) \colon N, x \nVdash \sim\phi, \phi \in \text{Fmla}(L)\} = \{M(\phi) \colon N, x^{*_N} \Vdash \phi, \phi \in \text{Fmla}(L)\} = f(x^{*_N})$, so $y^{*_\delta} Z_1 x^{*_N}$ as desired.

To establish (2) from Definition 3.1 suppose that xZ_1y, i.e., $x \subseteq f(y)$ and $R^N ybc$. Take $f(b)$ and $f(c)$. It is easy to see that $R^\delta x f(b) f(c)$. For assume that $M(\phi) \to M(\psi) = M(\phi \to \psi) \in x$ while $M(\phi) \in f(b)$. Then $N, b \Vdash \phi$ whereas $M(\phi \to \psi) \in f(y)$, which means that $N, y \Vdash \phi \to \psi$, so $N, c \Vdash \psi$ given that $R^N ybc$. Thus, $M(\psi) \in f(c)$ as desired. Moreover, $bZ_2 f(b)$ while $f(c)Z_1c$. On the other hand, assume that xZ_2y, i.e., $f(x) \subseteq y$ and $R^\delta ybc$. Consider the set Δ defined as

$$\{ST_x(\phi) \colon M^\delta, b \Vdash \phi, \phi \in \text{Fmla}(L)\} \cup \{\neg ST_x(\psi) \colon M^\delta, c \nVdash \psi, \psi \in \text{Fmla}(L)\}.$$

Take any finite $\Delta_0 \subseteq \Delta$. We may assume it is $\{ST_x(\phi_0), \ldots, ST_x(\phi_j)\} \cup \{\neg ST_x(\psi_0), \ldots, \neg ST_x(\psi_k)\}$. Now, $M^\delta, y \nVdash \bigwedge_{i<j+1} \phi_i \to \bigvee_{i<k+1} \psi_i$, so $M(\bigwedge_{i<j+1} \phi_i \to \bigvee_{i<k+1} \psi_i) \notin y$, so $M(\bigwedge_{i<j+1} \phi_i \to \bigvee_{i<k+1} \psi_i) \notin f(x)$. The latter means then that $N, x \nVdash \bigwedge_{i<j+1} \phi_i \to \bigvee_{i<k+1} \psi_i$, hence, there must be b_0, c_0 such that $R^N x b_0 c_0$ and $N, b_0 \Vdash \bigwedge_{i<j+1} \phi_i$ while $N, c_0 \nVdash \bigvee_{i<k+1} \psi_i$. By the ω-saturation of N we must have b', c' such that $R^N x b'c'$ while also b' satisfies $\{ST_x(\phi) \colon M^\delta, b \Vdash \phi, \phi \in \text{Fmla}(L)\}$ (i.e., $M^\delta, b \Rrightarrow_L N, b'$) and c' satisfies $\{\neg ST_x(\psi) \colon M^\delta, c \nVdash \psi, \psi \in \text{Fmla}(L)\}$ (i.e., $N, c' \Rrightarrow_L M^\delta, c$). Finally, $b \subseteq f(b')$, i.e., bZ_2b' since $M(\phi) \in b$ implies that $M^\delta, b \Vdash \phi$ which means that $N, b' \Vdash \phi$, so indeed $M(\phi) \in f(b')$, and $f(c') \subseteq c$, i.e., $c'Z_2c$ similarly since $N, c' \Rrightarrow_L M^\delta, c$.

To check (3) from Definition 3.1 first assume that xZ_1y, i.e., $x \subseteq f(y)$. Now, $M^\delta, x \Vdash p$ implies that $M(p) \in x$, so by assumption, $M(p) \in f(y)$ as well. The latter means that $N, y \Vdash p$ by definition of f. On the other hand, if xZ_2y, i.e., $f(x) \subseteq y$ and $N, x \Vdash p$, then $M(p) \in f(x)$. Consequently, $M(p) \in y$, by our assumption, and that means that $M^\delta, y \Vdash p$ as desired.

Clause (4) from Definition 3.1 is established along similar lines to (2) above. For (5), from Definition 3.1 let us assume that xZ_2y, i.e., $f(x) \subseteq y$ and that $x \in O^N$. Hence, $N, x \Vdash \mathbf{t}$, so $M(\mathbf{t}) = O^M \in f(x)$, so $O^M \in y$, which means that $y \in O_{M^\delta}$. The case of Z_1 follows simply by reversing this argument.

Finally, we show that $\langle Z_1, Z_2 \rangle$ is surjective w.r.t. both O^N and O_{M^δ}. So assume first that $x \in O^N$, then $f(x)$ is such that $f(x)Z_1x$ trivially and also $f(x) \in O_{M^\delta}$ given that $N, x \Vdash \mathbf{t}$ implies that $M(\mathbf{t}) \in f(x)$, which in turn means that $f(x) \in O_{M^\delta}$. On the other hand, if $x \in O_{M^\delta}$, then it suffices to find $y \in O^N$ such that $M^\delta, x \nVdash \phi$ implies that $N, y \nVdash \phi$ for all formulas ϕ of L. This will show that yZ_2x. Consider the set $\{ST_x(\mathbf{t})\} \cup \{\neg ST_x(\phi) \colon M^\delta, x \nVdash \phi, \phi \in \text{Fmla}(L)\}$ and take some finite subset

$\{ ST_x(\mathbf{t}) \} \cup \{ \neg ST_x(\phi_0), \ldots, \neg ST_x(\phi_n) \}$. We know that $M^\delta, x \not\Vdash \bigvee j < n + 1\phi_j$, so $M^\delta \not\Vdash \bigvee_{j<n+1} \phi_j$, so $M \not\Vdash \bigvee_{j<n+1} \phi_j$ and by the hypothesis of the proposition, $N \not\Vdash \bigvee_{j<n+1} \phi_j$. Consequently, there is $z \in O^N$ such that $N, z \not\Vdash \bigvee_{j<n+1} \phi_j$, so z satisfies $\{ ST_x(\mathbf{t}) \} \cup \{ \neg ST_x(\phi_0), \ldots, \neg ST_x(\phi_n) \}$. By the ω-saturation of N, there must be y satisfying $\{ ST_x(\mathbf{t}) \} \cup \{ \neg ST_x(\phi) \colon M^\delta, x \not\Vdash \phi, \phi \in \mathrm{Fmla}(L) \}$ as desired. $\qquad\square$

Lemma 4.4. *Let K be a class of Routley–Meyer $\boldsymbol{B}^{\circ t}$-models. If K is closed under surjective (w.r.t. O) relevant directed bisimulations and both K and $\overline{K}$ are closed under definable prime extensions, then both K and its complement are closed under ultrapowers.*

Proof. First suppose $M \in K$ and $\prod M/U$ is an ultrapower of M. Consider next an ultrapower $N = \prod(\prod M/U)/D$ obtained by Keisler's method which is ω-saturated. Applying Łoś's theorem a couple of times, we have that $M \Vdash \phi$ iff $\prod(\prod M/U)/D \Vdash \phi$. Thus, using Proposition 4.3, there is a surjective (w.r.t. O^N) relevant directed bisimulation between M^δ and M.

Now $M^\delta \in K$ by the closure of K under definable prime extensions. By the closure under surjective (w.r.t. O) relevant directed bisimulations, we also see that $N \in K$. Again applying Proposition 4.3 with N and $\prod M/U$ this time, there must be a relevant directed bisimulation surjective w.r.t. $O^{\prod M/U}$ between N and $(\prod M/U)^\delta$, so indeed the latter is in K and since $\overline{K}$ is closed under definable prime extensions, we must have that $\prod M/U \in K$, as desired.

On the other hand, if $M \in \overline{K}$, since also $M^\delta \in \overline{K}$, then by Proposition 4.3, we have that $\prod(\prod M/U)/D \in \overline{K}$ by the closure of K under surjective (w.r.t. O) relevant directed bisimulations. Moreover, $\prod M/U \in \overline{K}$ by the above established closure of K under ultrapowers. $\qquad\square$

Theorem 4.5. *Let K be a class of Routley–Meyer $\boldsymbol{B}^{\circ t}$-models. Then the following are equivalent:*

(i) K is relevantly definable, that is, $K = \mathrm{Mod}(\Theta)$ for some collection Θ of formulas of L.

(ii) K is closed under surjective (w.r.t. O) relevant directed bisimulations and disjoint unions while both K and its complement $\overline{K}$ are closed under prime filter extensions.

(iii) K is closed under surjective (w.r.t. O) relevant directed bisimulations and disjoint unions while both K and its complement $\overline{K}$ are closed under definable extensions.

(iv) K is closed under surjective (w.r.t. O) relevant directed bisimulations and disjoint unions while both K and its complement $\overline{K}$ are closed under ultrapowers.

Proof. (i) $\Rightarrow$ (ii): Closure under surjective (w.r.t. O) relevant directed bisimulations comes from Corollary 3.3. Closure under disjoint unions is a consequence of Proposition 3.9. Finally, the remaining closure properties follow from Proposition 3.14.

(ii) $\Rightarrow$ (iii): By Corollary 3.19, we see that if K is closed under prime filter extensions, then it is also closed under definable extensions.

(iii) $\Rightarrow$ (iv): By Lemma 4.4.

(iv) $\Rightarrow$ (i): Suppose that K and $\overline{K}$ are closed as indicated. Observe that K is closed under ultraproducts according to Lemma 4.1. Consider

$$\mathrm{Th}_{rel}(K) = \{\, \phi \in L \colon M \Vdash \phi, \text{ for all } M \in K \,\}.$$

All we need to do is show that $K = \mathrm{Mod}(\mathrm{Th}_{rel}(K))$. The direction $K \subseteq \mathrm{Mod}(\mathrm{Th}_{rel}(K))$ is obvious, so let $M \in \mathrm{Mod}(\mathrm{Th}_{rel}(K))$ to establish that $M \in K$. We may assume that M is an ω-saturated model (for if we can establish that some ω-saturated ultrapower of M is in K then since $\overline{K}$ is closed under ultrapowers, M will be in K).

Expand L^{corr} to a language L^{corr+} by adding a constant c_a for each $a \in O^M$. Next consider the union Δ of the following sets of formulas of L^{corr+}:

$$\{\, \neg ST_x^{c_a/x}(\phi) \colon M \not\Vdash \phi, \phi \in \mathrm{Fmla}(L) \,\} \cup \{\, ST_x^{c_a/x}(\mathbf{t}) \,\} \quad (a \in O^M).$$

Now we can see that Δ is finitely satisfiable in K, which in conjunction with K's closure under ultraproducts, lets us conclude, by appealing to Lemma 4.2, that indeed Δ is satisfiable in K. So take any finite $\Delta_0 \subseteq \Delta$. Without loss of generality, we may assume that Δ_0 is of the following form:

$$\{\, \neg ST_x^{c_{a_0}/x}(\phi) \colon M \not\Vdash \phi_{0j}, \phi_{0j} \in \mathrm{Fmla}(L), j < l_0 \,\} \cup \{\, ST_x^{c_{a_0}/x}(\mathbf{t}) \,\} \cup \cdots$$
$$\cup \{\, \neg ST_x^{c_{a_n}/x}(\phi) \colon M \not\Vdash \phi_{nj}, \phi_{nj} \in \mathrm{Fmla}(L), j < l_n \,\} \cup \{\, ST_x^{c_{a_n}/x}(\mathbf{t}) \,\}.$$

Since $a_p \leqslant^M a_p$ $(p < n+1)$, we must have that $M, a_p \not\Vdash \bigvee_{j<l_p} \phi_{pj}$, so $\bigvee_{j<l_p} \phi_{pj} \notin \mathrm{Th}_{rel}(K)$. The latter implies, by definition, that there is a model $N_p \in K$ such that for some $b_p \in O_p^N$, $N_p, b_p \not\Vdash \bigvee_{j<l_p} \phi_{pj}$. Now take the disjoint union $\bigoplus_{p<n+1} N_p$ which is in K by its closure under this construction. $\bigoplus_{p<n+1} N_p$ can be expanded in the obvious way to a model of Δ_0, that is, the constant c_{a_p} gets assigned the element b_p.

Given a model $N \in K$ of Δ, we can assume it is ω-saturated by the closure of K under ultrapowers. Now we have that for each $a \in O^M$, there is $a' \in O^N$ such that $N, a' \Rightarrow_L M, a$. But the relation $\Rightarrow_L$ can be used to define a relevant directed bisimulation between ω-saturated models according to Proposition 3.5, and in this case we get indeed a surjective directed bisimulation with respect to O^M. Thus, we must have that $M \in K$ as desired by the closure properties of K. $\qquad\square$

Observe that, as a referee points out, even though on the face of it, this proof might seem to show that we only needed closure under finite disjoint unions, this is in fact not true. The reason is that we appealed to Lemma 4.1, where closure under arbitrary disjoint unions was required.

5   Concluding Remarks

We have seen that a class K of Routley–Meyer models is relevantly definable iff K is closed under surjective (w.r.t. O) relevant directed bisimulations and disjoint unions while both K and its complement $\overline{K}$ are closed under prime filter extensions. Moreover, prime filter extensions could be replaced by definable prime filter extensions. This provides a complete characterization of definability in relevant languages at the level of models.

One final word on our inclusion of $\bot$ in our language (we do not discuss $\top$ for, in fact, it is definable as $\sim\bot$). Note that if we look at the empty class of models $\varnothing$ and L has only the set of connectives $\{\sim, \wedge, \vee, \rightarrow, \circ, \mathbf{t}\}$, then $\varnothing$ is simply not definable in L for a model with a trivial world (in the sense that it satisfies all the formulas of L) included among the distinguished worlds cannot really be ruled out. But $\varnothing$ is trivially closed under any model-theoretic relations or constructions, so we could not get a characterization like the one we have presented here. One way to solve this would be to add to the language a connective $\bot_O$ with the semantics: $M, w \not\Vdash \bot_O$ if $w \in O^M$. We could have gone this way rather than added $\bot$, but we have decided in favor of the latter for simplicity and to work in the same setting as [22], where we have gotten our duality theory from.

Finally, this paper is part of a larger project by the author to fill in some gaps in the literature on the Routley–Meyer semantics and bring it up to date with the state of the research on the Kripkean semantics for modal logic. We are under the, perhaps misguided, impression that the late 1970s, the 1980s and the 1990s were the golden days of the Routley–Meyer framework in terms of the number of people actively working in the field and results appearing. In this sense, following the current political climate, we are seeking to make the Routley–Meyer semantics great again.

Acknowledgments

We are grateful to an anonymous referee who provided a great number of very useful corrections. Their careful reading of a previous version of this paper helped to make it a much better final text. We also wish to thank the audience at the *Third Workshop* held in Edmonton, in particular, Kit Fine, Ed Mares, Michael Dunn, and Arnon Avron. Without the support of Katalin Bimbó and Zach Weber we wouldn't have been able to attend such an exciting event. A Marsden Fund Grant awarded to Zach Weber by the Royal Society of New Zealand partially funded research for this paper. Finally, we also acknowledge the support by the Austrian Science Fund (FWF): project I 1923-N25 (*New perspectives on residuated posets*).

References

[1] John L. Bell and Alan B. Slomson. *Models and Ultraproducts: An Introduction.* North-Holland Publishing Company, Amsterdam, 1971. (Dover, Mineola, NY, 2006).

[2] M. Bílková, R. Horcík, and J. Velebil. Distributive substructural logics as coalgebraic logics over posets. In *Advances in Modal Logic*, volume 9, pages 119–142. College Publications, 2012.

[3] Katalin Bimbó and J. Michael Dunn. *Generalized Galois Logics. Relational Semantics of Nonclassical Logical Calculi*, volume 188 of *CSLI Lecture Notes*. CSLI Publications, Stanford, CA, 2008.

[4] Patrick Blackburn, Maarten de Rijke, and Yde Venema. *Modal Logic*, volume 53 of *Cambridge Tracts in Theoretical Computer Science*. Cambridge University Press, Cambridge, UK, 2001.

[5] Ross T. Brady, editor. *Relevant Logics and Their Rivals. A Continuation of the Work of R. Sylvan, R. Meyer, V. Plumwood and R. Brady*, volume II. Ashgate, Burlington, VT, 2003.

[6] C. C. Chang and H. J. Keisler. *Model Theory*. North-Holland, Amsterdam, 1992.

[7] R. Goldblatt. Axiomatic classes of intuitionistic models. *Journal of Universal Computer Science*, 11(12):1945–1962, 2005.

[8] R. Goldblatt and S. K. Thomason. Axiomatic classes in propositional modal logic. In J. Crossley, editor, *Algebra and Logic*, pages 163–173. Springer, 1974.

[9] Robert Goldblatt. Varieties of complex algebras. *Annals of Pure and Applied Logic*, 44:173–242, 1989.

[10] Wilfrid Hodges. *Model Theory*, volume 42 of *Encyclopedia of Mathematics and its Applications*. Cambridge University Press, Cambridge, UK, 1993.

[11] D. Kaplan. Review of "Semantical analysis of modal logic I" by S. Kripke. *Journal of Symbolic Logic*, 31:120–122, 1966.

[12] A. Kurz and J. Rosický. The Goldblatt–Thomason theorem for coalgebras. In *Proceedings of CALCO 2007*, number 4624 in LNCS, pages 342–355, 2007.

[13] Edwin D. Mares. Halldén-completeness and modal relevant logic. *Logic et Analyse*, 181:59–76, 2003.

[14] Greg Restall. *An Introduction to Substructural Logics*. Routledge, London, UK, 2000.

[15] M. de Rijke. Modal model theory. Technical Report CS–R9517, Computer Science/Department of Software Technology, University of Amsterdam, Amsterdam, 1995.

[16] P. H. Rodenburg. *Intuitionistic Correspondence Theory*. PhD Thesis, University of Amsterdam, 1986.

[17] Richard Routley and Robert K. Meyer. The semantics of entailment – II. *Journal of Philosophical Logic*, 1:53–73, 1972.

[18] Richard Routley and Robert K. Meyer. The semantics of entailment – III. *Journal of Philosophical Logic*, 1:192–208, 1972.

[19] Richard Routley and Robert K. Meyer. The semantics of entailment. In H. Leblanc, editor, *Truth, Syntax and Modality. Proceedings of the Temple University Conference on Alternative Semantics*, pages 199–243, Amsterdam, 1973. North-Holland.

[20] Richard Routley, Robert K. Meyer, Val Plumwood, and Ross T. Brady. *Relevant Logics and Their Rivals*, volume 1. Ridgeview Publishing Company, Atascadero, CA, 1982.

[21] Takahiro Seki. General frames for relevant modal logics. *Notre Dame Journal of Formal Logic*, 44(2):93–109, 2004.

[22] Alasdair Urquhart. Duality for algebras of relevant logics. *Studia Logica*, 56: 263–276, 1996.

[23] Y. Venema. Model definability, purely modal. In J. Gerbrandy, M. Marx, M. de Rijke, and Y. Venema, editors, *JFAK. Essays Dedicated to Johan van Benthem on the Occasion of his 50th Birthday*. Amsterdam University Press, Amsterdam, 1999.

Received 12 June 2016

KRIPKE–GALOIS FRAMES AND THEIR LOGICS

CHRYSAFIS HARTONAS
Department of Computer Science and Engineering,
University of Applied Sciences of Thessaly (TEI of Thessaly), Thessaly, Greece
<hartonas@teilar.gr>

Abstract

This article introduces Kripke–Galois frames and, more specifically, τ-frames for a similarity type τ, and studies their logics. τ-logics include a number of well-known logic systems and we focus here on providing a semantic treatment of familiar substructural logics, ranging from the Full Lambek and Lambek–Grishin calculi, to Linear Logic (with, or without exponentials) and to non-distributive Relevance Logic.

Keywords: gaggle theory, Lambek calculus, linear logic, non-distributive logics, relational semantics, relevance logic

1 Introduction

Kripke–Galois relational semantics is proposed as an alternative and improvement (in the opinion of this author at least) over generalized Kripke frames [11], bi-approximation semantics [27], or approaches [10, 1] building on Urquhart's representation of lattices [28], including TiRS graphs [7]. Whereas [11] builds on the canonical extensions theory of Gehrke and Harding [12], where the latter builds on this author's shared work with Dunn on lattice representation [24], the framework presented here is developed in the context of this author's Stone duality for lattice expansions [16] and it may be also seen as an elaboration of Dunn's theory of Generalized Galois Logics (gaggle theory) [8], a comprehensive presentation of which can be found in Bimbó and Dunn [6].

Kripke–Galois frames are a simple generalization of Kripke frames and the focus of the approach lies, first, with delineating appropriate classes of first-order definable frames for the logics of interest and second with re-capturing the classical interpretation of familiar operators, such as possibility, or cotenability (fusion), despite the lack of distribution of conjunctions over disjunctions and conversely. The significance of this latter goal is that it makes it possible to study applied logics, such as temporal, epistemic, or dynamic logics over a non-distributive propositional basis [22, 20]. For example, tense logic is interpreted in linear orders and the tense operators 'sometimes/always in the future/past' are used to capture properties of points in time. In the generalized Kripke frames approach of [11], the possibility operator is interpreted over 2-sorted frames by the clause

$$X \ni x \Vdash \Diamond\varphi \text{ iff } \forall y \in Y \left(\forall z \in X(z \Vdash \varphi \implies yR_\Diamond z) \implies x \leq_R y \right)$$

and there appears to be no intelligible way to understand the meaning of diamond as a future operator. Similarly, processes are typically interpreted as binary relations and then graphs are the appropriate frame structures for the interpretation of modal operators in dynamic logic. However, again, the semantic condition above makes it hard to understand the meaning of $\langle\alpha\rangle\varphi$ in the approach of [11], not to mention the fact that 2-sorted frames are used in [11].

It is this author's opinion that there is no reason why the absence of distribution of conjunctions over disjunctions and conversely should affect the way we understand and interpret other operators in the logic. It, therefore, becomes important to seek a solution where the standard interpretation of familiar operators is recaptured and this has been the objective of this author's recent research. Naturally, recapturing the meaning that operators have in a distributive context cannot apply to disjunction, which can no longer be interpreted as union. But we have demonstrated in [20, 21] that non-distributive disjunction can be interpreted modally and this sheds some light into its semantics.

Though 'Kripke–Galois semantics' is a term we introduce in this article, published results and applications of the framework to the semantics of non-distributive logics have appeared in [19, 22, 18, 20] and in [21], where the groundwork for the framework of this article first appeared (called 'order-dual relational semantics' in [20, 21]). Our approach is based on a Stone-type representation and duality result for bounded lattices with operators published several years ago [16], with minor improvements, first explicitly presented in [21].

The main objective of the present article is to demonstrate the wide application scope of the Kripke–Galois semantics approach and its ability to handle, in a uniform way, a variety of logical systems lacking distribution. Having studied modal and temporal extensions of non-distributive propositional logic in [22, 20], we turn here to studying a number of well-known substructural logic systems.

Following an introduction to Kripke–Galois frames, to generalized image operators and τ-frames, as well as to τ-logics for some similarity type τ (Section 2), applications of the proposed framework are considered in detail, providing a semantic treatment of both the associative and the non-associative Full Lambek Calculus

(**FL**) with, or without exchange, contraction and weakening (Section 3). An extension of the calculus with the Grishin dual operators is also considered and completeness results for both the minimal Full Lambek–Grishin calculus (**FLG**$_\varnothing$) as well as with Grishin's interaction axioms are proven. **FLG**$_\varnothing$ includes non-commutative, contraction, weakening and negation-free Relevance Logic, or Linear Logic without negation and without exponentials. We then also turn to a study of applications of Kripke–Galois semantics to relational semantics for the logic of De Morgan Lattices and Monoids (Section 5.1), for (non-distributive) **RL** and for **LL** without exponentials (Sections 5.2, 5.3 and 5.4), and, finally, for full Linear Logic (Section 5.5).

Notational Conventions: We use a, b, c, d, e for lattice elements and x, y, z, u, v for lattice filters. $x_a = a \uparrow$ designates the principal filter generated by the lattice element a, while Γ, Δ are used to designate closure operators, typically on subsets of some set X and we simplify notation by writing Γx for the more accurate $\Gamma(\{x\})$, for $x \in X$, and similarly for Δ. Furthermore, we overload the use of $\leq$ whose primary use is for the lattice order and write $x \leq y$ for filter inclusion, $x \leq U$, where $x \in X, U \subseteq X$ as an abbreviation for $\forall u \in U \ x \leq u$. Similarly for $U \leq x$. Also, we let $a \leq x$, for a lattice element a and a filter x, designate the fact that a is a lower bound of the elements in x (i.e., $\forall b \in x \ a \leq b$). Note that $a \leq x$ iff $x \leq x_a$.

2 Kripke–Galois Frames and Models

2.1 From Kripke Frames to Kripke–Galois Frames

By a Kripke–Galois frame we mean any relational structure $(X, R, (R_i)_{i \in I})$, where $R \subseteq X \times X$ is called the *Galois relation* of the frame and each R_i is a relation of some specified arity on the carrier set X of the frame. Whereas any subset of X is classically viewed as a proposition, we restrict to Galois stable subsets, i.e., subsets $A \subseteq X$ such that $A = \lambda \rho A$ and where λ, ρ is the Galois connection generated by the relation R as shown in equations (1, 2) below.

$$\lambda U \ = \ \{x : URx\} = \{x : \forall u \ (u \in U \implies uRx)\} \tag{1}$$
$$\rho V \ = \ \{y : yRV\} = \{y : \forall v \ (v \in V \implies yRv)\} \tag{2}$$

$\mathcal{G}_\lambda(X)$ hereafter designates the set of stable subsets $A = \lambda \rho A = \Gamma A$ of the carrier set of a Kripke–Galois frame $\mathfrak{F} = (X, R, (R_i)_{i \in I})$. The co-stable sets are the sets $B = \rho \lambda B = \Delta B$ and $\mathcal{G}_\rho(X)$ designates the dual family of co-stable sets. The Galois connection restricts to a dual isomorphism of the complete lattices of stable and, respectively, co-stable sets. We refer to members of $\mathcal{G}_\rho(X)$ as the *co-propositions* of the frame. A stable set interpreting a sentence φ will be designated by $[\![\varphi]\!]$, while the co-interpretation of φ will be written as $(\!(\varphi)\!)$, and we shall write $x \Vdash \varphi$ (x satisfies φ) iff $x \in [\![\varphi]\!]$ and $x \Vdash^{\partial} \varphi$ (x co-satisfies, or refutes, φ) iff $x \in (\!(\varphi)\!)$.

It is a straightforward observation that given a binary relation R and where $\overline{R}$ is its complement, if λ, ρ is the Galois connection generated by R as in equations (1, 2) and $\blacksquare, \diamond$ is the residuated pair generated by $\overline{R}$ as in equation (3)

$$\blacksquare U = \{x : \forall x' \, (x'\overline{R}x \implies x' \in U)\} \qquad \diamond V = \{x : \exists x' \, (x\overline{R}x' \text{ and } x' \in V)\} \qquad (3)$$

then the Galois connection and the residuated pair generate the same closure operator, in other words $\lambda\rho = \blacksquare\diamond$. Therefore, if $[\![\varphi]\!]$ stands for the interpretation of a sentence, then $\lambda\rho([\![\varphi]\!] \cup [\![\psi]\!]) = [\![\varphi \vee \psi]\!] = \blacksquare(\diamond[\![\varphi]\!] \cup \diamond[\![\psi]\!])$. Hence, disjunction can be modeled either as the closure of a union, or equivalently, it can be interpreted modally by the clause $x \Vdash \varphi \vee \psi$ iff $\forall y \, (y\overline{R}x \implies \exists z \, (y\overline{R}z \text{ and } (z \Vdash \varphi \text{ or } z \Vdash \psi)))$. The reader is referred to [20] for further details on the modal representation of lattices, a result which was implicit in the lattice representation and duality of [16].

An orthoframe $(X, \perp)$ [14] is a well-known example of what is called here a Kripke–Galois frame (where $\perp$ is symmetric and then $\lambda = \rho$), with no additional relations. TiRS graphs [7] are also instances of what we call Kripke–Galois frames.

An ordinary Kripke frame is a Kripke–Galois frame where $\mathcal{G}_\lambda(X) = \wp(X) = \mathcal{G}_\rho(X)$, i.e., every subset of the carrier set X is both Galois stable and co-stable. This is the case when the Galois relation is chosen to be the non-identity relation xRy iff $x \neq y$, as the reader can easily verify, since in that case each of the maps of the Galois connection is set-complementation and $U = --U$ for any $U \subseteq X$.

Therefore, Kripke–Galois frames, to be used in providing semantics to non-distributive lattice-based logics, are a generalization of classical Kripke frames, used for the semantics of distributive, or Boolean logics. The generalization resides precisely in abandoning set-complementation as the Galois connection with respect to which stable sets are determined, and replacing it by an arbitrary Galois connection.

2.2 τ-Frames and Generalized Image Operators

This section introduces τ-frames, as a subclass of the general Kripke–Galois frame class. Having fixed the semantic structures, our interest is with studying the logics of τ-frames. τ-frames are specified in Definition 2.2, following a preliminary definition of distribution and similarity types and of generalized image operators (Definition 2.1). τ-languages, i.e., the languages of τ-frames are introduced in Definition 2.3. In [21], we studied the minimal logic $\Lambda_0(\tau)$ of τ-frames and we turn here to a number of substructural logical systems which are naturally regarded as τ-logics for appropriate classes of τ-frames.

A *distribution type* is an element δ of the set $\{1, \partial\}^{n+1}$, for some $n \geq 0$, typically to be written as $\delta = (i_1, \ldots, i_n; i_{n+1})$ and where $i_{n+1} \in \{1, \partial\}$ will be referred to as the *output type* of δ. A *similarity type* τ is a finite sequence of distribution types, $\tau = \langle \delta_1, \ldots, \delta_k \rangle$. To each distribution type $\delta = (i_1, \ldots, i_n; i_{n+1})$ Kripke–Galois semantics associates a pair of frame relations $R_\delta, R_\delta^\partial \subseteq X \times X^n$ from which generalized image operators are defined (see Definition 2.1) on $\mathcal{G}_\lambda(X)$ and $\mathcal{G}_\rho(X)$, respectively.

The reason we associate to a distribution type a pair of relations $R_\delta, R_\delta^\partial$, rather than a single relation, is grounded on an essential feature of every lattice representation theorem, where two dually isomorphic concrete meet-semilattices $\mathcal{S} \simeq \mathcal{K}^{op}$ are shown to be isomorphic and dually isomorphic to the original lattice $\mathcal{L}$, see [24, 16, 28, 25]. Thereby, a normal n-ary lattice operator $\mathfrak{f}$ is also both represented as an operator $\odot_f$ and dually represented as an operator $\odot_f^\partial$ in each of $\mathcal{S}$ and $\mathcal{K}$, respectively, and so that $\odot_f(e_1, \ldots, e_n) = \lambda \odot_f^\partial(\rho e_1, \ldots, \rho e_n)$. Each of the relations $R_\delta, R_\delta^\partial$ is used to generate its respective operator $\odot_f$ (on $\mathcal{S}$) and $\odot_f^\partial$ (on $\mathcal{K}$).

Definition 2.1 (Generalized Image Operators). Let (X, R) be a frame with $R \subseteq X \times X$, λ, ρ the generated Galois connection, $\delta = (i_1, \ldots, i_n; i_{n+1})$ a distribution type and let $R_\delta, R_\delta^\partial \subseteq X \times X^n$, for some n depending on δ, be $(n+1)$-ary relations on the set X. When $\delta = (i_1, \ldots, i_n; 1)$ is of output type 1, we designate the relations by $R_\oplus, R_\oplus^\partial$, rather than $R_\delta, R_\delta^\partial$. Similarly, if $\delta = (i_1, \ldots, i_n; \partial)$ is of output type ∂, we use the notation $R_\ominus, R_\ominus^\partial$ for $R_\delta, R_\delta^\partial$. In other words, R_δ is either $R_\oplus$, or $R_\ominus$, depending on the output type of δ, and similarly for R_δ^∂. The relations $R_\oplus, R_\oplus^\partial$ (and similarly for $R_\ominus, R_\ominus^\partial$) are used to define a pair of order-dual operators $\oplus, \oplus^\partial$ ($\ominus, \ominus^\partial$, respectively) and we think of a relation R_δ^∂ as the 'dual' of the relation R_δ. Equations (4, 5) define the *generalized image operators* on $\mathcal{P}(X)$ generated by the

$$\oplus(U_1, \ldots, U_n) = \left\{ x : \exists u_1, \ldots, u_n \left(x R_\oplus u_1 \cdots u_n \wedge \bigwedge_{j=1\cdots n}^{i_j=1} (u_j \in U_j) \wedge \bigwedge_{r=1\cdots n}^{i_r=\partial} (u_r \in \rho U_r) \right) \right\} \quad (4)$$

$$\oplus^\partial(U_1, \ldots, U_n) = \left\{ x : \forall u_1, \ldots, u_n \left(\bigwedge_{j=1\cdots n}^{i_j=1} (u_j \in \lambda U_j) \wedge \bigwedge_{r=1\cdots n}^{i_r=\partial} (u_r \in U_r) \implies x R_\oplus^\partial u_1 \cdots u_n \right) \right\} \quad (5)$$

relations, when $\delta = (i_1, \ldots, i_n; 1)$ is of output type 1, while equations (6, 7)

$$\ominus(U_1, \ldots, U_n) = \left\{ x : \forall u_1, \ldots, u_n \left(\bigwedge_{j=1\cdots n}^{i_j=1} (u_j \in U_j) \wedge \bigwedge_{r=1\cdots n}^{i_r=\partial} (u_r \in \rho U_r) \implies x R_\ominus u_1 \cdots u_n \right) \right\} \quad (6)$$

$$\ominus^\partial(U_1, \ldots, U_n) = \left\{ x : \exists u_1, \ldots, u_n \left(x R_\ominus^\partial u_1 \cdots u_n \wedge \bigwedge_{j=1\cdots n}^{i_j=1} (u_j \in \lambda U_j) \wedge \bigwedge_{r=1\cdots n}^{i_r=\partial} (u_r \in U_r) \right) \right\} \quad (7)$$

define them when $\delta = (i_1, \ldots, i_n; \partial)$ is of output type ∂.

The definition of the image operator $\oplus$ in (4) is a generalization of the Jónsson–Tarski additive image operators in a mere distributive setting (lacking a complementation operator), resulting by the addition of the extra conditions that $u_r \in \rho U_r$, whenever $i_r = \partial$, a case that is captured in a Boolean context by composition with classical negation. The reader may wish to consider the case where the operators are defined on a plain Kripke frame, i.e., the relation R is the non-identity relation xRy iff $x \neq y$ and then $\lambda U = \rho U = -U$.

Definition 2.2 (τ-Frames). Let $\tau = \langle \delta_1, \ldots, \delta_k \rangle$ be a similarity type. A *Kripke–Galois τ-frame* (or simply τ-frame) $\mathfrak{F}_\tau = (X, R, (R_\delta, R_\delta^\partial)_{\delta \in \tau})$ is a frame (X, R) together with a pair of relations $R_\delta, R_\delta^\partial \subseteq X \times X^n$, for each $\delta \in \tau$, where $n + 1 = \ell(\delta)$ is the length of $\delta = (i_1, \ldots, i_n; i_{n+1})$. If $\tau = \langle \rangle$ is the empty sequence, then we refer to the frame as a *lattice frame*.

For each $\delta_\vee = (i_1, \ldots, i_n; 1) \in \tau$ of output type 1, and where $(R_\oplus, R_\oplus^\partial)$ is its corresponding relation pair, let $\oplus, \oplus^\partial$ be the generalized image operators generated by $R_\oplus, R_\oplus^\partial$, respectively, defined by equations (4, 5). Similarly, for each $\delta_\wedge = (i_1, \ldots, i_n; \partial) \in \tau$ of output type ∂ and where $(R_\ominus, R_\ominus^\partial)$ is its corresponding relation pair, let $\ominus, \ominus^\partial$ be the generalized image operators generated by $R_\ominus, R_\ominus^\partial$, respectively, defined by equations (6, 7). The following requirements are placed on the operators of the frame.

1. $\mathcal{G}_\lambda(X)$ is closed under the operators $\oplus, \ominus$, while $\mathcal{G}_\rho(X)$ is closed under the operators $\oplus^\partial, \ominus^\partial$.

2. The operators $\oplus, \oplus^\partial$ and the operators $\ominus, \ominus^\partial$ are order-dual, i.e., they are interdefinable by means of the Galois connection generated by the binary relation R of the frame. More specifically, for any sets $A_1, \ldots, A_n \in \mathcal{G}_\lambda(X)$ and any $D_1, \ldots, D_n \in \mathcal{G}_\rho(X)$ the following two (equivalent) conditions hold: $\oplus(A_1, \ldots, A_n) = \lambda(\oplus^\partial(\rho A_1, \ldots, \rho A_n)), \oplus^\partial(D_1, \ldots, D_n) = \rho(\oplus(\lambda D_1, \ldots, \lambda D_n))$. Similarly, the following two (equivalent) conditions hold: $\ominus(A_1, \ldots, A_n) = \lambda(\ominus^\partial(\rho A_1, \ldots, \rho A_n))$ and $\ominus^\partial(D_1, \ldots, D_n) = \rho(\ominus(\lambda D_1, \ldots, \lambda D_n))$.

A *general τ-frame* $\mathfrak{G}_\tau = (X, R, (R_\delta, R_\delta^\partial)_{\delta \in \tau}, \mathfrak{P}_\lambda)$ is a frame with a distinguished sublattice $\mathfrak{P}_\lambda \subseteq \mathcal{G}_\lambda(X)$ such that $\oplus, \ominus$ restrict to operators of the respective distribution type on $\mathfrak{P}_\lambda$ and similarly for $\oplus^\partial, \ominus^\partial$ and $\mathfrak{P}_\rho = \{\rho A : A \in \mathfrak{P}_\lambda\}$.

2.3 (Co)Interpretation of Propositional τ-Languages

If $\mathcal{L} = (L, \wedge, \vee, 0, 1)$ is a bounded lattice, **PLL** designates Positive Lattice Logic, whose language is generated by the schema $\varphi := p \; (p \in P) \mid \top \mid \bot \mid \varphi \wedge \varphi \mid \varphi \vee \varphi$, where P is a countable set of propositional variables.

Definition 2.3 (Propositional τ-Languages). Given a similarity type τ, the propositional τ-language is the extension $\mathbf{PLL}_\tau$ of the language of $\mathbf{PLL}$ with an n-ary operator symbol $\bigcirc_\delta$ for each $\delta \in \tau$. Explicitly, sentences are generated by the grammar $\varphi := p \; (p \in P) \mid \top \mid \bot \mid \varphi \wedge \varphi \mid \varphi \vee \varphi \mid \bigcirc_\delta(\varphi, \ldots, \varphi) \; (\delta \in \tau)$.

Given a lattice frame $\mathfrak{F} = (X, R)$, where $R \subseteq X \times X$, a *lattice model* $\mathfrak{M} = (\mathfrak{F}, V)$ is a frame together with an *admissible valuation* $V = (V_1, V_2)$ consisting of a pair of valuations $V_1 : P \longrightarrow \mathcal{G}_\lambda(X)$ and $V_2 : P \longrightarrow \mathcal{G}_\rho(X)$ such that $V_1(p) = \lambda V_2(p)$ and then also $V_2(p) = \rho V_1(p)$.

An *interpretation* $[\![\;]\!]$ and *co-interpretation* (or *refutation*) $(\!(\;)\!)$ is a pair of functions extending V_1, V_2, respectively, to all sentences of the language and subject to the conditions in Table 1, together with the constraint that for all φ, $[\![\varphi]\!] = \lambda(\!(\varphi)\!)$ and $(\!(\varphi)\!) = \rho([\![\varphi]\!])$. A *model* on a general lattice frame $\mathfrak{G} = (\mathfrak{F}, \mathfrak{P}_\lambda)$ is a pair $\mathfrak{M} = (\mathfrak{G}, V)$ where V is an admissible valuation as previously detailed, but with the additional requirement that for every propositional variable p, $V_1(p) \in \mathfrak{P}_\lambda$ and then also $V_2(p) \in \mathfrak{P}_\rho$. The satisfaction $\Vdash$ and co-satisfaction $\Vdash^\partial$ relations are defined by $x \Vdash \varphi$ iff $x \in [\![\varphi]\!]$ and $x \Vdash^\partial \varphi$ iff $x \in (\!(\varphi)\!)$.

$x \Vdash p$	iff	$x \in V_1(p)$	$x \Vdash^\partial p$	iff	$x \in V_2(p)$
$x \Vdash \top$		always	$x \Vdash^\partial \bot$		always
$x \Vdash \bot$	iff	$x \in \varnothing_\lambda$	$x \Vdash^\partial \top$	iff	$x \in \varnothing_\rho$
$x \Vdash \varphi \wedge \psi$	iff	$x \Vdash \varphi$ and $x \Vdash \psi$	$x \Vdash^\partial \varphi \vee \psi$	iff	$x \Vdash^\partial \varphi$ and $x \Vdash^\partial \psi$
$x \Vdash \varphi \vee \psi$	iff	$\forall y \, (y\overline{R}x$ implies	$x \Vdash^\partial \varphi \wedge \psi$	iff	$\forall y \, (x\overline{R}y$ implies
		$\exists z \, (y\overline{R}z$ and $(z \Vdash \varphi$ or $z \Vdash \psi))$			$\exists z \, (z\overline{R}y$ and $(z \Vdash^\partial \varphi$ or $z \Vdash^\partial \psi))$
	iff	$\forall y \, (y \Vdash^\partial \varphi \vee \psi$ implies $yRx)$		iff	$\forall y \, (y \Vdash \varphi \wedge \psi$ implies $xRy)$

$$x \Vdash \oplus_\delta(\varphi_1, \ldots, \varphi_n) \text{ iff } \exists u_1, \ldots, u_n (xR_\oplus u_1 \cdots u_n \wedge \bigwedge_{\substack{j=1\cdots n}}^{i_j=1} (u_j \Vdash \varphi_j) \wedge \bigwedge_{\substack{r=1\cdots n}}^{i_r=\partial} (u_r \Vdash^\partial \varphi_r)) \quad (8)$$

$$x \Vdash^\partial \oplus_\delta(\varphi_1, \ldots, \varphi_n) \text{ iff } \forall u_1, \ldots, u_n (\bigwedge_{\substack{j=1\cdots n}}^{i_j=1} (u_j \Vdash \varphi_j) \wedge \bigwedge_{\substack{r=1\cdots n}}^{i_r=\partial} (u_r \Vdash^\partial \varphi_r) \implies xR_\oplus^\partial u_1 \cdots u_n) \quad (9)$$

$$x \Vdash \ominus_\delta(\varphi_1, \ldots, \varphi_n) \text{ iff } \forall u_1, \ldots, u_n (\bigwedge_{\substack{j=1\cdots n}}^{i_j=1} (u_j \Vdash \varphi_j) \wedge \bigwedge_{\substack{r=1\cdots n}}^{i_r=\partial} (u_r \Vdash^\partial \varphi_r) \implies xR_\ominus u_1 \cdots u_n) \quad (10)$$

$$x \Vdash^\partial \ominus_\delta(\varphi_1, \ldots, \varphi_n) \text{ iff } \exists u_1, \ldots, u_n (xR_\ominus^\partial u_1 \cdots u_n \wedge \bigwedge_{\substack{j=1\cdots n}}^{i_j=1} (u_j \Vdash \varphi_j) \wedge \bigwedge_{\substack{r=1\cdots n}}^{i_r=\partial} (u_r \Vdash^\partial \varphi_r)) \quad (11)$$

Table 1: Interpretation and Dual Interpretation

A model $\mathfrak{M} = (\mathfrak{G}, V)$ on a general τ-frame $\mathfrak{G}$ is a lattice model (on the underlying general lattice frame) where the satisfaction and co-satisfaction relations are subject to the conditions of Table 1, where we make the convention to write $\circled{0}_\delta$ for $\circled{0}_\delta$ when $\delta = (i_1, \ldots, i_n; 1)$ is of output type 1 and we write $\ominus_\delta$, respectively, when $\delta = (i_1, \ldots, i_n; \partial)$ is of output type ∂. A sentence φ is *(dually) satisfied* in a model $\mathfrak{M}$ if there is a world $x \in X$ such that $x \Vdash \varphi$ (respectively, $y \Vdash^\partial \varphi$, for some $y \in X$). It is *(dually) valid* in $\mathfrak{M}$ iff it is satisfied (respectively, dually satisfied) at all worlds $x \in X$ (respectively, at all $y \in X$).

A symmetric sequent $\varphi \vdash \psi$ is valid in a model $\mathfrak{M}$ iff for every world x of $\mathfrak{M}$, if $x \Vdash \varphi$, then $x \Vdash \psi$. Equivalently, the sequent is valid in the model iff for every world y, if $y \Vdash^\partial \psi$, then $y \Vdash^\partial \varphi$. The sequent is valid in a general frame $\mathfrak{G}$ if it is valid in every model $\mathfrak{M}$ based on the frame $\mathfrak{G}$. Finally, we say that the sequent is valid in a class $\mathbb{G}$ of general frames iff it holds in every frame in $\mathbb{G}$.

The operators semantically specified by (8–11) include all cases of unary and binary operators of various logical calculi, such as the Full Lambek and Lambek–Grishin calculus, Orthologic, Relevance and Linear Logic etc., see Sections 3–5.5.

Example 2.4. We present some cases of interest for the operators:

- If $\delta = (1; 1)$, then (8) specializes to the clause

$$x \Vdash \circled{0}^{1;1}\varphi \ \text{iff} \ \exists u \, (xR_{\circled{0}^{1;1}}u \ \text{and} \ u \Vdash \varphi)$$

so that $\circled{0}^{1;1}$ is a unary diamond operator $\Diamond$. Similarly, $\circled{0}^{1,\ldots,1;1} = \Diamond$ is an n-ary diamond operator with the familiar satisfaction clause. In particular, the distribution type $\delta = (1, 1; 1)$ corresponds to the binary diamond operator known as the *fusion* operator in substructural and relevance logics.

- If $\delta = (\partial; 1)$, then $\circled{0}^{\partial;1}$ is a falsifiability operator. Indeed, the satisfaction clause (8) provided above becomes

$$x \Vdash \circled{0}^{\partial;1}\varphi \ \text{iff} \ \exists u \, (xR_{\circled{0}^{\partial;1}}u \ \text{and} \ u \Vdash^\partial \varphi)$$

In words, φ is falsifiable at x iff it is refuted at some successor state u of x.

- If $\delta = (1; \partial)$, then $\ominus^{1;\partial}$ is an impossibility operator, i.e., a modally interpreted negation operator $\sim$. This can be seen from the respective clause (10) instantiated below, where $\overline{R}_{\ominus^{1;\partial}}$ is the complement of $R_{\ominus^{1;\partial}}$

$$x \Vdash \ominus^{1;\partial}\varphi \ \text{iff} \ \forall u \, (u \Vdash \varphi \implies xR_{\ominus^{1;\partial}}u) \ \text{iff} \ \forall u \, (x\overline{R}_{\ominus^{1;\partial}}u \implies u \not\Vdash \varphi)$$

- If $\delta = (\partial; \partial)$ then the respective clause (10) reads as follows

$$x \Vdash \ominus^{\partial;\partial}\varphi \ \text{iff} \ \forall u \, (u \Vdash^\partial \varphi \implies xR_{\ominus^{\partial;\partial}}u) \ \text{iff} \ \forall u \, (x\overline{R}_{\ominus^{\partial;\partial}}u \implies u \not\Vdash^\partial \varphi)$$

which is precisely an irrefutability operator.

- If $\delta = (1, \partial; \partial)$, then $\ominus^{1,\partial;\partial} = \multimap$ is an implication operator, with satisfaction clause instantiating (10),

$$x \Vdash \varphi \multimap \psi \quad \text{iff} \quad \forall u, v \, (u \Vdash \varphi \text{ and } v \Vdash^{\partial} \psi \implies x R_{\multimap} uv)$$
$$\text{iff} \quad \forall u, v \, (x \overline{R}_{\multimap} uv \text{ and } u \Vdash \varphi \implies v \Vdash^{\partial} \psi)$$

which we treated in [20] and as noted there it resembles the clause for Relevant implication [2, 3], except for replacing satisfaction of the conclusion at v by its non-refutation. Co-satisfaction is specified by the following clause, instantiating (11)

$$x \Vdash^{\partial} \varphi \multimap \psi \quad \text{iff} \quad \exists u, v \, (x R_{\multimap} uv \text{ and } u \Vdash \varphi, \text{ but } v \Vdash^{\partial} \psi)$$

which is the natural analogue of a clause for negated implication.

- For $\delta = (\partial, 1; \partial)$, the operator $\ominus^{\delta} = \ominus^{\partial,1;\partial}$ is a reverse implication $\multimapinv$, as the reader can easily see by instantiating clause (10).

- The case $\delta = (\partial, \partial; \partial)$ corresponds to a binary non co-refutability operator, with semantic clause (instantiating (10) and after contraposition and writing $\oslash$ for $\ominus^{\partial,\partial;\partial}$)

$$x \Vdash \varphi \oslash \psi \quad \text{iff} \quad \forall u, v \, (x \overline{R}_{\oslash} uv \implies (u \Vdash^{\partial} \varphi \text{ or } v \Vdash^{\partial} \psi))$$
$$\text{iff} \quad \forall u, v \, (x \overline{R}_{\oslash} uv \text{ and } u \Vdash^{\partial} \varphi \implies v \Vdash^{\partial} \psi)$$

In [21], we defined the minimal logic $\Lambda_0(\tau)$ for a similarity type τ as the logic on the τ-language $\mathbf{PLL}_{\tau}$ that includes the usual lattice axioms and rules, as well as monotonicity and distribution axioms for each operator $\bigcirc_{\delta}$ corresponding to its distribution type δ, but no interaction axioms between the operators. The following general soundness and completeness result was proved in [21].

Theorem 2.5 (Completeness, [21]). *Let* $\tau = \langle \delta_1, \ldots, \delta_k \rangle$ *be a similarity type and* $\Lambda_0(\tau)$ *the corresponding minimal propositional logic for this type. Then* $\Lambda_0(\tau)$ *is sound and complete in the class of general τ-frames of Definition 2.2.*

The objective of this article is to apply the Kripke–Galois semantic framework and extend the completeness Theorem 2.4 to various substructural logic systems, which are naturally viewed as τ-logics for appropriate classes of τ-frames.

3   The Full Lambek Calculus

We consider here systems that arise from the Gentzen system **LJ** for intuitionistic logic by dropping a combination of the structural rules of exchange, contraction and weakening (perhaps, also association) and expanding the logical signature of the language to include the operator symbols ∘ (fusion, cotenability), ← (reverse implication) and a constant t. The algebraic semantics of these systems has been investigated by Hiroakira Ono, see [26]. Following Ono, we let **FL** be the system with all structural rules dropped, which is precisely the (associative) Full Lambek calculus, and for $r \subseteq \{c, e, w\}$ we designate by **FL**$_r$ the system resulting by adding to **FL** the structural rules in r (where c abbreviates 'contraction', e abbreviates 'exchange' and similarly for w and 'weakening'). With the exception of **FL**$_{ecw}$, which is precisely **LJ**, distribution of conjunctions over disjunctions and conversely does not hold, unless explicitly postulated in the axiomatization.

An **FL**-algebra is a structure $\langle L, \leq, \wedge, \vee, 0, 1, \leftarrow, \circ, \rightarrow, t \rangle$ where

1. $\langle L, \leq, \wedge, \vee, 0, 1 \rangle$ is a bounded lattice

2. $\langle L, \leq, \circ, t \rangle$ is a partially-ordered monoid (∘ is monotone and associative and t is a two-sided identity element $a \circ t = a = t \circ a$)

3. $\leftarrow, \circ, \rightarrow$ are residuated, i.e., $a \circ b \leq c$ iff $b \leq a \rightarrow c$ iff $a \leq c \leftarrow b$

4. for any $a \in L$, $a \circ 0 = 0 = 0 \circ a$

An **FL**-algebra is known as a *residuated lattice*. **FL**-algebras (residuated lattices) are precisely the algebraic models of the (associative) full Lambek calculus.

An **FL**$_{ew}$-algebra adds to the axiomatization the exchange (commutativity) axiom $a \circ b = b \circ a$ for the cotenability operator (in which case ← and → coincide), as well as the weakening axiom $b \circ a \leq a$, in which case combining with commutativity $a \circ b \leq a \wedge b$ follows. In addition, by $1 \circ t \leq t$, the identity t = 1 holds in **FL**$_{ew}$-algebras. **FL**$_{ew}$-algebras are also referred to in the literature as *full BCK-algebras*, corresponding to full **BCK**-logic, resulting from **BCK** whose purely implicational signature is expanded to include conjunction and disjunction connectives, alongside the cotenability logical operator and the constants 0, 1. Algebraically, they constitute the class of *commutative integral residuated lattices*.

The language of **FL** is displayed below, where P is a non-empty, countable set of propositional variables.

$$L \ni \varphi := p \, (p \in P) \mid \top \mid \bot \mid t \mid \varphi \wedge \varphi \mid \varphi \vee \varphi \mid \varphi \leftarrow \varphi \mid \varphi \circ \varphi \mid \varphi \rightarrow \varphi$$

Note that we have preferred to use the notation common in the substructural logics community for the residuals of the cotenability operator, rather than the typical notation of the Lambek calculus where $\varphi \backslash \psi$ is used instead of $\varphi \to \psi$ and similarly for ψ/φ and $\psi \leftarrow \varphi$.

Since we have no interest in this article in studying proof theoretic issues, we may as well assume that the proof system is presented as a symmetric consequence system, directly encoding the corresponding algebraic specification.

3.1 Kripke–Galois Frames for the Full Lambek Calculus

Cotenability is a binary diamond operator, generated by a relation $R_\otimes$ in a Kripke–Galois frame $\mathfrak{G} = (X, R, R_\otimes, R_\otimes^\partial, \mathfrak{P}_\lambda)$ by equation (4), for $n = 2$, instantiated below.

$$A \otimes C = \{x : \exists u, v \ (xR_\otimes uv \text{ and } u \in A \text{ and } v \in C)\}$$
$$B \otimes^\partial D = \{x : \forall u, v \ (u \in \lambda B \text{ and } v \in \lambda D \implies xR_\otimes^\partial uv)\}$$

The corresponding satisfaction clauses, resulting by instantiating clauses (8, 9), are displayed below.

$$x \Vdash \varphi \circ \psi \quad \text{iff} \quad \exists u, v \ (xR_\otimes uv \text{ and } u \Vdash \varphi \text{ and } v \Vdash \psi)$$
$$x \Vdash^\partial \varphi \circ \psi \quad \text{iff} \quad \forall u, v \ (u \Vdash \varphi \text{ and } v \Vdash \psi \implies xR_\otimes^\partial uv)$$

Observe that fusion (a binary diamond) is interpreted by the familiar clause from the distributive setting and it is this author's opinion that there is no reason why lack of distribution of conjunctions over disjunctions and conversely should force abandoning the way we semantically understand other operators in the logic.

If the Kripke–Galois frame is a plain Kripke frame $(X, \neq, R_\otimes, \overline{R}_\otimes)$, i.e., R is the non-identity relation, $\lambda U = \rho U = -U$, hence $\mathcal{G}_\lambda(X) = \mathcal{P}(X)$, and then $x \Vdash^\partial \varphi$ iff $x \nVdash \varphi$ and the relation $R_\otimes^\partial = \overline{R}_\otimes$ is the complement of $R_\otimes$, then neither R, nor $R_\otimes^\partial$ need to be mentioned and the frame is simply $(X, R_\otimes)$.

What does force a difference of Kripke–Galois semantics from classical Kripke semantics is lack of classical complementation, which dictates the use of Galois-stable sets as propositions, for a non-trivial Galois connection λ, ρ, and where unlike the classical case $\rho A \neq -A$ and $\lambda B \neq -B$. Non-triviality of the Galois connection results in a refutation (co-satisfaction, or dual satisfaction) relation $x \Vdash^\partial \varphi$ which is distinct from $x \nVdash \varphi$ and, similarly, its negation $x \nVdash^\partial \varphi$ is distinct from satisfaction $x \Vdash \varphi$. This becomes relevant when considering the semantics of some operators, such as implication.

Forward implication $\to$ has the distribution type $(1, \partial; \partial)$ (it takes joins in the first and meets in the second argument place to meets) and backwards implication

657

$\leftarrow$ has the distribution type $(\partial, 1; \partial)$. Kripke–Galois frames for forward implication are of the form $(X, R, R_{1\partial;\partial}, R^{\partial}_{1\partial;\partial}, \mathfrak{P}_\lambda)$ and they are equipped with the indicated relations, which generate set operators by equations (6, 7), instantiated below.

$$A \Rightarrow C = \{x : \forall u, v\ (u \in A \text{ and } v \in \rho C \implies x R_{1\partial;\partial} uv)\} \tag{12}$$

$$B \Rightarrow^{\partial} D = \{x : \exists u, v\ (x R^{\partial}_{1\partial;\partial} uv \text{ and } u \in \lambda B, \text{ but } v \in D)\} \tag{13}$$

Correspondingly, thinking that $A = [\varphi], C = [\psi]$ and then $B = (\!(\varphi)\!), D = (\!(\psi)\!)$, the appropriate semantic clauses are obtained by instantiating clauses (10, 11):

$$x \Vdash \varphi \to \psi \quad \text{iff} \quad \forall u, v\ (u \Vdash \varphi \text{ and } v \Vdash^{\partial} \psi \implies x R_{1\partial;\partial} uv)$$

$$x \Vdash^{\partial} \varphi \to \psi \quad \text{iff} \quad \exists u, v\ (x R^{\partial}_{1\partial;\partial} uv \text{ and } u \Vdash \varphi, \text{ but } v \Vdash^{\partial} \psi)$$

We consider Kripke–Galois general frames $(X, \leq, (R_i, R^{\partial}_i)_{i \in I}, \mathfrak{P}_\lambda)$, where R_i, R^{∂}_i are appropriate relations on X and the partial order $\leq$ is the Galois relation of the frame. The partial order generates the Dedekind–MacNeille Galois connection λ, ρ and $\mathcal{G}_\lambda(X), \mathcal{G}_\rho(X)$ designate the stable and co-stable sets, respectively. By choice of the Galois relation, each upper set $x \!\uparrow$ is Γ-stable $x \!\uparrow = \Gamma(x\!\uparrow) = \Gamma(\{x\})$ (which we hereafter designate simply by Γx) and similarly $x \!\downarrow = \Delta(x\!\downarrow) = \Delta x$ is a co-stable set. We set things up in a way that ensures first-order definability of frames. For this purpose, we further require that there is a subset $M \subseteq X$ such that every $A \in \mathfrak{P}_\lambda$ is $A = \Gamma x$ for some $x \in M$ and then it follows that every $B \in \mathfrak{P}_\rho$ is of the form $B = \Delta x = \rho(\Gamma x)$, for some $x \in M$. Thus, frames $(X, \leq, R_{11;1}, R^{\partial}_{11;1}, M)$ can be described in the first-order frame language $L^1(\leq, R_{11;1}, R^{\partial}_{11;1}, M)$, where $M \subseteq X$, since $\mathfrak{P}_\lambda$ is completely determined by M. For use in the sequel we define the properties $\Psi(x, u, v), \Psi^{\partial}(x, u, v)$ by conditions (14, 15):

$$\Psi(x, u, v) \equiv \exists u', v'\ (u \leq u' \wedge v \leq v' \wedge x R_{11;1} u'v') \tag{14}$$

$$\Psi^{\partial}(x, u, v) \equiv \forall u', v'\ (u \leq u' \wedge v \leq v' \longrightarrow x R^{\partial}_{11;1} u'v') \tag{15}$$

Note that, given (4), $\Psi(x, u, v)$ holds iff $x \in \Gamma u \otimes \Gamma v$ and similarly $\Psi^{\partial}(x, u, v)$ holds iff $x \in \Delta u \boxtimes \Delta v$. As usual, for any property Φ we define $\exists! x\ \Phi(x)$ as shorthand for $\exists x\ (\Phi(x) \wedge \forall y\ (\Phi(y) \longrightarrow y = x))$.

Modeling the Cotenability (Fusion) Operator: Groupoid and monoid frames are next defined. Conditions 1(a,b) of Definition 3.1 ensure that each of $\mathfrak{P}_\lambda, \mathfrak{P}_\rho$ is closed under binary intersections. Conditions 1(c,d) ensure that $\mathfrak{P}_\lambda, \mathfrak{P}_\rho$ are closed under $\otimes$ and $\boxtimes = \otimes^{\partial}$, respectively. Condition 2 enforces $A \otimes C = \lambda(\rho A \boxtimes \rho C)$ and $B \boxtimes D = \rho(\lambda B \otimes \lambda D)$, for $A, C \in \mathfrak{P}_\lambda$ and $B, D \in \mathfrak{P}_\rho$. Condition 3 equips $\mathfrak{P}_\lambda$ with a bottom element which is a zero element for $\otimes$. Likewise, Condition 4 equips $\mathfrak{P}_\lambda$ with an identity element $\mathfrak{T} = \Gamma(\mathfrak{e})$. Condition 5 forces distribution of $\otimes$ over joins in $\mathfrak{P}_\lambda$, in each argument place. All of the above are verified in the proof of Lemma 3.2.

Definition 3.1 (Groupoid frames). A structure $(X, \leq, R_{11;1}, R_{11;1}^\partial, M)$, with $M \subseteq X$ and such that $R_{11;1}, R_{11;1}^\partial \subseteq X \times X^2$ and $\leq$ is a partial order on X, will be called a *Groupoid-frame* (**G**-frame) provided that

1. the following closure conditions hold:

    (a) $\forall x, y \exists z [x \in M \wedge y \in M \longrightarrow z \in M \wedge \forall u ((x \leq u \wedge y \leq u) \longleftrightarrow z \leq u)]$

    (b) $\forall x, y \exists z [x \in M \wedge y \in M \longrightarrow z \in M \wedge \forall u ((u \leq x \wedge u \leq y) \longleftrightarrow u \leq z)]$

    (c) $\forall x, y \exists z [x \in M \wedge y \in M \longrightarrow z \in M \wedge \forall u (\Psi(u, x, y) \longleftrightarrow z \leq u)]$

    (d) $\forall x, y \exists z [x \in M \wedge y \in M \longrightarrow z \in M \wedge \forall u (\Psi^\partial(u, x, y) \longleftrightarrow u \leq z)]$

2. For all $x, y, z \in M$ the following equivalences hold:

    (a) $\Psi(z, x, y) \longleftrightarrow \forall z' [\Psi^\partial(z', x, y) \longrightarrow z' \leq z]$

    (b) $\Psi^\partial(z, x, y) \longleftrightarrow \forall z' [\Psi(z', x, y) \longrightarrow z \leq z']$

3. $\exists x (x \in M \wedge \forall y (y \leq x \wedge \forall x', u ((x' R_{11;1} xu \vee x' R_{11;1} ux) \longrightarrow x' = x)))$
    Thus X has an upper bound, which belongs to M and we shall freely use the name ω for it in the sequel (extending the signature of the frame language with the constant ω). Part of the above condition then is equivalent to the statement $\forall x, u ((x R_{11;1} \omega u \vee x R_{11;1} u\omega) \longrightarrow x = \omega)$.

4. $\exists! w \in M \; \forall u, z \; (u \in M \longrightarrow ((\Psi(z, u, w) \longleftrightarrow u \leq z) \wedge (\Psi(z, w, u) \longleftrightarrow u \leq z)))$
    For ease of reference we again extend the signature of the frame language, introducing the constant $\mathfrak{e}$ for the unique $w \in M \subseteq X$ of this axiom.

5. For all $x, y, z \in M$ and all $u \in X$

    (a) $\Psi^\partial(u, x, y) \wedge \Psi^\partial(u, x, z) \longrightarrow \forall u', v' (x \leq u' \wedge y \leq v' \wedge z \leq v' \longrightarrow u R_{11;1}^\partial u'v')$

    (b) $\Psi^\partial(u, x, z) \wedge \Psi^\partial(u, y, z) \longrightarrow \forall u', v' (x \leq u' \wedge y \leq u' \wedge z \leq v' \longrightarrow u R_{11;1}^\partial u'v')$

A *Monoid frame* (**M**-frame) is a frame where the associativity condition (M) holds.

(M) for any points $x \in X$ and $u, v, z \in M$, the following equivalence holds

$$\exists w \in M \; (w R_{11;1} vz \wedge x R_{11;1} uw) \longleftrightarrow \exists w \in M \; (w R_{11;1} uv \wedge x R_{11;1} wz)$$

An *E, C, W-frame* is a monoid frame such that, in addition, the following corresponding condition holds:

(E) For all x, u, v in X, $xR_{11;1}uv$ holds, iff $xR_{11;1}vu$ does.

(C) For all $u \in X$ and any $x, y \in M$, $x \leq u \wedge y \leq u \longrightarrow \Psi(u, x, y)$.

(W) For all $u \in X$ and any $x, y \in M$, $\Psi(u, x, y)$ implies $y \leq u$.

Lemma 3.2. *Let $\mathfrak{G} = (X, \leq, R_{11;1}, R_{11;1}^{\partial}, M)$ be a G-frame and $\otimes$ the binary operator on subsets of X generated according to equation (4) by the relation $R_{11;1}$ and $\boxtimes = \otimes^{\partial}$ the dual operator generated by the relation $R_{11;1}^{\partial}$ by equation (5). Then*

1. *$\mathfrak{P}_\lambda$ is closed under $\otimes$, $\mathfrak{P}_\rho$ is closed under $\boxtimes$ and both are closed under binary intersections, where $\mathfrak{P}_\lambda$ is the set $\{\Gamma x : x \in M\}$ and, similarly, $\mathfrak{P}_\rho = \{\Delta x : x \in M\}$. Furthermore, for any stable sets $A, C \in \mathfrak{P}_\lambda$ and co-stable sets $B, D \in \mathfrak{P}_\rho$ we have $A \otimes C = \lambda(\rho A \boxtimes \rho C)$ and $B \boxtimes D = \rho(\lambda B \otimes \lambda D)$. Hence, $\mathfrak{G}$ is a τ-frame for the similarity type $\tau = \langle (1, 1; 1) \rangle$.*

2. *$(\mathfrak{P}_\lambda, \cap, \vee, \otimes, \mathfrak{T}, \mathfrak{O})$ is a lattice-ordered groupoid with two-sided identity element $\mathfrak{T}$ and zero element $\mathfrak{O}$, where the latter are defined by $\mathfrak{T} = \Gamma(\mathfrak{e})$ and $\mathfrak{O} = \Gamma(\omega)$, while joins in $\mathfrak{P}_\lambda$ are defined by $A \vee C = \lambda(\rho A \cap \rho C)$.*

3. *If condition (M) holds as well, then $(\mathfrak{P}_\lambda, \cap, \vee, \otimes, \mathfrak{T})$ is a lattice-ordered monoid.*

4. *If the frame is an EW-frame, then $(\mathfrak{P}_\lambda, \cap, \vee, \otimes, \mathfrak{T}, \mathfrak{O})$ is a commutative integral lattice-ordered monoid.*

5. *If the frame is an EC-frame, $A \cap C \subseteq A \otimes C$ holds in $\mathfrak{P}_\lambda$.*

Proof. For 1), closure of $\mathfrak{P}_\lambda$ under $\otimes$, of $\mathfrak{P}_\rho$ under $\boxtimes$ and of both under binary intersections follows from Conditions 1(c), 1(d) and 1(a,b), respectively. Now $z \in A \otimes C = \Gamma x \otimes \Gamma y$, for some $x, y \in M$ iff (using Definition (4)) there exist $u \in A, v \in C$, i.e., $x \leq u$ and $y \leq v$, such that $zR_{11;1}uv$. By the second frame condition, given the definition (5) this is equivalent to $z \in \lambda(\Delta x \boxtimes \Delta y) = \lambda(\rho \Gamma x \boxtimes \rho \Gamma y)$. The proof of the second part is similar, now using the second equivalence of the second frame condition.

For 2), it is straightforward to show that $(\mathfrak{P}_\lambda, \cap, \vee, \otimes, \mathfrak{T}, \mathfrak{O})$ is a lattice with bottom element $\Gamma(\omega) = \{\omega\}$. By a simple calculation it is verified that the third condition in the definition of **G**-frames entails that for any set $A \in \mathfrak{P}_\lambda$, it holds that $A \otimes \mathfrak{O} = \mathfrak{O} = \mathfrak{O} \otimes A$ and therefore $\mathfrak{O}$ is a zero element. Next, Condition 4 enforces that $z \in (\Gamma u \otimes \mathfrak{T})$ iff $u \leq z$ and $z \in (\mathfrak{T} \otimes \Gamma u)$ iff $u \leq z$, which is equivalent to $z \in \Gamma u$

and this establishes that $\mathfrak{T}$ is an identity element for $\otimes$. It remains to show that $\otimes$ distributes over binary joins in $\mathfrak{P}_\lambda$, in each argument place. Given part 1 of this lemma, the claim is equivalent to showing that the dual operator $\boxtimes$ distributes over binary intersections in $\mathfrak{P}_\rho$. One direction follows from the monotonicity properties verified in Lemma 2.7 of [21]. Condition 5(a) ensures inclusion in the other direction $(\Delta x \boxtimes \Delta y) \cap (\Delta x \boxtimes \Delta z) \subseteq \Delta x \boxtimes (\Delta y \cap \Delta z)$ holds. Similarly, for Condition 5(b) and distribution in the first argument place.

For 3), using condition (M) of Definition 3.1 and closure of $\mathfrak{P}_\lambda$ under $\otimes$, we have

$$x \in \Gamma u \otimes (\Gamma v \otimes \Gamma z)$$

iff $\quad \exists u', w \; (w \in M \;\wedge\; u \leq u' \;\wedge\; (\exists v', z' \; (v \leq v' \wedge z \leq z' \wedge w R_{11;1} v' z')) \;\wedge\; x R_{11;1} u' w)$

iff $\quad \exists u', v', z' \; (u \leq u' \wedge v \leq v' \wedge z \leq z' \;\wedge\; \exists w \in M \; (w R_{11;1} v' z' \;\wedge\; x R_{11;1} u' w))$

iff $\quad \exists u', v', z' \; (u \leq u' \wedge v \leq v' \wedge z \leq z' \;\wedge\; \exists w \in M \; (w R_{11;1} u' v' \;\wedge\; x R_{11;1} w z'))$

iff $\quad \exists w, z' \; (w \in M \;\wedge\; \exists u', v' \; (u \leq u' \;\wedge\; v \leq v' \;\wedge\; w R_{11;1} u' v') \;\wedge\; z \leq z' \;\wedge\; x R_{11;1} w z')$

iff $\quad x \in (\Gamma u \otimes \Gamma v) \otimes \Gamma z,$

hence, $\otimes$ is associative on $\mathfrak{P}_\lambda$.

For 4), the (E) condition implies directly that the $\otimes$ operator is commutative, while if the (W) condition holds then it immediately follows that for any $A, C \in \mathfrak{P}_\lambda$ the inclusion $A \otimes C \subseteq C$ obtains. The (C) condition directly enforces inclusion of intersections $A \cap C$ into their product $A \otimes C$, for any $A, C \in \mathfrak{P}_\lambda$. $\qquad\square$

Modeling Implication: The following properties will be useful in this section, defined in the first-order frame language $L^1(\leq, R_{1\partial;\partial}, R^\partial_{1\partial;\partial}, R_{\partial 1;\partial}, R^\partial_{\partial 1;\partial}, M)$.

$$\Phi(x, u, v) \qquad \text{iff} \qquad \forall u', v' \; (u \leq u' \;\wedge\; v' \leq v \;\longrightarrow\; x R_{1\partial;\partial} u' v') \tag{16}$$

$$\Phi^\partial(x, u, v) \qquad \text{iff} \qquad \exists u', v' \; (u \leq u' \;\wedge\; v' \leq v \;\wedge\; x R^\partial_{1\partial;\partial} u' v') \tag{17}$$

$$\Theta(x, u, v) \qquad \text{iff} \qquad \forall u', v' \; (u \leq u' \;\wedge\; v' \leq v \;\longrightarrow\; x R_{\partial 1;\partial} v' u') \tag{18}$$

$$\Theta^\partial(x, u, v) \qquad \text{iff} \qquad \exists u', v' \; (u \leq u' \;\wedge\; v' \leq v \;\wedge\; x R^\partial_{\partial 1;\partial} v' u') \tag{19}$$

To fix our notation, we adopt the conventions below:

• An operator for the distribution type $(1, \partial; \partial)$ (the distribution type of forward implication) will be designated by $\Rightarrow$. It is defined using equation (6), instantiated below for subsets of the form $\Gamma u, \Gamma v$

$$\Gamma u \Rightarrow \Gamma v \;=\; \{x : \Phi(x, u, v)\} \tag{20}$$

• Its dual, $\Rightarrow^\partial$, will be designated by $\Downarrow$. It is defined using equation (7), instantiated below for the case of interest

$$\Delta u \Downarrow \Delta v \;=\; \{x : \Phi^\partial(x, u, v)\} \tag{21}$$

- An operator for the distribution type $(\partial, 1; \partial)$ (the distribution type of backwards implication) will be designated by $\Leftarrow$. It is also defined using equation (6), instantiated below for the case of interest

$$\Gamma v \Leftarrow \Gamma u = \{x : \Theta(x, u, v)\} \tag{22}$$

- Its dual, $\Leftarrow^{\partial}$, will be designated by $\Uparrow$, defined also by equation (7), instantiated below for the case of interest

$$\Delta v \Uparrow \Delta u = \{x : \Theta^{\partial}(x, u, v)\} \tag{23}$$

We next introduce a definition for biimplicative frames. Conditions 1(a, b) are the same as the corresponding conditions in the definition of groupoid frames and their significance is that they enforce that each of $\mathfrak{P}_\lambda, \mathfrak{P}_\rho$ is closed under binary intersections. The remaining Conditions 1(c, d) enforce closure of $\mathfrak{P}_\lambda$ under the implication operators $\Leftarrow$ and $\Rightarrow$, as well as closure of $\mathfrak{P}_\rho$ under the dual operators $\Downarrow$ and $\Uparrow$. Condition 2 ensures that dual operators are interdefinable using the Galois connection, i.e., $A \Rightarrow C = \lambda(\rho A \Downarrow \rho C)$ and $B \Downarrow D = \rho(\lambda B \Rightarrow \lambda D)$, for $A, C \in \mathfrak{P}_\lambda$ and $B, D \in \mathfrak{P}_\rho$, and similarly for $\Leftarrow$ and $\Uparrow$. Conditions 3–4 enforce the distribution properties of $\Rightarrow$ and $\Leftarrow$, while Condition 5 enforces the residuation condition of biimplicative lattices. All of these claims are proven in Lemma 3.4.

Definition 3.3 (Biimplicative frames). A *biimplicative-frame* is a structure $(X, R, (R_{1\partial;\partial}, R^{\partial}_{1\partial;\partial}), (R_{\partial 1;\partial}, R^{\partial}_{\partial 1;\partial}), M)$ such that

1. the following closure conditions hold:

    (a) $\forall x, y \exists z [x \in M \wedge y \in M \longrightarrow z \in M \wedge (\forall u \, (x \leq u \wedge y \leq u) \longleftrightarrow z \leq u)]$

    (b) $\forall x, y \exists z [x \in M \wedge y \in M \longrightarrow z \in M \wedge (\forall u \, (u \leq x \wedge u \leq y) \longleftrightarrow u \leq z)]$

    (c) $\forall x, y \exists z [x \in M \wedge y \in M \longrightarrow z \in M \wedge \forall u \, (\Upsilon(u, x, y) \longleftrightarrow z \leq u)]$,
    for each case $\Upsilon \in \{\Phi, \Theta\}$

    (d) $\forall x, y \exists z [x \in M \wedge y \in M \longrightarrow z \in M \wedge \forall u \, (\Upsilon^{\partial}(u, x, y) \longleftrightarrow u \leq z)]$,
    for each case $\Upsilon^{\partial} \in \{\Phi^{\partial}, \Theta^{\partial}\}$.

2. For all $x, y, z \in M$ the following hold, for each case $\Upsilon \in \{\Phi, \Theta\}$ with its corresponding $\Upsilon^{\partial} \in \{\Phi^{\partial}, \Theta^{\partial}\}$:

    (a) $\Upsilon(z, x, y) \longleftrightarrow \forall z' \, [\Upsilon^{\partial}(z', x, y) \longrightarrow z' \leq z]$

    (b) $\Upsilon^{\partial}(z, x, y) \longleftrightarrow \forall z' \, [\Upsilon(z', x, y) \longrightarrow z \leq z']$.

3. For all $x, y, z \in M$ and any $w \in X$,

(a) $\Phi(w,x,z) \wedge \Phi(w,y,z) \longrightarrow \forall u_1, v_1 \ (v_1 \leq z \ \wedge \ \forall u' \ (u' \leq x \wedge u' \leq y \longrightarrow u' \leq u_1) \longrightarrow wR_{1\partial;\partial}u_1v_1)$

(b) $\Phi(w,x,y) \wedge \Phi(w,x,z) \longrightarrow \forall u_1, v_1 \ (x \leq u_1 \ \wedge \ \forall v' \ (y \leq v' \wedge z \leq v' \longrightarrow v_1 \leq v') \longrightarrow wR_{1\partial;\partial}u_1v_1)$.

4. For all $x, y, z \in M$ and any $w \in X$,

(a) $\Theta(w,x,y) \wedge \Theta(w,x,z) \longrightarrow \forall u_1, v_1 \ (u_1 \leq x \wedge \forall v' \ (v' \leq y \wedge v' \leq z \longrightarrow v' \leq v_1) \longrightarrow wR_{\partial 1;\partial}u_1v_1)$

(b) $\Theta(w,x,z) \wedge \Theta(w,y,z) \longrightarrow \forall u_1, v_1 \ (z \leq v_1 \ \wedge \ \forall u' \ (x \leq u' \wedge y \leq u' \longrightarrow u_1 \leq u') \longrightarrow wR_{\partial 1;\partial}u_1v_1)$.

5. The following two conditions are equivalent, forall $x, y, z \in M$.

(a) $\forall w \ (y \leq w \longrightarrow \Phi(w,x,z))$

(b) $\forall w \ (x \leq w \longrightarrow \Theta(w,z,y))$

Lemma 3.4. *Let $\mathfrak{G} = (X, \leq, R_{1\partial;\partial}, R^{\partial}_{1\partial;\partial}, R_{\partial 1;\partial}, R^{\partial}_{\partial 1;\partial}, M)$ be a biimplicative frame and $\Rightarrow, \Leftarrow, \Downarrow, \Uparrow$ the binary operators on subsets of X generated according to equations (20, 21, 22, 23) by the frame relations. Then*

1. *$\mathfrak{P}_\lambda$ is closed under $\Rightarrow, \Leftarrow$, $\mathfrak{P}_\rho$ is closed under $\Downarrow, \Uparrow$ and both are closed under binary intersections, where $\mathfrak{P}_\lambda$ is the set $\{\Gamma x : x \in M\}$ and, similarly, $\mathfrak{P}_\rho = \{\Delta x : x \in M\}$. Furthermore, for any stable sets $A, C \in \mathfrak{P}_\lambda$ and co-stable sets $B, D \in \mathfrak{P}_\rho$ we have $A \Rightarrow C = \lambda(\rho A \Downarrow \rho C)$ and $B \Downarrow D = \rho(\lambda B \Rightarrow \lambda D)$ and similarly for $\Leftarrow$ and $\Uparrow$. Therefore, $\mathfrak{G}$ is a τ-frame for the similarity type $\tau = \langle (1, \partial; \partial), (\partial, 1; \partial) \rangle$.*

2. *$(\mathfrak{P}_\lambda, \cap, \vee, \Leftarrow, \Rightarrow)$ is a biimplicative lattice, where joins in $\mathfrak{P}_\lambda$ are defined by $A \vee C = \lambda(\rho A \cap \rho C)$.*

Proof. Closure under intersections and under the respective operators is immediate by conditions 1(a–d). For interdefinability of the operators, we do one case as an example, leaving the other cases to the interested reader.

$$z \in (\Gamma x \Rightarrow \Gamma y) \quad \text{iff} \quad \Phi(z,x,y)$$
$$\text{iff} \quad \forall z' \ [\Phi^{\partial}(z',x,y) \longrightarrow z' \leq z]$$
$$\text{iff} \quad \forall z' \ (z' \in (\Delta x \Downarrow \Delta y) \longrightarrow z' \leq z)$$
$$\text{iff} \quad z \in \lambda(\Delta x \Downarrow \Delta y) = \lambda(\rho\Gamma x \Downarrow \rho\Gamma y)$$

It follows that $\mathfrak{G}$ is a τ-frame.

For the distribution properties, one direction of the relevant identities follows from the monotonicity properties of the operators, verified in Lemma 2.7 of [21]. For the other direction, Condition 3(a) directly enforces the inclusion $(A \Rightarrow C) \cap (B \Rightarrow C) \subseteq (A \vee B) \Rightarrow C$, for $A, B, C \in \mathfrak{P}_\lambda$, and similarly, for left implication $\Leftarrow$, now using Condition 4(a). For distribution over intersections in the consequent place, again one direction of the inclusion follows by monotonicity, see Lemma 2.7 of [21]. For the other direction, Condition 3(b) directly enforces that $(A \Rightarrow B) \cap (A \Rightarrow C) \subseteq A \Rightarrow (B \cap C)$, and similarly, for $\Leftarrow$, using Condition 4(b), and where $A, B, C \in \mathfrak{P}_\lambda$.

Finally, the equivalence $\Gamma y \subseteq \Gamma x \Rightarrow \Gamma z$ iff $\Gamma x \subseteq \Gamma z \Leftarrow \Gamma y$ is directly derivable from the fifth frame condition. $\qquad \square$

Remark 3.5 (Implicative Frames). Frames with a single relation pair only for the distribution type $(1, \partial; \partial)$ (corresponding to right implication $\Rightarrow$) are defined by simplifying Definition 3.3 in the obvious way. More specifically,

- Conditions 1(a, b) and 2 of Definition 3.3 now have a single instance $\Upsilon = \Phi$ and $\Upsilon^\partial = \Phi^\partial$;

- Condition 4, relating to the distribution properties of $\Leftarrow$ is now redundant;

- Condition 5 now simply needs to capture that $\Gamma x \subseteq \Gamma y \Rightarrow \Gamma z$ iff $\Gamma y \subseteq \Gamma x \Rightarrow \Gamma z$ and this is achieved by replacing the occurrence of Θ in 5(b) by Φ.

Lemma 3.4 can be restated for the case of a single implication and the proof only needs a minor adjustment to the single implication situation.

Modeling Fusion–Implication Residuation:

Definition 3.6 (FL frames). An FL^--frame $\mathfrak{G} = (X, \leq, (R_{11;1}, R^\partial_{11;1}), (R_{1\partial;\partial}, R^\partial_{1\partial;\partial}), (R_{\partial 1;\partial}, R^\partial_{\partial 1;\partial}), M)$ is a structure $\mathfrak{G}$, where

1. $(X, \leq, R_{11;1}, R^\partial_{11;1}, M)$ is a groupoid frame (Definition 3.1)

2. $(X, \leq, R_{1\partial;\partial}, R^\partial_{1\partial;\partial}, R_{\partial 1;\partial}, R^\partial_{\partial 1;\partial}, M)$ is a biimplicative frame (Definition 3.3) and, in addition,

3. the condition $\forall w\ (\Psi(w, x, y) \longrightarrow z \leq w)$, for any $x, y, z \in M$ and where Ψ is defined by (14), is equivalent to each of the two conditions 5(a) and 5(b) of biimplicative frames.

If the associativity condition (M) for groupoid frames holds, then the frame will be called an **FL**-*frame*. An **FL$_e$**-*frame* is an **FL**-frame where, in addition,

1. the Exchange condition (E) for monoid frames holds (Definition 3.1);

2. the frame is now reduced to an implicative frame $(X, \leq, R_{1\partial;\partial}, R_{1\partial;\partial}^{\partial}, M)$, as specified in Remark 3.5;

3. the residuation Condition 3 above assumes that both of the above mentioned Conditions 5(a), 5(b) for biimplicative frames use Φ (see Remark 3.5).

Lemma 3.7. *If $\mathfrak{G}$ is an $\mathbf{FL}^-$-frame, then its dual algebra $(\mathfrak{P}_\lambda, \cap, \varnothing_\lambda, X, \mathfrak{O}, \mathfrak{T}, \Leftarrow,$ $\otimes, \Rightarrow)$ is a non-associative $\mathbf{FL}$-algebra. It is an $\mathbf{FL}$-algebra (i.e., $\otimes$ is associative) if condition (M) holds in the frame and it is an $\mathbf{FL}_e$-algebra if, in addition, condition (E) holds. If, furthermore, condition (W) holds, then it is an $\mathbf{FL}_{ew}$-algebra and if conditions (E) and (C) hold, then it is an $\mathbf{FL}_{ec}$-algebra, i.e., it satisfies $A \cap C \subseteq A \otimes C$.*

Proof. Immediate, left to the interested reader. □

Models and Soundness: The definition of models over general τ-frames for each of the cases of the logic of lattice-ordered groupoids, monoids, or abelian monoids, or biimplicative lattices, as well as over τ-frames for the associative, or non-associative Full Lambek calculus, with or without Exchange and Weakening, is a specialization of the general definition of models over τ-frames (see Section 2.3). For concreteness, we list the related satisfaction and refutation clauses.

$x \Vdash \varphi \circ \psi$	iff	$\exists u, v \, (u \Vdash \varphi \wedge v \Vdash \psi \wedge x R_{11;1} uv)$	iff	$x \in [\![\varphi]\!] \otimes [\![\psi]\!]$
$x \Vdash^{\partial} \varphi \circ \psi$	iff	$\forall u, v \, (u \Vdash^{\partial} \varphi \wedge v \Vdash^{\partial} \psi \implies x R_{11;1}^{\partial} uv)$	iff	$x \in (\!(\varphi)\!) \boxtimes (\!(\psi)\!)$
$x \Vdash \varphi \rightarrow \psi$	iff	$\forall u, v \, (x \overline{R}_{1\partial;\partial} uv \wedge u \Vdash \varphi \implies v \Vdash^{\partial} \psi)$	iff	$x \in [\![\varphi]\!] \Rightarrow [\![\psi]\!]$
$x \Vdash^{\partial} \varphi \rightarrow \psi$	iff	$\exists u, v \, (x R_{1\partial;\partial}^{\partial} uv \wedge u \Vdash \varphi \wedge v \Vdash^{\partial} \psi)$	iff	$x \in (\!(\varphi)\!) \Downarrow (\!(\psi)\!)$
$x \Vdash \varphi \leftarrow \psi$	iff	$\forall u, v (x \overline{R}_{\partial 1;\partial} uv \wedge v \Vdash \psi \implies u \Vdash^{\partial} \varphi)$	iff	$x \in [\![\varphi]\!] \Leftarrow [\![\psi]\!]$
$x \Vdash^{\partial} \varphi \leftarrow \psi$	iff	$\exists u, v \, (x R_{\partial 1;\partial}^{\partial} uv \wedge u \Vdash^{\partial} \varphi \wedge v \Vdash \psi)$	iff	$x \in (\!(\varphi)\!) \Uparrow (\!(\psi)\!)$
$x \Vdash \mathsf{t}$	iff	$x \in \mathfrak{T}$	$x \Vdash^{\partial} \mathsf{t}$ iff	$x \in \rho(\mathfrak{T})$

Theorem 3.8 (Soundness). *The following hold:*

1. *The logic of lattice-ordered groupoids is sound in models over G-frames, and similarly for the logic of lattice-ordered monoids and M-frames, as well as for the logic of lattice-ordered abelian monoids and M-frames with condition (E).*

2. *The logic of (bi)implicational lattices is sound in the class of models over (bi)implicative frames.*

3. *The non-associative Full Lambek Calculus is sound in models over FL⁻-frames and, similarly for the associative Full Lambek calculus and models over FL-frames, as well as the associative-commutative calculus over FL_e-frames.*

Proof. The proof is a straightforward consequence of Lemmas 3.2, 3.4 and 3.7. □

3.2 Canonical Model Construction and Completeness

In this section, we construct a canonical τ-frame and model, based on [21], itself extending and improving work first presented in [16], where a full functorial, Stone-type duality was presented for lattice-ordered algebras (lattice expansions).

Let $(X, \leq)$ be the partially ordered set of filters, including the improper filter ω (the whole lattice) of the Lindenbaum–Tarski algebra of the logic. The partial order will be the Galois relation of the frame. To define relation pairs $R_\delta, R_\delta^\partial$ for each of the operators of corresponding distribution type δ we first define operators on points (filters) of the carrier set X, assuming the logic is equipped with the respective logical connective.

$$u\widehat{\otimes}v = \bigwedge\{(a \circ b)\uparrow : a \leq u \text{ and } b \leq v\} = \{e : \forall a, b \, (a \leq u \text{ and } b \leq v \implies a \circ b \leq e)\} \tag{24}$$

$$u\widehat{\rightarrow}v = \bigvee\{(a \rightarrow b)\uparrow : a \leq u \text{ and } b \in v\} = \{e : \exists a, b \, (a \leq u \text{ and } b \in v \text{ and } a \rightarrow b \leq e)\} \tag{25}$$

$$u\widehat{\leftarrow}v = \bigvee\{(a \leftarrow b)\uparrow : a \in v \text{ and } b \leq u\} = \{e : \exists a, b \, (a \in v \text{ and } b \leq u \text{ and } a \leftarrow b \leq e)\} \tag{26}$$

The definitions follow a general pattern first introduced in [16] and presented in a simpler form in [21], to which we refer the reader for intuitions and details. Define also relations on filters by (27), as shown below:

$$xR_\otimes uv \text{ iff } u\widehat{\otimes}v \leq x \qquad xR_\rightarrow uv \text{ iff } u\widehat{\rightarrow}v \leq x \qquad xR_\leftarrow vu \text{ iff } v\widehat{\leftarrow}u \leq x \tag{27}$$

Let $\otimes, \Rightarrow, \Leftarrow$ be the operators generated by the relations $R_\otimes, R_\rightarrow, R_\leftarrow$ by equations (4, 6) corresponding to the respective distribution types $(1, 1; 1), (1, \partial; \partial)$ and $(\partial, 1; \partial)$. Define also the dual relations by (28).

$$xR_\otimes^\partial uv \text{ iff } x \leq u\widehat{\otimes}v \qquad xR_\rightarrow^\partial uv \text{ iff } x \leq u\widehat{\rightarrow}v \qquad xR_\leftarrow^\partial vu \text{ iff } x \leq v\widehat{\leftarrow}u \tag{28}$$

and let $\boxtimes, \Downarrow, \Uparrow$ be the dual operators, generated by the relations $R_\otimes^\partial, R_\rightarrow^\partial, R_\leftarrow^\partial$, respectively, according to equations (5, 7). Furthermore, let $M \subseteq X$ be the set of principal filters and let $\mathfrak{T} = \Gamma(\mathfrak{e}) = \Gamma x_{\mathfrak{t}}$ and $\mathfrak{O} = \Gamma x_0 = \Gamma\omega$.

Lemma 3.9. *Let $\otimes$ be any of $\leftarrow, \circ, \rightarrow$, of respective distribution type $(i_1, i_2; i_3) \in \{(\partial, 1; \partial), (1, 1; 1), (1, \partial; \partial)\}$. Let also $\widehat{\otimes}$ be any of the defined filter operators $\widehat{\leftarrow}, \widehat{\circ}, \widehat{\rightarrow}$.*

1. *$\otimes, \widehat{\otimes}$ have the same monotonicity type. In other words, if for all $j = 1, 2$, if $x_j \leq u_j$ whenever $i_j = 1$ and $u_j \leq x_j$ whenever $i_j = \partial$, then $\widehat{\otimes}(x_1, x_2) \leq^{i_3} \widehat{\otimes}(u_1, u_2)$, where $\leq^{i_3}$ is $\leq$ if $i_3 = 1$ and it is $\geq$ if $i_3 = \partial$*

2. *$\widehat{\otimes}(x_a, x_b) = \otimes(a, b)\uparrow$, for any lattice elements a, b*

Proof. The lemma is a special case of Lemma 4.3 of [21], to which we refer the reader. □

Lemma 3.10 (Canonical Frame Lemma). *Let* $\mathfrak{G} = (X, \le, R_\otimes, R_\otimes^\partial, R_\rightarrow, R_\rightarrow^\partial, R_\leftarrow, R_\leftarrow^\partial,$ $\mathfrak{O}, \mathfrak{T}, M)$, *where* X *is the filter space of the Lindenbaum–Tarski algebra of the Full Lambek calculus. Then* $\mathfrak{G}$ *is an* **FL⁻***-frame and it is an* **FL***-frame if the associativity condition (M) on the relation* $R_{11;1}$ *is further assumed.*

Proof. The proof is given in a sequence of claims.

Claim 3.11. *The structure* $(X, \le, R_\otimes, R_\otimes^\partial, \mathfrak{O}, \mathfrak{T}, M)$ *is a groupoid frame. If the logic assumes associativity, then it is an M-frame. If the logic further assumes the Exchange and Weakening rules, then the generated set operator* $\otimes$ *is commutative and for* $A, C \in \mathfrak{P}_\lambda$ *the inclusion* $A \otimes C \subseteq C$ *obtains.*

Proof. Recall that a stable set $A \in \mathfrak{P}_\lambda$ of the canonical structure is an upper set Γx_a over a principal filter x_a, for some a in the Lindenbaum–Tarski algebra of the logic. By $\Gamma x_a \cap \Gamma x_b = \Gamma(x_{a \vee b})$ and $\Delta x_a \cap \Delta x_b = \Delta(x_{a \wedge b})$ it follows that each of $\mathfrak{P}_\lambda$ and $\mathfrak{P}_\rho$ is an intersection semilatice. Since λ, ρ restricts to a duality on $\mathfrak{P}_\lambda$ and $\mathfrak{P}_\rho$, each is a lattice, with joins in $\mathfrak{P}_\lambda$ and $\mathfrak{P}_\rho$ defined by $A \vee C = \lambda(\rho A \cap \rho C)$ and $B \vee D = \rho(\lambda B \cap \lambda D)$, respectively. If $\omega = x_0$ is the improper filter (the whole lattice), then $\Gamma \omega = \{\omega\}$ and for any filter x, $\omega \in \Gamma x$, hence $\Gamma \omega = \mathfrak{O}$ is the bottom element of $\mathfrak{P}_\lambda$, while $\Delta \omega = X$ is the top element of $\mathfrak{P}_\rho$. The trivial filter $x_1 = \{1\}$ generates the whole space by $\Gamma x_1 = X$, while $\Delta x_1 = \{x_1\}$, so that these elements are the top and bottom elements of $\mathfrak{P}_\lambda$ and $\mathfrak{P}_\rho$, respectively. Hence $(\mathfrak{P}_\lambda, \cap, \vee, \{x_0\}, X)$ is a bounded lattice and so is its dual isomorphic copy $(\mathfrak{P}_\rho, \cap, \vee, \{x_1\}, X)$.

By the definition of $\otimes$ using the canonical relation,

$$\Gamma u \otimes \Gamma v = \{z : \exists u', v' \ (u \le u' \text{ and } v \le v' \implies u' \widehat{\otimes} v' \le z)\} = \Gamma(u \widehat{\otimes} v)$$

(where Lemma 3.9 was used) and so $\mathcal{G}_\lambda(X)$ is closed under $\otimes$. In particular, given Lemma 3.9, it follows that $\mathfrak{P}_\lambda$ is closed under $\otimes$ as well. Similarly, both $\mathcal{G}_\lambda(X)$ and $\mathfrak{P}_\lambda$ are closed under $\Rightarrow$ and $\Leftarrow$.

By definition of the dual operator $\boxtimes$, using the dual canonical relation and recalling Lemma 3.9 again we obtain

$$\Delta u \boxtimes \Delta v = \{z : \forall u', v' \ (u \le u' \text{ and } v \le v' \text{ and } z \le u' \boxtimes v')\} = \Delta(u \widehat{\otimes} v)$$

so that each of $\mathcal{G}_\rho(X)$ and $\mathfrak{P}_\rho$ is closed under $\boxtimes$. By similar argument, left to the reader, closure of $\mathfrak{P}_\rho$ under $\Downarrow, \Uparrow$ follows.

Since $\Delta x = \rho(\Gamma x)$ and $\Gamma x = \lambda(\Delta x)$ it immediately follows that

$$\Gamma u \otimes \Gamma v = \lambda(\Delta u \boxtimes \Delta v) = \lambda(\rho \Gamma u \boxtimes \rho \Gamma v) \quad \Gamma u \Rightarrow \Gamma v = \lambda(\rho \Gamma u \Downarrow \rho \Gamma v)$$
$$\Gamma u \Leftarrow \Gamma v = \lambda(\rho \Gamma u \Uparrow \rho \Gamma v).$$

It follows from the above that the canonical structure is a τ-frame for $\tau = \langle(1,1;1)\rangle$ and the first and second conditions of G-frames (Definition 3.1) obtain.

For the third condition, the upper bound in the canonical structure is the improper filter ω and since $\omega = x_0$ and M is the set of principal filters we get $\omega \in M$. Note that $\omega \widehat{\otimes} u = \{e : \forall a, b \; (a \leq \omega, \; b \leq u \implies a \circ b \leq e)\}$. If $a \leq \omega$, then $a = 0$, hence $a \circ b = 0 \circ b = 0 \leq e$ for any e and thereby $\omega \widehat{\otimes} u = \omega$, and similarly $u \widehat{\otimes} \omega = \omega$. By definition, $x R_\otimes u \omega$ iff $u \widehat{\otimes} \omega \leq x$ and therefore the hypothesis implies that $\omega \leq x$, i.e., $x = \omega$. Hence, if $x R_\otimes u \omega$ or $x R_\otimes \omega u$, then $x = \omega$ and therefore the third condition on G-frames also holds in the canonical structure.

For the fourth condition, we have set $\mathfrak{e} = x_{\mathsf{t}}$ and $\mathfrak{T} = \Gamma \mathfrak{e} = \Gamma x_{\mathsf{t}}$. For any $x_a \in M$, it is immediate that $\Gamma x_a \otimes \Gamma x_{\mathsf{t}} = \Gamma x_a = \Gamma x_{\mathsf{t}} \otimes \Gamma x_a$, because $a \circ \mathsf{t} = a = \mathsf{t} \circ a$. Uniqueness of x_{t} follows by uniqueness of the identity element t.

For the fifth condition, relating to the distribution properties of $\otimes$, assuming $\Psi^\partial(u, x, y)$ and $\Psi^\partial(u, x, z)$, for $x = x_a, y = x_b, z = x_c \in M$ and given the defining clause for Ψ^∂, the hypothesis is equivalent to $u \in \Delta x_a \boxtimes \Delta x_b$ and $u \in \Delta x_a \boxtimes \Delta x_c$. In fact, either one of these hypotheses suffices to derive the conclusion. We let u', v' be any filters such that $x_a \leq u', x_b \leq v'$ and $x_c \leq v'$ and demonstrate that $u \leq u' \widehat{\otimes} v'$. By $\Delta x_a \boxtimes \Delta x_b = \rho(\Gamma x_a \otimes \Gamma x_b) = \rho \Gamma(x_a \widehat{\otimes} x_b) = \Delta(x_a \widehat{\otimes} x_b)$, if $u \in \Delta x_a \boxtimes \Delta x_b$ and $x_a \leq u', x_b \leq v'$, then it follows that $u \leq x_a \widehat{\otimes} x_b \leq u' \widehat{\otimes} v'$.

By the above arguments, it has been established that the canonical structure is a G-frame.

Now suppose the underlying logic assumes association, in other words that $\circ$ is an associative operator in the Lindenbaum–Tarski algebra of the logic. Let $u = x_a, v = x_b$ and $z = x_c$ and x be any filter. Assume that $w = x_e$ satisfies the left-hand-side of the biconditional (M). Then it follows that $e \leq b \circ c$ and $a \circ e \in x$. Hence, since x is a filter, we obtain $a \circ e \leq a \circ (b \circ c) \in x$. By associativity, $(a \circ b) \circ c \in x$ and then letting $d = a \circ b$ and $w' = x_d$ the conclusion is immediate. The other direction is similar.

For the exchange condition (E) we assume that the underlying logic assumes the Exchange rule, hence $\circ$ is commutative. It is then obvious that $x_a \widehat{\otimes} x_b = x_b \widehat{\otimes} x_a$. In fact, from the definition of the filter operator $\widehat{\otimes}$ and the commutativity assumption for $\circ$, it immediately follows that $\widehat{\otimes}$ is commutative on all filters, which is equivalent to condition (E) in Definition 3.1.

If the logic assumes, in addition, the Weakening rule, then since condition (W) for the canonical structure is reduced to deriving $c \in x$ from $a \circ c \in x$ and $a \circ c \leq c$

holds in the Lindenbaum-Tarski algebra of the calculus, it is immediate that (W) obtains in the canonical structure. □

Claim 3.12. *The canonical structure* $(X, \leq, R_\to, R_\to^\partial, R_\leftarrow, R_\leftarrow^\partial, M)$ *is a biimplicative frame (Definition 3.3).*

Proof. The closure Conditions 1(a, b) are the same as for G-frames and they have already been shown to hold in the proof of Claim 3.11. The stronger closure property of $\mathcal{G}_\lambda(X)$, rather than just $\mathfrak{P}_\lambda$, under $\Leftarrow, \Rightarrow$ actually holds and similarly for $\mathcal{G}_\rho(X)$ and $\mathfrak{P}_\rho$. Since every stable set in the canonical frame is of the form Γx, for some filter x, and the definition of $\Gamma x \Rightarrow \Gamma y$ identifies the latter with $\Gamma(x \stackrel{\Rightarrow}{} y)$, it is immediately seen that Condition 1(c) obtains. The arguments for each of the Conditions 1(d, f) is similar.

For Condition 2, we only discuss 2(a). It is immediate that 2(a) is equivalent in the canonical frame to the equivalence of $z \in \Gamma x \Rightarrow \Gamma y$ and $z \in \rho(\Delta x \Downarrow \Delta y)$, which falls out directly from definitions. The cases 2(b–d) are similar.

For 3(a), the hypothesis is equivalent to $w \in (\Gamma x \Rightarrow \Gamma z) \cap (\Gamma y \Rightarrow \Gamma z)$, i.e., to $x \stackrel{\Rightarrow}{} z \leq w$ and $y \stackrel{\Rightarrow}{} z \leq w$. Let u_1, v_1 be filters such that $v_1 \leq z$ and for any u', if $u' \leq x$ and $u' \leq y$, then $u' \leq u_1$. The latter means that $u_1 = x \cap y$. It needs to be shown that $w R_{1\partial;\partial} u_1 v_1$, i.e., $(x \cap y) \stackrel{\Rightarrow}{} v_1 \leq w$. Under the hypothesis of the condition that $x, y, z \in M$, let $x = x_a, y = x_b, z = x_c$, so that $x \cap y = x_{a \lor b}$, $c \leq v_1$ and both $a \to c, b \to c \in w$, using Lemma 3.9. Since $x \cap y \stackrel{\Rightarrow}{} v_1 \leq x \cap y \stackrel{\Rightarrow}{} z$, by Lemma 3.9, it suffices to verify that $x \cap y \stackrel{\Rightarrow}{} z \leq w$. But this is immediate since the left hand side of the inclusion is the principal filter $x_{a \lor b \to c}$ and $a \lor b \to c = (a \to c) \land (b \to c) \in w$.

The Conditions 3(b) and each of 4(a, b) are similar and they can be safely left to the interested reader.

For the last condition, observe that it follows from definitions that the condition is equivalent to $\Gamma y \subseteq \Gamma x \Rightarrow \Gamma z$ iff $\Gamma x \subseteq \Gamma z \Leftarrow \Gamma y$, for principal filters $x = x_a, y = y_b$ and $z = x_c$. Hence the desired equivalence of the conditions is immediate, given Lemma 3.9 and this completes the proof of the claim. □

Returning to the proof of Lemma 3.10, it only remains to prove residuation of $\otimes$ with $\Leftarrow$ and $\Rightarrow$, i.e., the equivalence claimed in Condition 3 of Definition 3.6 with the equivalent, by the previous claim, Conditions 5(a), 5(b) of Definition 3.3. In the canonical frame, this amounts to showing that $\Gamma x \otimes \Gamma y \subseteq \Gamma z$ iff $\Gamma y \subseteq \Gamma x \Rightarrow \Gamma z$, for principal filters $x = x_a, y = y_b$ and $z = x_c$. Given definitions, this reduces directly to the residuation condition of the underlying logic, $a \circ b \leq c$ iff $b \leq a \to c$. Then the canonical frame is an **FL**$^-$-frame and it is an **FL**-frame when the association rule is assumed, by the proof of Claim 3.11. Hence, the proof of Lemma 3.10 is complete. □

By the arguments in the proof of Claim 3.11, provided in the course of the proof of Lemma 3.10, the following is a direct conclusion.

Corollary 3.13. *If the logic assumes in addition both the exchange and the weakening rule, then the canonical frame is an FL_{ew}-frame.*

Lemma 3.14 (Canonical Interpretation Lemma). *The canonical interpretation and co-interpretation $[\![\varphi]\!] = \{x \in X : [\varphi] \in x\}$ and $(\!(\varphi)\!) = \{x \in X : [\varphi] \leq x\}$ satisfy the recursive conditions of Section 2.3.*

Proof. Set $V_1(p) = \{x : [p] \in x\}$ and $V_2(p) = \{x : [p] \leq x\}$. Then $\rho V_1(p) = \{z : \forall u \, (u \in V_1(p) \text{ implies } z \leq u)\}$. Therefore, for $z \in \rho V_1(p)$, since the principal filter $x_{[p]} \in V_1(p)$, it holds that $z \leq x_{[p]}$, which is equivalent to $[p] \leq z$. Thus $\rho V_1(p) = V_2(p)$. By a similar argument, which is left to the reader, $\lambda V_2(p) = V_1(p)$.

Clearly, $[\![T]\!] = \{x : 1 \in x\} = X$ and $[\![\bot]\!] = \{x : 0 \in x\} = \{\omega\} = \varnothing_\lambda$ in the canonical frame, while $(\!(\top)\!) = \{x : 1 \leq x\} = \{1\} = \varnothing_\rho$ in the canonical frame and $(\!(\bot)\!) = \{x : 0 \leq x\} = X$. For conjunctions, obviously $[\![\varphi \wedge \psi]\!] = [\![\varphi]\!] \cap [\![\psi]\!] = \{x : a \in x \text{ and } b \in x\}$ and $(\!(\varphi \wedge \psi)\!) = \{x : a \wedge b \leq x\}$, where we set $a = [\varphi]$ and $b = [\psi]$. Then $\lambda(\!(\varphi \wedge \psi)\!) = \{z : \forall x \, (a \wedge b \leq x \text{ implies } x \leq z)\}$. Since $a \wedge b \leq x_{a \wedge b}$, it follows $a \wedge b \in z$, for any $z \in \lambda(\!(\varphi \wedge \psi)\!)$, hence $\lambda(\!(\varphi \wedge \psi)\!) \subseteq [\![\varphi \wedge \psi]\!]$ and then also $\rho([\![\varphi \wedge \psi]\!]) \subseteq (\!(\varphi \wedge \psi)\!)$. For the converse inclusions, if $x \in [\![\varphi \wedge \psi]\!]$, i.e., $a \wedge b \in x$, then since x is a filter any y with $a \wedge b \leq y$ will be contained in x, hence $[\![\varphi \wedge \psi]\!] \subseteq \lambda(\!(\varphi \wedge \psi)\!)$. A similar argument establishes that $[\![\varphi \wedge \psi]\!] \subseteq \lambda(\!(\varphi \wedge \psi)\!)$. Hence,

$$
\begin{aligned}
x \Vdash^\partial \varphi \wedge \psi \quad &\text{iff} \quad x \in (\!(\varphi \wedge \psi)\!) = \rho([\![\varphi \wedge \psi]\!]) \\
&\text{iff} \quad \forall y \, (y \in [\![\varphi \wedge \psi]\!] \text{ implies } x \leq y) \\
&\text{iff} \quad \forall y \, (x \not\leq y \text{ implies } y \notin [\![\varphi]\!] \cap [\![\psi]\!]) \\
&\text{iff} \quad \forall y \, (x \not\leq y \text{ implies } y \notin \lambda(\rho[\![\varphi]\!] \cup \rho[\![\psi]\!])) \\
&\text{iff} \quad \forall y \, (x \not\leq y \text{ implies } \exists z \, (z \not\leq y \text{ and } (z \Vdash^\partial \varphi \text{ or } z \Vdash^\partial \psi)))
\end{aligned}
$$

which is precisely the co-satisfaction clause for conjunction, given that the Galois relation of the canonical frame is the partial order $\leq$ of filter inclusion. A similar argument applies to the satisfaction clause for disjunction. The satisfaction and co-satisfaction clauses for the unit element t are immediately seen to hold in the canonical model, by definition of the unit element $\mathfrak{T} = \Gamma x_t$ for the $\otimes$ operator on stable sets.

For the cotenability operator, the proof is provided in the following claim, where recall that we set $a = [\varphi]$, $b = [\psi]$.

Claim 3.15. *For all a, b and any filter x,*

1. $a \circ b \in x$ *iff* $\exists u, v \, (x R_\otimes uv \text{ and } a \in u \text{ and } b \in v)$.

2. $a \circ b \leq x$ iff $\forall u, v \ (a \leq u \ and \ b \leq v \ implies \ xR_{\otimes}^{\partial} uv)$.

Proof. The claim is an instance of Claims 4.9, 4.11 of [21], where we handled the case of any operator of distribution type $(i_1, \ldots, i_n; 1)$. □

By definition of the canonical interpretation and by the proof of the above claim it follows that the satisfaction and co-satisfaction clauses for $\varphi \circ \psi$ are respected by the canonical model.

Claim 3.16. *For all a, b and any filter x,*

1. $a \to b \in x$ iff $\forall u, v \ (a \in u \ and \ b \leq v \ implies \ xR_{\to} uv)$

2. $a \to b \leq x$ iff $\exists u, v \ (xR_{\to}^{\partial} \ and \ a \in u \ and \ b \leq v)$

Proof. Again the claim is an instance of Claims 4.10, 4.12 of [21]. □

The definition of the canonical interpretation and the above claim entail that the satisfaction and co-satisfaction clauses for $\varphi \to \psi$ are respected by the canonical model.

Claim 3.17. *For all a, b and any filter x,*

1. $a \leftarrow b \in x$ iff $\forall u, v \ (a \in u \ and \ b \leq v \ implies \ xR_{\leftarrow} vu)$

2. $a \leftarrow b \leq x$ iff $\exists u, v \ (xR_{\leftarrow} vu^{\partial} \ and \ a \in u \ and \ b \leq v)$

Proof. The proof is similar to that of the previous claim. □

Hence the satisfaction and co-satisfaction clauses for $\varphi \leftarrow \psi$ are respected by the canonical model and the proof of the canonical interpretation lemma is complete. □

Theorem 3.18 (Completeness). *The associative and non-associative Full Lambek calculus, without, or with Exchange and Weakening (Full BCK), as well as its reducts, the logic of groupoids, or monoids and the logic of (bi)implicative bounded lattices are sound and complete in the class of τ-frames, where τ is the corresponding similarity type of the logic.*

Proof. Immediate, by the Canonical Frame and Interpretation Lemmas 3.10, 3.14. □

4 The Full Lambek–Grishin Calculus

In this section, we consider the Lambek–Grishin calculus [15], in fact its full extension with conjunctions and disjunctions, without distribution, to which we refer as the Full Lambek–Grishin calculus (**FLG**). The minimal system, without interaction axioms between the two sets of dual operators will be referred to as **FLG**$_\varnothing$. Algebraically, an **FLG**$_\varnothing$-algebra is a structure $\langle L, \leq, \wedge, \vee, 0, 1, \leftarrow, \circ, \rightarrow, \mathfrak{t}, \leftharpoonup, *, \rightharpoonup, \mathfrak{f} \rangle$ such that the reduct without the constant $\mathfrak{f}$ and the operators $\leftharpoonup, *, \rightharpoonup$ is an **FL**-algebra and in addition, the following residuation axioms (in L^∂) hold: $a*b \geq c$ iff $b \geq a \rightharpoonup c$ iff $a \geq c \leftharpoonup b$. By residuation, $*$ distributes over joins of L^∂, in other words it distributes over meets of L: $a*(b \wedge c) = (a*b) \wedge (a*c)$ and similarly for the first argument place. Hence $*$ behaves like a binary box and its distribution type is $(\partial, \partial; \partial)$. Similarly, $(a \wedge b) \rightharpoonup c = (a \rightharpoonup c) \vee (b \rightharpoonup c)$ and $a \rightharpoonup (b \vee c) = (a \rightharpoonup b) \vee (a \rightharpoonup c)$, hence the distribution type of $\rightharpoonup$ is $(\partial, 1; 1)$ and that of $\leftharpoonup$ is $(1, \partial; 1)$.

The **FLG**$_\varnothing$ language is generated by the schema (where $(p \in P)$)

$$L \ni \varphi ::= p \mid \top \mid \bot \mid \mathfrak{t} \mid \mathfrak{f} \mid \varphi \wedge \varphi \mid \varphi \vee \varphi \mid \varphi \leftarrow \varphi \mid \varphi \circ \varphi \mid \varphi \rightarrow \varphi \mid \varphi \leftharpoonup \varphi \mid \varphi * \varphi \mid \varphi \rightharpoonup \varphi$$

The Grishin dual operator $*$ of cotenability $\circ$ may be semantically understood as a *non-corefutability* operator, witness the definition of respective set operators

$$
\begin{aligned}
A \otimes C &= \{u : \exists u', v' \, (xR_\otimes u'v' \text{ and } u' \in A \text{ and } v' \in C)\} \\
A \oplus C &= \{u : \forall u', v' (u' \in \rho A \wedge v' \in \rho C \implies uR_\oplus u'v')\} \\
&= \{u : \forall u', v' \, (u\overline{R}_\oplus u'v' \implies (u' \notin \rho A \text{ or } v' \notin \rho C))\}
\end{aligned}
$$

Non-corefutability is a binary box-like operator and in plain Kripke frames where $\rho A = -A$ the definition reduces to that of a binary box image operator, generated by the relation $\overline{R}_\otimes$.

4.1 Kripke–Galois Frames for the Full Lambek–Grishin calculus

Kripke–Galois frames for **FLG**$_\varnothing$ are next specified, where the properties defined below will be of use:

$$
\begin{aligned}
\Psi_*(u, x, y) &\equiv \forall u', v' \, (u' \leq x \wedge v' \leq y \implies uR_{\partial\partial;\partial}u'v') \\
\Psi_*^\partial(u, x, y) &\equiv \exists u', v' \, (uR_{\partial\partial;\partial}u'v' \wedge x \leq u' \wedge y \leq v') \quad &(29)\\
\Phi_*(u, x, y) &\equiv \exists u', v' \, (uR_{\partial 1;1}u'v' \wedge u \leq x \wedge y \leq v) \\
\Phi_*^\partial(u, x, y) &\equiv \forall u', v' \, (x \leq u' \wedge v' \leq y \implies uR_{\partial 1;1}^\partial u'v') \quad &(30)\\
\Theta_*(u, x, y) &\equiv \exists u', v' \, (uR_{1\partial;1} \wedge x \leq u' \wedge v' \leq y) \\
\Theta_*^\partial(u, x, y) &\equiv \forall u', v' \, (u' \leq x \wedge y \leq v' \implies uR_{1\partial;1}^\partial u'v') \quad &(31)
\end{aligned}
$$

The properties Ψ_*, Φ_*, Θ_* define respective set operators $\oplus, \rhd, \lhd$, while their duals $\Psi_*^\partial, \Phi_*^\partial, \Theta_*^\partial$ define the dual operators, which we designate, respectively, by $\oplus\!\!\!\!\!\!/\,, \triangledown, \triangle$.

Definition 4.1 (FLG$_\varnothing$-frames). An **FLG$_\varnothing$-frame** is a τ-frame $(X, \leq, (R_{11;1}, R_{11;1}^\partial),$ $(R_{1\partial;\partial}, R_{1\partial;\partial}^\partial), (R_{\partial1;\partial}, R_{\partial1;\partial}^\partial), (R_{\partial\partial;\partial}, R_{\partial\partial;\partial}^\partial), (R_{1\partial;1}, R_{1\partial;1}^\partial), (R_{\partial1;1}, R_{\partial1;1}^\partial), M)$, where $\tau = \langle (1,1;1), (1,\partial;\partial), (\partial,1;\partial), (\partial,\partial;\partial), (1,\partial;1), (\partial,1;1) \rangle$ and

1. $(X, \leq, (R_{11;1}, R_{11;1}^\partial), (R_{1\partial;\partial}, R_{1\partial;\partial}^\partial), (R_{\partial1;\partial}, R_{\partial1;\partial}^\partial), M)$ is an **FL**-frame (Definition 3.6)

2. (closure conditions)

    (a) $\forall x, y \; \exists z \; [x \in M \wedge y \in M \longrightarrow (z \in M \wedge \forall u \; (\Upsilon(u, x, y) \longleftrightarrow z \leq u))]$, for each case $\Upsilon \in \{\Psi_*, \Phi_*, \Theta_*\}$

    (b) $\forall x, y \; \exists z \; [x \in M \wedge y \in M \longrightarrow (z \in M \wedge \forall u \; (\Upsilon^\partial(u, x, y) \longleftrightarrow u \leq z))]$, for each case $\Upsilon^\partial \in \{\Psi_*^\partial, \Phi_*^\partial, \Theta_*^\partial\}$

3. (dual operator interdefinability conditions) For all $x, y, z \in M$ the following equivalences hold, for each case $\Upsilon \in \{\Psi_*, \Phi_*, \Theta_*\}$ with its corresponding $\Upsilon^\partial \in \{\Psi_*^\partial, \Phi_*^\partial, \Theta_*^\partial\}$

    (a) $\Upsilon(u, x, y) \longleftrightarrow \forall z \; (\Upsilon^\partial(z, x, y) \longrightarrow z \leq u)$

    (b) $\Upsilon^\partial(u, x, y) \longleftrightarrow \forall z \; (\Upsilon(z, x, y) \longrightarrow u \leq z)$

4. (existence of left-right identity for $*$ condition)
    $\exists! w \in M \; \forall u, z \; (u \in M \longrightarrow ((\Psi_*(z, u, w) \longleftrightarrow u \leq z) \wedge (\Psi_*(z, w, u) \longleftrightarrow u \leq z)))$
    The unique $w \in M$ of this condition will be designated by the constant $\mathfrak{f}$, extending again the signature of the frame language.

5. ($*$ distribution conditions) For all $x, y, z \in M$ and all $u \in X$,

    (a) $\Psi_*(u, x, y) \wedge \Psi_*(u, x, z) \longrightarrow \forall u', v' \; (x \leq u' \wedge y \leq v' \wedge z \leq v' \longrightarrow u R_{\partial\partial;\partial} u' v')$

    (b) $\Psi_*(u, x, z) \wedge \Psi_*(u, y, z) \longrightarrow \forall u', v' \; (x \leq u' \wedge y \leq u' \wedge z \leq v' \longrightarrow u R_{\partial\partial;\partial} u' v')$

To assist the reader in keeping track of symbols for the various operators and their duals, we display our notational choices in a table below. The last row of the table displays the Lambek and Grishin notation for the operators.

δ	$* : (\partial,\partial;\partial)$	$\rightharpoonup : (\partial,1;1)$	$\leftharpoonup : (1,\partial;1)$	$\circ : (1,1;1)$	$\rightarrow : (\partial,1;\partial)$	$\leftarrow : (1,\partial;\partial)$
1	$\oplus$ (6)	$\rhd$ (4)	$\lhd$ (4)	$\otimes$ (4)	$\Rightarrow$ (6)	$\Leftarrow$ (6)
∂	$\oplus = \oplus^{\partial}$ (7)	$\triangledown = \rhd^{\partial}$ (5)	$\triangle = \lhd^{\partial}$ (5)	$\boxtimes = \otimes^{\partial}$ (5)	$\Downarrow = \Rightarrow^{\partial}$ (7)	$\Uparrow = \Leftarrow^{\partial}$ (7)
LG	$\oplus$	$\oslash$	$\oslash$	$\otimes$	$\backslash$	$/$

For later use we list the following definition.

Definition 4.2 ($\mathbf{FLG}_e^{\rightarrow}$-frames). An $\mathbf{FLG}_e^{\rightarrow}$-*frame* is a τ-frame

$$(X, \leq, (R_{11;1}, R_{11;1}^{\partial}), (R_{1\partial;\partial}, R_{1\partial;\partial}^{\partial}), (R_{\partial\partial;\partial}, R_{\partial\partial;\partial}^{\partial}), M),$$

where $\tau = \langle (1,1;1), (1,\partial;\partial), (\partial,\partial;\partial), \rangle$ and

1. $(X, \leq, (R_{11;1}, R_{11;1}^{\partial}), (R_{1\partial;\partial}, R_{1\partial;\partial}^{\partial}), M)$ is an $\mathbf{FL}_e$-frame (Definition 3.6);

2. the closure Conditions 2(a, b) of Definition 4.1 hold with the restriction of Υ to Ψ, Φ and of Υ^{∂} to $\Psi^{\partial}, \Phi^{\partial}$, respectively;

3. the dual operator interdefinability Conditions 3(a, b) of Definition 4.1 hold, with the restriction again of Υ to Ψ, Φ and of Υ^{∂} to $\Psi^{\partial}, \Phi^{\partial}$, respectively;

4. Condition 4 of Definition 4.1 for the existence of a left-right identity for $*$ holds;

5. Condition 5 of Definition 4.1 for the distribution properties of $*$ holds.

Lemma 4.3. *Let $\mathfrak{G}$ be a frame, let $\mathfrak{P}_{\lambda}$ be the family of upper closures Γx of members $x \in M$ and $\mathfrak{P}_{\rho}$ the family of images ρA of $A \in \mathfrak{P}_{\lambda}$. Let also $F = \Gamma(\mathfrak{f}), I = \Gamma(\mathfrak{t}) \in \mathfrak{P}_{\lambda}$.*

1. *If $\mathfrak{G}$ is an $\mathbf{FLG}_{\varnothing}$-frame, then*

   (a) *$\mathfrak{P}_{\lambda}$ is closed under all operators $\Leftarrow, \otimes, \Rightarrow$ (Lemma 3.7) and $\lhd, \oplus, \rhd$, while $\mathfrak{P}_{\rho}$ is closed under the dual operators $\Uparrow, \boxtimes, \Downarrow$ (Lemma 3.7) as well as $\triangle, \oplus, \triangledown$. Furthermore, for any stable sets $A, C \in \mathfrak{P}_{\lambda}$ and co-stable sets $B, D \in \mathfrak{P}_{\rho}$ the following interdefinability identities hold, in addition to those of Lemma 3.7:*
   $A \oplus C = \lambda(\rho A \oplus \rho B), \qquad A \rhd C = \lambda(\rho A \triangledown \rho B), \qquad A \lhd C = \lambda(\rho A \triangle \rho B).$
   Hence, $\mathfrak{G}$ is a τ-frame (Definition 2.2) for the specific similarity type τ.

   (b) *$(\mathfrak{P}_{\lambda}, \cap, \vee, \Leftarrow, \otimes, \Rightarrow, I)$ is a residuated lattice,*

   (c) *$(\mathfrak{P}_{\lambda}, \cap, \vee, \lhd, \oplus, \rhd, F)$ is a dual residuated lattice (i.e., residuation is with respect to the opposite order), $A \oplus A' \supseteq C$ iff $A' \supseteq A \rhd C$ iff $A \supseteq C \lhd A'$ with identity element $F \oplus A = A = A \oplus F$.*

674

2. *If $\mathfrak{G}$ is an $\mathbf{FLG}_e^{\rightarrow}$-frame, then claims 1(a, b) above hold, with the obvious restrictions to the operators $\otimes, \Rightarrow$ and $\oplus$ only.*

Proof. The frame conditions are completely analogous to those for **FL**-frames and the proof arguments are, correspondingly, completely similar, as the interested reader can easily verify. $\qquad\square$

Definition 4.4 ($\mathbf{FLG}_\varnothing$-models). An $\mathbf{FLG}_\varnothing$-*model* $\mathfrak{M} = (\mathfrak{G}, V)$ is an instance of a τ-model (see Section 2.3) for the specific similarity type $\tau = \langle(1, 1; 1), (1, \partial; \partial), (\partial, 1; \partial), (\partial, \partial; \partial), (1, \partial; 1), (\partial, 1; 1)\rangle$. In particular, instantiating the general semantic definitions (8–11) we obtain the following semantic conditions for the satisfaction and co-satisfaction (refutation) relations, setting $R_* = \overline{R}_{\partial\partial;\partial}, R_\rightarrow = R_{\partial 1;1}, R_\leftharpoonup = R_{1\partial;1}$ and similarly for their duals.

$$x \Vdash \varphi \leftharpoonup \psi \quad \text{iff} \quad \exists u, v \, (xR_\leftharpoonup uv \wedge u \Vdash^\partial \varphi \wedge v \Vdash \psi) \tag{32}$$

$$x \Vdash^\partial \varphi \leftharpoonup \psi \quad \text{iff} \quad \forall u, v \, (u \Vdash \varphi \wedge v \Vdash^\partial \psi \implies xR_\leftharpoonup^\partial uv) \tag{33}$$

$$x \Vdash \varphi \rightarrow \psi \quad \text{iff} \quad \exists u, v \, (xR_\rightarrow uv \wedge u \Vdash \varphi \wedge v \Vdash^\partial \psi) \tag{34}$$

$$x \Vdash^\partial \varphi \rightarrow \psi \quad \text{iff} \quad \forall u, v \, (u \Vdash^\partial \varphi \wedge v \Vdash \psi \implies xR_\rightarrow^\partial uv) \tag{35}$$

$$x \Vdash \varphi * \psi \quad \text{iff} \quad \forall u, v \, (xR_* uv \implies (u \Vdash^\partial \varphi \vee v \Vdash^\partial \psi)) \tag{36}$$

$$x \Vdash^\partial \varphi * \psi \quad \text{iff} \quad \exists u, v \, (x\overline{R}_*^\partial uv \wedge u \Vdash^\partial \varphi \wedge v \Vdash^\partial \psi) \tag{37}$$

$$x \Vdash \mathtt{f} \quad \text{iff} \quad x \in F \tag{38}$$

$$x \Vdash^\partial \mathtt{f} \quad \text{iff} \quad x \in \rho(F) \tag{39}$$

The definitions of satisfaction and validity of a sentence or of a sequent are the same as in Section 2.3.

The reader can easily verify that $[\![\varphi * \psi]\!] = [\![\varphi]\!] \oplus [\![\psi]\!]$, $(\![\varphi * \psi]\!) = (\![\varphi]\!) \oplus (\![\psi]\!)$ and similarly for $\rightarrow, \leftharpoonup$. As a consequence, the proof of soundness follows, by arguments analogous to those developed for the **FL**-calculus.

Theorem 4.5 (Soundness). $\mathbf{FLG}_\varnothing$ *is sound in the class of $\mathbf{FLG}_\varnothing$-frames.*

4.2 Canonical Model Construction and Completeness

For completeness, we extend the canonical frame construction of Section 3.2. The definitions of the filter operators corresponding to the new logical operators are obtained as a special case from the corresponding general filter operator definitions of [16, 21], to which we refer the reader for intuitions and details.

$$u \widehat{*} v \;=\; \{e : \exists a, b \,(a \in u \text{ and } b \in v \text{ and } a * b \le e)\} \tag{40}$$

$$u \widehat{\rightharpoonup} v \;=\; \{e : \forall a, b \,(a \in u \text{ and } b \le v \implies a \rightharpoonup b \le e)\} \tag{41}$$

$$u \widehat{\leftharpoonup} v \;=\; \{e : \forall a, b \,(a \le u \text{ and } b \in v \implies a \leftharpoonup b \le e)\} \tag{42}$$

and we define $F = \Gamma(\mathfrak{f}) = \Gamma x_{\mathfrak{f}}$. The canonical accessibility relations are also defined as special instances of a general schema, see [21], instantiated below for the cases of interest.

$$xR_{\oplus}uv \quad \text{iff} \quad u \widehat{*} v \le x \qquad\qquad xR_{\oplus}^{\partial}uv \quad \text{iff} \quad \le u \widehat{*} v \tag{43}$$

$$xR_{\rightharpoonup}uv \quad \text{iff} \quad u \widehat{\rightharpoonup} v \le x \qquad\qquad xR_{\rightharpoonup}^{\partial}uv \quad \text{iff} \quad x \le u \widehat{\rightharpoonup} v \tag{44}$$

$$xR_{\leftharpoonup}uv \quad \text{iff} \quad u \widehat{\leftharpoonup} v \le x \qquad\qquad xR_{\leftharpoonup}^{\partial}uv \quad \text{iff} \quad x \le u \widehat{\leftharpoonup} v \tag{45}$$

Lemma 4.6. *Let $\circledcirc$ be any of $\leftharpoonup, *, \rightharpoonup$, of respective distribution type $(i_1, i_2; i_3) \in \{(1, \partial; 1), (\partial, \partial; \partial), (\partial, 1; 1)\}$. Let also $\widehat{\circledcirc}$ be any of the defined filter operators $\widehat{\leftharpoonup}, \widehat{*}, \widehat{\rightharpoonup}$.*

1. $\circledcirc, \widehat{\circledcirc}$ have the same monotonicity type. In other words, if for all $j = 1, 2$, if $x_j \le u_j$ whenever $i_j = 1$ and $u_j \le x_j$ whenever $i_j = \partial$, then $\widehat{\circledcirc}(x_1, x_2) \le^{i_3} \widehat{\circledcirc}(u_1, u_2)$, where $\le^{i_3}$ is $\le$ if $i_3 = 1$ and it is $\ge$ if $i_3 = \partial$.

2. $\widehat{\circledcirc}(x_a, x_b) = \circledcirc(a, b){\uparrow}$, for any lattice elements a, b.

Proof. The lemma is analogous to Lemma 3.9 and it is again a special case of Lemma 4.3 of [21] (restricted to the three additional operators of interest in this section). $\qquad\square$

To drive the completeness proof to a conclusion, what is needed is to prove a canonical frame and a canonical interpretation lemma.

Lemma 4.7 (Canonical Frame Lemma). *The canonical frame*

$$(X, \le, R_{\otimes}, R_{\otimes}^{\partial}, R_{\rightarrow}, R_{\rightarrow}^{\partial}, R_{\leftarrow}, R_{\leftarrow}^{\partial}, R_{\oplus}, R_{\oplus}^{\partial}, R_{\rightharpoonup}, R_{\rightharpoonup}^{\partial}, R_{\leftharpoonup}, R_{\leftharpoonup}^{\partial}, F, \mathfrak{T}, \mathfrak{O}, M)$$

where X is the filter space of the Lindenbaum–Tarski algebra of $\mathbf{FLG}_{\varnothing}$, $M \subseteq X$ is the set of principal filters and the relations and special sets are as previously defined, is an $\mathbf{FLG}_{\varnothing}$-frame (Definition 4.3).

Proof. Much of the proof (corresponding to the first condition in the Definition 4.3 of $\mathbf{FLG}_{\varnothing}$-frames) has been already given, see the canonical frame lemma for $\mathbf{FL}$, Lemma 3.10. What remains to be proved is that $(\mathfrak{P}_{\lambda}, \cap, \vee, \triangleleft, \oplus, \triangleright, F)$ is a dual residuated lattice. Members of $\mathfrak{P}_\lambda$ are the upper sets over principal filters Γx_a, where a is the equivalence class of a sentence. As in the case of Lemma 3.10 it is readily verified, given the definitions of the filter operators and of the canonical accessibility relations that, for u, v any filters

$$\Gamma u \oplus \Gamma v = \Gamma(u\widehat{*}v) \qquad \Gamma u \rhd \Gamma v = \Gamma(u\widehat{\rightharpoonup}v) \qquad \Gamma u \lhd \Gamma v = \Gamma(u\widehat{\leftharpoonup}v)$$

and similarly for the dual operators

$$\Delta u \oplus \Delta v = \Delta(u\widehat{*}v) \qquad \Delta u \vartriangle \Delta v = \Delta(u\widehat{\rightharpoonup}v) \qquad \Delta u \triangledown \Delta v = \Delta(u\widehat{\leftharpoonup}v)$$

Hence Condition 2 of Definition 4.3 holds.

Lemma 4.6 ensures that for $u, v \in M$ (where recall that M is the set of principal filters) the filter operators return a principal filter $x_a\widehat{*}x_b = x_{a*b}$, $x_a\widehat{\rightharpoonup}x_b = x_{a\to b}$, $x_a\widehat{\leftharpoonup}x_b = x_{a\leftarrow b}$. Therefore, $\mathfrak{P}_\lambda$ is closed under the operators $\lhd, \oplus, \rhd$ and similarly $\mathfrak{P}_\rho$ is closed under the dual operators $\vartriangle, \oplus, \triangledown$ and it is readily verified, from definitions, that

$$\Gamma u \oplus \Gamma v = \lambda(\Delta u \oplus \Delta v) = \lambda(\rho\Gamma u \oplus \rho\Gamma v) \qquad \Gamma u \rhd \Gamma v = \lambda(\Delta u \vartriangle \Delta v)$$
$$\Gamma u \lhd \Gamma v = \lambda(\Delta u \triangledown \Delta v)$$

We may conclude that Condition 3 of Definition 4.3 holds.

Verifying that F is an identity for $\oplus$ restricted to members of $\mathfrak{P}_\lambda$ is elementary, so Condition 4 of Definition 4.3 also holds. Condition 5 refers to the distribution properties of $\oplus$ on $\mathfrak{P}_\lambda$. Recall that we proved the corresponding condition for distribution of $\otimes$ over joins by proving that its dual distributes over intersections. The observant reader will have noticed the analogy of the distribution conditions in Definition 4.3 with the definition of the conditions for the case of $\otimes$. We shall leave it up to the interested reader to verify the required distribution property of $\oplus$ over intersections, by imitating the corresponding argument given for $\otimes$ and we shall conclude that Condition 5 of Definition 4.3 also holds and thereby the proof of the canonical frame lemma is complete. $\square$

Lemma 4.8 (Canonical Interpretation Lemma). *The canonical interpretation $[\![\varphi]\!] = \{x : [\varphi] \in x\}$ and co-interpretation $(\!(\varphi)\!) = \{x : [\varphi] \le x\}$ satisfy the recursive conditions (32–33).*

Proof. The cases for conjunction, disjunction, the constants $\top, \bot$ and the operators $\leftarrow, \circ, \rightarrow$ and $\mathbf{t}$ have been dealt with in Lemma 3.14, hence they are not mentioned in the present proof. As mentioned in the proof of Claims 3.15 and 3.16 in the course of the proof of the canonical interpretation lemma for **FL** the proof is actually a specialization to the particular operators of a general argument for n-ary logical operators of any distribution type that was developed in a recent article by this author [21]. The same applies to the new operators $\leftharpoonup, *, \rightharpoonup$. $\qquad\square$

Claim 4.9. *For all a, b and any filter x*

1. *$a*b \in x$ iff $\forall u, v\, (x\overline{R}_{\partial\partial;\partial}uv \implies (a \nleq u \text{ or } b \nleq v))$*

2. *$a*b \leq x$ iff $\exists u, v\, (xR^{\partial}_{\partial\partial;\partial}uv \text{ and } a \leq u \text{ and } b \leq v)$*

Proof. Consult the proof of Claims 4.10 and 4.12 of [21]. $\qquad\square$

Claim 4.10. *For all a, b and any filter x*

1. *$a \rightarrow b \in x$ iff $\exists u, v\, (xR_{\partial 1;\partial}uv \text{ and } a \in u \text{ and } b \leq v)$*

2. *$a \rightarrow b \leq x$ iff $\forall u, v\, (a \leq u \text{ and } b \in v \implies xR^{\partial}_{\partial 1;\partial}uv)$*

Proof. Consult the proof of Claims 4.9 and 4.11 of [21]. $\qquad\square$

Claim 4.11. *For all a, b and any filter x*

1. *$a \leftharpoonup b \in x$ iff $\exists u, v; (xR_{1\partial;1}uv \text{ and } a \leq u \text{ and } b \in v)$*

2. *$a \leftharpoonup b \leq x$ iff $\forall u, v\, (a \in u \text{ and } b \leq v \implies xR^{\partial}_{1\partial;1}uv)$*

Proof. Consult the proof of Claims 4.9 and 4.11 of [21]. $\qquad\square$

By the above, the proof of the canonical interpretation lemma (Lemma 4.8) is complete.

Thereby, the proof of completeness has been concluded.

Theorem 4.12 (Completeness). *The minimal Full Lambek–Grishin calculus $\mathbf{FLG}_\varnothing$ is (sound and) complete in the class of τ-frames of Definition 4.1.*

4.3 Grishin's Interaction Axioms

Grishin further proposed the strengthening of the system by the addition of interaction axioms from the following two dual axiom groups.

(G1)	$(a \rightarrow b) \circ c \leq a \rightarrow (b \circ c)$	(G3)	$c \circ (b \leftharpoonup a) \leq (c \circ b) \leftharpoonup a$	
(G2)	$c \circ (a \rightarrow b) \leq a \rightarrow (c \circ b)$	(G4)	$(b \leftharpoonup a) \circ c \leq (b \circ c) \leftharpoonup a$	(46)
(G1')	$(c*b) \leftharpoonup a \leq c*(b \leftharpoonup a)$	(G3')	$a \rightarrow (b*c) \leq (a \rightarrow b)*c$	
(G2')	$(b*c) \leftharpoonup a \leq (b \leftharpoonup a)*c$	(G4')	$a \rightarrow (c*b) \leq c*(a \rightarrow b)$	(47)

In this article, Kripke–Galois frames $(X, \leq, (R_\delta, R_\delta^\partial)_{\delta \in \tau}, M)$ of similarity type τ, with $M \subseteq X$, take the set $\mathfrak{P}_\lambda$ of propositions to be the set Γx with $x \in M$, where Γx is apparently a stable set under the Dedekind–MacNeille Galois connection generated by the partial order. The Grishin axioms translate directly to inclusion requirements, like $(\Gamma x \rhd \Gamma y) \otimes \Gamma z \subseteq \Gamma x \rhd (\Gamma y \otimes \Gamma z)$, corresponding to (G1). The properties Ψ and Φ_* define, respectively, the sets generated by $\otimes$ and $\rhd$, given also the requirement of closure of $\mathfrak{P}_\lambda$ under the operators. It is then a straightforward exercise to translate the inclusion to a frame requirement captured by a sentence in the first-order frame language. In particular for (G1), this is the requirement that condition (A) below implies condition (B), for all $w \in X$ and any $x, y, z \in M$:

(A) $\exists u, v \, (\Phi_*(u, x, y) \wedge z \leq v \wedge w R_\otimes uv)$

(B) $\exists u', v' \, [x \leq u' \wedge \forall y', z' \, (y' \leq y \wedge z' \leq z \longrightarrow v R_\otimes^\partial y' z') \wedge x R_\rightarrow u' v']$

The implication $\forall x, y, z \, (x \in M \wedge y \in M \wedge z \in M \longrightarrow (\text{A} \longrightarrow \text{B}))$ is equivalent to the inclusion requirement $(\Gamma x \rhd \Gamma y) \otimes \Gamma z \subseteq \Gamma x \rhd (\Gamma y \otimes \Gamma z)$ and this is what is used to prove soundness of the G1 axiom.

For completeness, only the canonical frame lemma is affected and the proof must be extended to verify that the canonical frame satisfies the above condition. By definition of $\otimes$, $\rhd$ we have $(\Gamma x \rhd \Gamma y) \otimes \Gamma z = \Gamma(x \circeq y) \otimes \Gamma z = \Gamma((x \circeq y) \widehat{\otimes} z)$ and when x, y, z are principal filters x_a, x_b, x_c this is just the upper closure of the principal filter $x_{(a \rightarrow b) \circ c}$, by Lemmas 3.9, 4.6. Similarly, the right hand side is the upper closure of the principal filter $x_{a \rightarrow (b \circ c)}$ and $\Gamma x_{(a \rightarrow b) \circ c} \subseteq \Gamma x_{a \rightarrow (b \circ c)}$ iff $x_{a \rightarrow (b \circ c)} \leq x_{(a \rightarrow b) \circ c}$ iff $(a \rightarrow b) \circ c \leq a \rightarrow (b \circ c)$ hence the frame is canonical. The same type of argument applies to any of the other axioms and we may then safely conclude with the following theorem.

Theorem 4.13. *The Full Lambek–Grishin calclus axiomatized by also including any combination of G or G' axioms (see 46, 47) is sound and complete in the corresponding class of τ-frames.*

5   Relevance and Linear Logic

We next provide Kripke–Galois semantics for (non-distributive) Relevance Logic (**RL**) and Linear Logic (**LL**), its original [13] commutative version where the left and right residuals $\circ\!-$, $-\!\circ$ of intensional conjunction coincide. Both **RL** and **LL** are equipped with a De Morgan negation $\varphi^\perp$ converting multiplicatives to additives and vice versa, $(\varphi \vee \psi)^\perp \equiv \varphi^\perp \wedge \psi^\perp$, $(\varphi \circ \psi)^\perp \equiv \varphi^\perp * \psi^\perp$ etc., and where $\varphi \multimap \psi \equiv \varphi^\perp * \psi$. It is useful to study separately the logic of abelian residuated De Morgan monoids.

5.1 The Logic of Abelian Residuated De Morgan Monoids

τ-frames for negation are structures $(X, \leq, \bot, \bot^{\partial}, (R_{\delta}, R_{\delta}^{\partial})_{\delta \in \tau_1}, M)$, where $M \subseteq X$, since $\mathfrak{P}_{\lambda} = \{\Gamma x : x \in M\}$ is completely determined by M, with further conditions to be placed on $M, \bot, \bot^{\partial}$. This section relies partly on definitions and results from [19], to which we shall refer the reader for details and proofs. Partially ordered spaces for ortholattices and De Morgan lattices (assuming distribution) have been also considered by Bimbó in [5], where the partial order structure allows for an extension of Goldblatt's [14] representation of ortholattices to a full duality.

Definition 5.1. An $\mathbf{FL}_e^{\bot}$-frame $\mathfrak{G} = (X, \leq, (\bot, \bot^{\partial}), (R_{\delta}, R_{\delta}^{\partial})_{\delta \in \tau_1}, M)$ is a structure where (1) $\bot, \bot^{\partial}$ are binary relations on X, (2) $M \subseteq X$, (3) $\tau_1 = \langle(1, 1; 1), (1, \partial; \partial)\rangle$, with $\tau = \langle(1; \partial)\rangle^\frown \tau_1$ and the following axioms in the frame language are assumed:

1. $(X, \leq, (R_{\delta}, R_{\delta}^{\partial})_{\delta \in \tau_1}, M)$ is an $\mathbf{FL}_e$ frame (Definition 3.6).

2. For each $x \in M$ the set $\{y : x \perp y\}$ is generated as the upper closure of a single point $z \in M$, which also generates the set $\{y : x \perp^{\partial} y\}$ as its down closure in the partial order. In addition, the set $\{u : z \perp u\}$, for this z, is contained in the upper closure of x, while the set $\{u : z \perp^{\partial} u\}$ is contained in the down closure of x. In symbols,

$$\forall x \left[x \in M \longrightarrow \exists z \left(z \in M \wedge \forall y \left[(x \perp y \longleftrightarrow z \leq y) \wedge \forall u \left(z \perp u \longrightarrow x \leq u \right) \right] \right) \wedge \right.$$
$$\left. \forall y' \left[(x \perp^{\partial} y' \longleftrightarrow y' \leq z) \wedge \forall u' \left(z \perp^{\partial} u' \longrightarrow u' \leq x \right) \right] \right]$$

3. (symmetry and increasingness axioms for the binary relation $\perp$)

   (a) $\forall x \forall y \, (x \in M \wedge y \in M \longrightarrow (x \perp y \longrightarrow y \perp x))$

   (b) $\forall x \forall y \forall z \, (x \perp y \wedge x \leq z \longrightarrow z \perp y)$

4. (symmetry and decreasingness axioms for the dual binary relation $\perp^{\partial}$)

   (a) $\forall x \forall y \, (x \in M \wedge y \in M \longrightarrow (x \perp^{\partial} y \longrightarrow y \perp^{\partial} x))$

   (b) $\forall x \forall y \forall z \, (x \perp^{\partial} y \wedge z \leq x \longrightarrow z \perp^{\partial} y)$.

Note that, comparing with the partially ordered orthoframes of [19] the irreflexivity condition on $\perp$ is not included in the above defined frames, since irreflexivity is precisely the condition that validates the characteristic axiom of intuitionistic negation $\varphi \wedge \neg\varphi \vdash \bot$. Note also that we have not yet included duality conditions to validate equivalences such as $\neg(\varphi \circ \psi) = (\neg\varphi) * (\neg\psi)$. We shall do this in the sequel.

For $U \subseteq X$, let $U^{\perp}$ and $U^{\perp^{\partial}}$ be defined as in equations (6, 7), instantiated below

$$U^{\perp} = \{z : \forall u \, (u \in U \implies x \perp u)\} \qquad U^{\perp^{\partial}} = \{z : \exists u \, (z \perp^{\partial} u \text{ and } u \in \lambda U)\}.$$

For the proof of the following claim the reader may wish to consult [19].

Lemma 5.2. *Let $\mathfrak{G}$ be an $FL_e^{\perp}$-frame. Then,*

1. *for all $x \in X$, $\{x\}^{\perp} = (\Gamma x)^{\perp}$ and $\{x\}^{\perp^{\partial}} = (\Delta x)^{\perp^{\partial}}$;*

2. *for each $x \in M$, there is a point $x^* \in M$ such that*

   (a) *$\{x\}^{\perp} = \Gamma x^*$ and $\{x\}^{\perp^{\partial}} = \Delta x^*$*

   (b) *$*$ is an antitone operator and an involution on M: $x \leq y \implies y^* \leq x^*$ and $x = x^{**}$, for $x, y \in M$.*

The following is an immediate consequence of Lemma 5.2.

Corollary 5.3. *The following hold:*

1. *$\mathfrak{P}_{\lambda}$ is closed under $(\ )^{\perp}$ and $\mathfrak{P}_{\rho}$ is closed under $(\ )^{\perp^{\partial}}$. Furthermore, $(\ )^{\perp}$ has the Galois property on $\mathfrak{P}_{\lambda}$, i.e., $A \subseteq C^{\perp}$ iff $C \subseteq A^{\perp}$, for $A, C \in \mathfrak{P}_{\lambda}$ and similarly for $(\ )^{\perp^{\partial}}$ and $\mathfrak{P}_{\rho}$.*

2. *for all $x \in M$, $(\Gamma x)^{\perp\perp} = \Gamma x$ and $(\Delta x)^{\perp^{\partial}\perp^{\partial}} = \Delta x$.*

3. *for all $x \in M$, $\lambda((\rho\Gamma x)^{\perp^{\partial}}) = (\Gamma x)^{\perp}$ and $(\Delta x)^{\perp^{\partial}} = \rho((\lambda\Delta x)^{\perp})$.*

Note that closure of both $\mathfrak{P}_{\lambda}, \mathfrak{P}_{\rho}$ under intersections (imposed by Conditions 1(a, b) of frames (see Definition 3.1) imply that $(\ )^{\perp}$ is a De Morgan negation on $\mathfrak{P}_{\lambda}$ and similarly for $(\ )^{\perp^{\partial}}$ and $\mathfrak{P}_{\rho}$. Consequently, the above suffices in order to prove that $\mathbf{FL}_e^{\perp}$, i.e., $\mathbf{FL}_e$ enriched with a De Morgan negation, is sound in the class of $\mathbf{FL}_e^{\perp}$-frames of Definition 5.1, where models on the frames are extended by adding to the clauses for fusion and implication the following:

$$x \Vdash \neg\varphi \text{ iff } \forall y \, (y \Vdash \varphi \implies x \perp y) \qquad x \Vdash^{\partial} \neg\varphi \text{ iff } \exists y \, (y \Vdash \varphi \text{ and } x \perp^{\partial} y)$$

Proposition 5.4 (Soundness). *$\mathbf{FL}_e^{\perp}$ (i.e., the logic of residuated De Morgan abelian monoids) is sound in the class of frames of Definition 5.1.*

For completeness, let X be the set of filters of the Lindenbaum–Tarski algebra of the logic, including the improper filter ω. In Section 3, we proved completeness for variants of $\mathbf{FL}$, including its associative-commutative version $\mathbf{FL}_e$. Since we assume no interaction of the negation operator with the monoidal operator and its residual, we only need to extend the proof of the canonical frame and interpretation lemmas of Section 3 to cover the case of negation as well. Define a filter star operator, following the representation approach of [16, 18, 21, 19], as in equation (48).

$$x^* = \{a : \neg a \leq x\} \tag{48}$$

The reader can easily verify that x^* is a filter, when x is one.

Lemma 5.5. *The filter operator* * *has the following properties, where we use* x_e *for the principal filter* $e\uparrow$ *generated by the lattice element* e:

1. $(x_a)^* = x_{\neg a}$

2. $x \leq y \implies y^* \leq x^*$

3. $x \leq x^{**}$

Proof. For the proof, the reader may consult any of [16, 18, 21, 19]. □

It follows from the Lemma that $(\ )^*$ is a Galois connection on the set X of filters and that it has the properties of De Morgan negation on the set $M \subseteq X$ of principal filters. For any stable set $A \in \mathfrak{P}_\lambda = \{\Gamma x : x \in M\}$, define the operator $(\ )^\perp$ by setting $A^\perp = (\Gamma x)^\perp = \Gamma x^*$. It is then straightforward to verify that $(\ )^\perp$ is a De Morgan negation on $\mathfrak{P}_\lambda$, since $(\Gamma x_a)^\perp = \Gamma x_a^* = \Gamma x_{a^\perp}$. We leave it to the reader to verify that all conditions for $\mathbf{FL}_e^\perp$-frames (Definition 5.1) obtain in the canonical frame and [19] can be consulted for this purpose.

For the canonical $\perp$ relation, we may define $x \perp z$ iff $x^* \leq z$ and $x \perp^\partial z$ iff $z \leq x^*$, following the general approach of [16, 20, 21]. It is typical to model negation by the clause $x \perp_G z$ iff $\exists a\ (a \in x$ and $\neg a \in z)$, see [14, 9], but it is easy to verify that $x_a \perp_G z$ iff $\neg a \in z$ iff $\exists e\ (e \in x_a$ and $\neg e \in z)$, so that $\perp$ and $\perp_G$ coincide on the set of principal filters. Symmetry of $\perp, \perp^\partial$ on M is immediate and so is increasingness (decreasingness) of $\perp$ (respectively, $\perp^\partial$) on M. The following lemma has been proved in [20].

Lemma 5.6. *For any element* a *of the Lindenbaum–Tarski algebra of the logic and filter* x *we have*

1. $\sim a \in x$ *iff* $\forall y\ (a \in y \implies y \perp x)$

2. $\neg a \leq x$ *iff* $\forall z\ (a \leq z \implies x \perp^\partial z)$

Proof. See the proofs of Lemmas 3.6 and 3.7 in [20]. □

It follows from the lemma that the canonical interpretation satisfies the recursive clauses for satisfaction and co-satisfaction:

$$x \Vdash \varphi^\perp \text{ iff } \forall y \, (x \Vdash \varphi \implies x \perp y) \qquad x \Vdash^\partial \varphi^\perp \text{ iff } \exists y \, (x \perp^\partial y \text{ and } y \Vdash^\partial \varphi)$$

This suffices to establish a completeness theorem for $\mathbf{FL}_e^\perp$, listed below.

Theorem 5.7 (Completeness for $\mathbf{FL}_e^\perp$). *The logic of De Morgan residuated abelian monoids is complete in the class of $\mathbf{FL}_e^\perp$-frames, as specified in Definition 5.1.*

5.2 Exponential-free Linear Logic (MALL)

The language of **LL** without exponentials (Multiplicative–Additive Linear Logic, **MALL**) is generated by the following grammar

$$\varphi := p, p^\perp \ (p \in P) \mid \top \mid \bot \mid \mathsf{t} \mid \mathsf{f} \mid \varphi \wedge \varphi \mid \varphi \vee \varphi \mid \varphi \circ \varphi \mid \varphi * \varphi$$

where P is a non-empty, countable set of propositional variables and where negation is defined on all sentences by the syntactic identities of Table 2 and linear implication is then defined as $\varphi \multimap \psi = \varphi^\perp * \psi$.

$$
\begin{array}{ll}
\mathsf{t}^\perp = \bot & \bot^\perp = \mathsf{t} \\
\top^\perp = \mathsf{f} & \mathsf{f}^\perp = \top \\
(p)^\perp = p^\perp & (p^\perp)^\perp = p \\
(\varphi \wedge \psi)^\perp = \varphi^\perp \vee \psi^\perp & (\varphi \vee \psi)^\perp = \varphi^\perp \wedge \psi^\perp \\
(\varphi \circ \psi)^\perp = \varphi^\perp * \psi^\perp & (\varphi * \psi)^\perp = \varphi^\perp \circ \psi^\perp
\end{array}
$$

Table 2: Definition of negation in **LL**

For **LL**'s sequent system, the reader may consult [13]. It is a known fact that the Lindenbaum–Tarski algebra of **LL** is a bounded lattice $\mathcal{L} = (L, \wedge, \vee, 0, 1, \circ, \mathsf{t}, *, \mathsf{f}, \multimap, ()^\perp)$ where $0 = [\![\bot]\!]$, $1 = [\![\top]\!]$, $\mathsf{t} = [\![\mathsf{t}]\!]$, $(L, \circ, \multimap, \mathsf{t})$ is a residuated abelian monoid with identity element t ($a \circ b = b \circ a$, $a \circ \mathsf{t} = a = \mathsf{t} \circ a$ and $a \circ b \leq c$ iff $b \leq a \multimap c$), $()^\perp$ is a De Morgan complementation operator ($a \leq b^\perp$ iff $b \leq a^\perp$, $a^{\perp\perp} \leq a$) and $(a \circ b)^\perp = a^\perp * b^\perp$, while $a \multimap b = a^\perp * b$ and $\mathsf{t} = 0^\perp$. Similarly, $(L, *, \mathsf{f})$ is an abelian monoid with identity element $[\![\mathsf{f}]\!] = \mathsf{f} = 1^\perp$. An algebra $(L, \wedge, \vee, 0, 1, \circ, \to, *, ()^\perp, \mathsf{t}, \mathsf{f})$ such that the above conditions hold is referred to in the literature as a **MALL**-algebra.

Kripke–Galois frames for **LL** will be defined as τ-frames, where τ is the similarity type $\langle (1, 1; 1), (1, \partial; \partial), (\partial, \partial; \partial), (1; \partial) \rangle$, with appropriate conditions in the first-order frame language $L^1(\leq, (R_\delta, R_\delta^\partial)_{\delta \in \tau}, M)$.

683

Note that in $\mathbf{FL}_e^\perp$ the dual of multiplicative conjunction can be defined using De Morgan negation: $a * b = (a^\perp \circ b^\perp)^\perp$. What $\mathbf{MALL}$ adds to the axiomatization is that $a^\perp * b$ is the residual of (commutative) $\circ$, hence identical to the $\rightarrow$ operation of $\mathbf{FL}_e^\perp$. Alternatively, $\mathbf{MALL}$ can be seen as obtained from the associative-commutative Full Lambek–Grishin calculus (see Section 4), dropping the co-implication residuals of $*$ and augmented with a De Morgan negation. Both observations underly the definition of $\mathbf{MALL}$-frames below.

Definition 5.8 ($\mathbf{MALL}$-Frames). A $\mathbf{MALL}$-*frame*

$$\mathfrak{G} = (X, \leq, (\perp, \perp^\partial), (R_\otimes, R_\otimes^\partial), (R_\oplus, R_\oplus^\partial), (R_\rightarrow, R_\rightarrow^\partial), M)$$

is a τ-frame where

1. $\mathfrak{G}_1 = (X, \leq, (\perp, \perp^\partial), (R_\otimes, R_\otimes^\partial), (R_\rightarrow, R_\rightarrow^\partial), M)$ is an $\mathbf{FL}_e^\perp$-frame (Definition 5.1)

2. $\mathfrak{G}_2 = (X, \leq, (R_\otimes, R_\otimes^\partial), (R_\oplus, R_\oplus^\partial), (R_\rightarrow, R_\rightarrow^\partial), M)$ is an $\mathbf{FLG}_e^{\rightarrow}$-frame (Definition 4.2)

3. For all $x, y \in M$ and any $u \in X$,

    (a) $\Psi_*(u, x, y) \longleftrightarrow \forall v \, [\Psi(v, x^*, y^*) \longrightarrow u \perp v]$

    (b) $\Phi(u, x, y) \longleftrightarrow \Psi_*(u, x^*, y)$

where Ψ, Φ, Ψ_* are defined respectively by (14), (16), (29) and where $(\ )^*$ is the operator on members of M whose existence, resting on Condition 2 of Definition 5.1, was proved in Lemma 5.2.

Lemma 5.9. *The dual algebra of a $\mathbf{MALL}$-frame is a $\mathbf{MALL}$-algebra.*

Proof. The last frame condition directly enforces that for $A, C \in \mathfrak{P}_\lambda$ both identities $A \oplus C = (A^\perp \otimes C^\perp)^\perp$ and $A \Rightarrow C = A^\perp \oplus C$ hold. The rest is a combination of previous results, proved in Lemmas 4.3 and 5.2. $\qquad\square$

Models for $\mathbf{MALL}$ are defined by only keeping the relevant satisfaction and refutation clauses for $\mathbf{FLG}_\varnothing$ models (listed below) and adding the clauses for De Morgan negation, which are the clauses specified for $\mathbf{FL}_e^\perp$ (Section 5.1).

$$x \Vdash \varphi * \psi \quad \text{iff} \quad \forall u, v \, (x R_* uv \implies (u \Vdash^\partial \varphi \lor v \Vdash^\partial \psi)) \tag{49}$$

$$x \Vdash^\partial \varphi * \psi \quad \text{iff} \quad \exists u, v \, (x \overline{R}_*^\partial uv \land u \Vdash^\partial \varphi \land v \Vdash^\partial \psi) \tag{50}$$

$$x \Vdash \mathbf{f} \quad \text{iff} \quad x \in F \tag{51}$$

$$x \Vdash^\partial \mathbf{f} \quad \text{iff} \quad x \in \rho(F) \tag{52}$$

The proof of soundness for $\mathbf{MALL}$ is a direct consequence of Lemma 5.9.

Theorem 5.10 (Soundness). *Multiplicative–Additive Linear Logic is sound in the class of Kripke–Galois frames specified in Definition 5.8.*

The completeness proof is given in Section 5.4.

5.3 Relevance Logic (without Distribution)

Non-distributive Relevance Logic and exponential-free Linear Logic differ only by their acceptance or rejection of contraction, or as Avron [4] puts it:

Linear Logic + contraction = Relevance Logic without distribution

Therefore, Kripke–Galois frames for (non-distributive) Relevance Logic are just frames for **MALL**, with the addition of condition (C) for the underlying monoid frame (see Definition 3.1). Soundness is immediate by Lemma 3.2 in combination with the soundness Theorem 5.10 for **MALL**.

Theorem 5.11 (Soundness for Non-Distributive **RL**). *Non-Distributive Relevance Logic is sound in **RL**-frames, i.e., **LL**-frames assuming, in addition, condition (C) in their underlying monoid frame.*

5.4 Completeness for MALL and RL

For **MALL**, Conditions 1 and 2 for **MALL**-frames (Definition 5.8) hold in the canonical frame, by the work presented in the canonical frame construction for **FLG**$_\varnothing$ (see Section 4) and we only need to verify that Condition 3 also holds. As pointed out in the proof of Lemma 5.9, the condition is equivalent to the claim that for $A, C \in \mathfrak{P}_\lambda$ both identities $A \oplus C = (A^\perp \otimes C^\perp)^\perp$ and $A \Rightarrow C = A^\perp \oplus C$ hold. But this is immediate in the canonical frame for any $A = \Gamma x_a$, $C = \Gamma x_b \in \mathfrak{P}_\lambda$. Hence the canonical frame is a **MALL**-frame. We verified above, when discussing completeness for **FL**$_e^\perp$ that the canonical interpretation satisfies the required recursive conditions (Lemma 5.6), hence we may conclude with completeness.

Theorem 5.12 (Completeness for **MALL**). *Multiplicative–Additive Linear Logic (MALL) is complete in the class of **MALL**-frames (Definition 5.8).*

Non-distributive **RL** is obtained from **MALL** by adding contraction. We have verified in Lemma 3.7 that for a groupoid frame the requirement that the contraction condition (C) holds is equivalent to the requirement that the inclusion $A \cap C \subseteq A \otimes C$ holds in the dual algebra of the frame. In the canonical frame for **RL**, verifying that the required inclusion holds is immediate, since $\Gamma x_a \cap \Gamma x_b = \Gamma(x_a \vee x_b) = \Gamma x_{a \wedge b} \subseteq \Gamma x_{a \circ b}$, since $a \wedge b \leq a \circ b$ holds in the logic. Hence we may conclude with the respective completeness result.

Theorem 5.13 (Completeness for Non-Distributive **RL**). *Non-Distributive Relevance Logic is complete in the respective class of **RL**-frames, i.e., **MALL**-frames assuming in addition that condition (C) holds in their underlying groupoid frame.*

5.5 Controlling Resources: Full Linear Logic

Full Linear Logic, as originally proposed by Girard [13], includes devices to control the use of resources (where sentences are viewed as representing resources), by controlling the use of contraction and weakening in the Gentzen system for the logic. Girard [13] introduced a single operator ! (of course, bang) with its dual (why not) $?\varphi = (!\varphi^\perp)^\perp$. The exponential operators $!, ?$ in [13] control both contraction and weakening. Though it is possible to separate control of weakening and contraction by distinct operators, whose composition delivers the original $!, ?$ (cf. [17]), we shall restrict ourselves in this article to the original exponential operators of **LL**. For full **LL**'s Gentzen system [13] may be consulted. Algebraically, ! is a monotone operator, $a \le b \implies !a \le !b$, and it satisfies both $!a \le a$ and $!!a = a$, hence it is an *interior operator*. It's dual is a closure operator $a \le b \implies ?a \le ?b$, $a \le ?a$ and $??a = ?a$. The distinctive property of ! is that it maps extensional to intentional conjunctions: $!(a \wedge b) = !a \circ !b$. Finally, ! satisfies the identity $!\top = \mathsf{t}$. The ! operator has been regarded as an S4-modality, a (monotone) box operator satisfying both the T and the S4 axioms, $\square a \le a$ and $\square a \le \square \square a$. Despite such appearances, we share the view of Bimbó and Dunn [6] that ! is best viewed as a diamond-like operator, rather than a box-like one. Definition 5.14 introduces τ-frames for Linear Logic.

Definition 5.14 (**LL** Frames). An **LL**-frame is a Kripke–Galois frame

$$\mathfrak{G} = (X, \le, (\perp, \perp^\partial), (R_\otimes, R_\otimes^\partial), (R_\oplus, R_\oplus^\partial), (R_\to, R_\to^\partial), (R_!, R_!^\partial), (R_?, R_?^\partial), M)$$

of similarity type $\tau = \langle (1; \partial), (1, 1; 1), (\partial, \partial; \partial), (1, \partial; \partial), (1; 1), (\partial; \partial) \rangle$, where

1. $\mathfrak{G} = (X, \le, (\perp, \perp^\partial), (R_\otimes, R_\otimes^\partial), (R_\oplus, R_\oplus^\partial), (R_\to, R_\to^\partial), M)$ is a **MALL**-frame (Definition 5.8)

2. $R_!, R_!^\partial, R_?, R_?^\partial \subseteq X \times X$ are binary relations on X generating the respective operators in (53, 54), instantiating equations (4–7) to the case of distribution types $(1; 1)$ for ! and $(\partial; \partial)$ for ?.

$$!A = \{x : \exists y\ (xR_! y \text{ and } y \in A)\} \qquad !^\partial B = \{x : \forall y; (y \in \lambda B \implies xR_!^\partial y)\} \quad (53)$$

$$?A = \{x : \forall y\ (y \in \rho A \implies xR_? y)\} \qquad ?^\partial B = \{x : \exists y\ (xR_?^\partial y \text{ and } y \in B)\} \quad (54)$$

3. For each $x \in X$, the set $R_! x = \{y : y R_! x\}$ is point-generated by some point above x. More specifically,

$$\forall x \, \exists ! w_x \, (x \le w_x \text{ and } w_x R_! w_x \text{ and } \forall y \, [(y R_! x \longleftrightarrow w_x \le y)$$
$$\text{and } (y \le x \longrightarrow w_x R_! y)]). \qquad (55)$$

For the convenience of referencing, we again extend the frame language by introducing a function symbol $\downarrow$ on X with the property prescribed above, letting $w_x = \downarrow x$.

4. For any $u \in X$ and any $x, y \in M$,

$$\Psi(u, \downarrow x, \downarrow y) \longleftrightarrow \forall z \, [\forall u' \, (x \le u' \wedge y \le u' \longrightarrow z \le u') \longrightarrow \downarrow z \le u] \qquad (56)$$

where Ψ is defined by (14).

5. Conditions (A) and (B) below are equivalent, for all $z \in X$ and any $x \in M$,

(A) $\forall u \, (\forall y \, (y \le x \longrightarrow u R_? y) \longrightarrow z \perp u)$

(B) $\exists v \, (z R_! v \wedge \forall v' \, (x \le v' \longrightarrow v \perp v'))$.

6. $\forall x \, (x \perp \omega \longleftrightarrow \exists y \, x R_! y)$

Models for **LL** are the **MALL**-models with the addition of the obvious clauses for $!, ?$ resulting from definitions (53, 54) of the corresponding set-operators.

Lemma 5.15. *The following hold in **LL**-frames, for all $z, z' \in X$ and any $x, y \in M$,*

1. $!(\Gamma z) = \Gamma(\downarrow z)$

2. $!(\Gamma z) \subseteq \Gamma z$ and $!(\Gamma z) = !!(\Gamma z)$

3. If $z' \le x$, then $!(\Gamma z) \subseteq !(\Gamma z')$

4. $!(\Gamma x \cap \Gamma y) = !(\Gamma x) \otimes !(\Gamma y)$

5. $!X = \{\omega\}^\perp$

Proof. The first three claims follow from the definition of $!$, see (53), and the third condition imposed on frames, see (55). The last claim follows from closure of $\mathfrak{P}_\lambda = \{\Gamma x : x \in M\}$ and of $\mathfrak{P}_\rho$ under binary intersections, the duality between $\mathfrak{P}_\lambda, \mathfrak{P}_\rho$ (which turns each to a lattice) and from the fourth frame condition, see (56). The claim that $!X = \{\omega\}^\perp$ follows directly from the last condition on **LL**-frames, where recall that ω is the $\le$-largest element of X and it belongs to M. $\qquad \square$

Lemma 5.16. *For any $x \in M$, $?(\Gamma x) = (!((\Gamma x)^{\perp}))^{\perp}$. Consequently, for each $z \in M$, there is a point $\Uparrow z$ such that $?(\Gamma z) = \Gamma(\Uparrow z)$, and therefore, $\mathfrak{P}_{\lambda}$ is closed under the set-operator $?$. Furthermore, $?$ is a closure operator on $\mathfrak{P}_{\lambda}$, i.e.,*

1. *if $z' \leq z$, then $?(\Gamma z) \subseteq ?(\Gamma z')$;*

2. *$\Gamma z \subseteq ?(\Gamma z)$ and $?(\Gamma z) = ??(\Gamma z)$.*

Proof. To show that $(?\Gamma x)^{\perp} = !((\Gamma x)^{\perp})$, just use the equivalence of Conditions 5(A) and 5(B). By Lemma 5.2, $(\Gamma x)^{\perp} = \Gamma x^{*}$, for $x \in M$, and by Lemma 5.15, $!\Gamma x = \Gamma(\Downarrow x)$. Hence, $?\Gamma x = \Gamma(\Downarrow(x^{*}))^{*}$ and we let $\Uparrow x = (\Downarrow(x^{*}))^{*}$, so that $?\Gamma x = \Gamma(\Uparrow x)$. From $\Downarrow\Downarrow u = \Downarrow u$, which follows from $!!\Gamma u = !\Gamma u$ and given the definition of $\Uparrow$, we obtain $\Uparrow\Uparrow x = \Uparrow x$, hence, $??\Gamma x = ?\Gamma x$, for $x \in M$. The monotonicity of $?$ and the inclusion $\Gamma x \subseteq ?\Gamma x$ are immediate, and left to the interested reader. $\square$

Corollary 5.17 (Soundness for **LL**). *Full Linear Logic is sound in the class of Kripke–Galois frames specified in Definition 5.14.*

Proof. Rather than proving rules to be sound we verify that all the inequalities of the Lindenbaum–Tarski algebra of **LL**, as these were summarized in the beginning of this section, actually hold. But this was precisely the content of Lemmas 5.15 and 5.16. $\square$

Most of the work needed for completeness of full Linear Logic has been already done in previous sections. For full **LL**, we have already verified that the canonical frame is a **MALL**-frame, hence Condition 1 in the definition of **LL**-frames holds. Define now operators $\Downarrow, \Uparrow$ on filters and canonical relations $R_!, R_?$ and their dual relations $R_!^{\partial}, R_?^{\partial}$, by (57, 58), so that Condition 2 of **LL**-frames is satisfied,

$$\Downarrow x = \{e : \forall b \, (b \leq x \implies !b \leq e)\} \qquad x R_! y \text{ iff } \Downarrow y \leq x \qquad x R_!^{\partial} y \text{ iff } x \leq \Downarrow y \quad (57)$$

$$\Uparrow x = \{e : \exists b \, (b \in x \text{ and } ?b \leq e)\} \qquad x R_? y \text{ iff } \Uparrow y \leq x \qquad x R_?^{\partial} y \text{ iff } x \leq \Uparrow y, \quad (58)$$

following again the approach of [16, 18, 20, 21]. The relations $R_!, R_?$ generate set-operators, by instantiating equations (4, 6) and dual set-operators, by instantiating equations (5, 7):

$$!U = \{x : \exists y \, (x R_! y \text{ and } y \in U)\} \qquad !^{\partial} U = \{x : \forall y \, (y \in \lambda U \implies x R_!^{\partial} y)\}$$

$$?U = \{x : \forall y \, (y \in \rho U \implies x R_? y)\} \qquad ?^{\partial} U = \{x : \exists y \, (x R_?^{\partial} y \text{ and } y \in U)\}$$

The reader is invited to verify that $\Downarrow x_a = x_{!a}$ and, similarly, $\Uparrow x_a = x_{?a}$, for any principal filter x_a.

Lemma 5.18 (Canonical Frame Lemma for full **LL**). *The canonical frame defined as above is an **LL**-frame in the sense of Definition 5.14.*

Proof. The proof of the following claim is needed. □

Claim 5.19. *The following hold for the filter operators $\downarrow, \uparrow$ and the set operators $!, ?$.*

1. *For all $x \in X$, $x \leq \downarrow x$, hence $!\Gamma x \subseteq \Gamma x$.*

2. *$\downarrow, \uparrow$ are monotone functions on X, i.e., $x \leq y$ implies $\downarrow x \leq \downarrow y$ and $\uparrow x \leq \uparrow y$, hence, $\Gamma y \subseteq \Gamma x$ implies $!\Gamma y \subseteq !\Gamma x$ and $?\Gamma y \subseteq ?\Gamma x$.*

3. *For all $x \in X$, $\downarrow x = \downarrow\downarrow x$, and consequently, $!!\Gamma x = !\Gamma x$.*

4. *For all $x \in M$, $\uparrow x = (\downarrow (x^*))^*$.*

5. *$\lambda(!^\partial(\rho\Gamma x)) = !\Gamma x$ and $!^\partial \Delta x = \rho(!(\lambda\Delta x))$.*

Proof. For 1), if $e \in x$ and $b \leq x$, then $b \leq e$, hence $!b \leq b \leq e$ and therefore $e \in \downarrow x$. It follows then that $!\Gamma x \subseteq \Gamma x$, using the fact proved above that $!\Gamma z = \Gamma(\downarrow z)$.

For 2), assume $x \leq y$ and let $e \in \downarrow x$ and $b \leq y$. The hypothesis implies $b \leq x$ and then $!b \leq e$ follows, since $e \in \downarrow x$, which shows that $e \in \downarrow y$. This implies also that if $\Gamma y \subseteq \Gamma x$, then $!\Gamma y \subseteq !\Gamma x$.

For 2) again, but for the $\uparrow$ operator, assuming $x \leq y$ and $e \in \uparrow x$, any $b \in x$ such that $?b \leq e$ is a $b \in y$ and $?b \leq e$, hence, $e \in \uparrow y$. It then also follows that $\Gamma x \subseteq \Gamma y$ implies $?\Gamma x \subseteq ?\Gamma y$.

For 3), it only needs to be verified that $\downarrow\downarrow x \leq \downarrow x$. Let $d \in \downarrow\downarrow x$ and assume $b \leq x$. To show that $d \in \downarrow x$ it suffices to verify that $!b \leq d$. From $b \leq x$ the filter inclusion $x \leq x_b$ is obtained and then by monotonicity of $\downarrow$ we get $\downarrow x \leq \downarrow x_b$. From the definition of the filter operator $\downarrow$ it easily follows that $\downarrow x_b = x_{!b}$, hence it follows by monotonicity that $\downarrow\downarrow x \leq \downarrow\downarrow x_b = x_{!!b} = x_{!b}$ and therefor $!b \leq \downarrow\downarrow x$ and since we assume that $d \in \downarrow\downarrow x$ the desired conclusion $!b \leq d$ follows. As a consequence, we also obtain the identity $!!\Gamma x = !\Gamma x$.

For 4), the result is immediate since $?a = (!(a^\perp))^\perp$ holds in the Lindenbaum–Tarski algebra of **LL** and given that M is the set of principal filters and that $\uparrow x_a = \{e : \exists b \, (a \leq b \text{ and } ?b \leq e)\} = \{e : ?a \leq e\} = x_{?a}$.

Case 5) is easy to prove, and it is left to the reader. □

Now, for any filter $z \in X$, we obtain by monotonicity of the filter operators

$$
\begin{aligned}
!\Gamma z &= \{x : \exists u \, (xR_! u \text{ and } u \in \Gamma z)\} &&= \{x : \exists u \, (z \leq u \text{ and } \downarrow u \leq x)\} \\
&&&= \{x : \downarrow z \leq x\} = \Gamma(\downarrow z)
\end{aligned}
$$

$$?\Gamma z = \{x : \forall u \, (u \in \rho(\Gamma z) \implies x R_? u)\} = \{x : \forall u \, (u \leq z \implies \mathord{\uparrow} u \leq x)\}$$
$$= \{x : \mathord{\uparrow} z \leq x\} = \Gamma(\mathord{\uparrow} x)$$

This shows, in particular, that if $A \in \mathfrak{P}_\lambda$, then also $!A, ?A \in \mathfrak{P}_\lambda$.

Condition 4 for **LL**-frames is easily seen to hold for any $x, y \in M$ (i.e., principal filters). This is shown by the following calculation: $!(\Gamma x_a \cap \Gamma x_b) = !\Gamma(x_a \vee x_b) = \Gamma \mathord{\downarrow}(x_{a \wedge b}) = \Gamma x_{!(a \wedge b)} = \Gamma x_{!a \circ !b} = !(\Gamma x_a) \otimes !(\Gamma x_b)$.

The equivalence of Conditions 5A and 5B, as noted in the proof of Lemma 5.16, is equivalent to the requirement that $(?\Gamma x)^\perp = !((\Gamma x)^\perp)$, for $x \in M$, which is equivalent to $(\mathord{\uparrow} x)^* = \mathord{\downarrow}(x^*)$, itself equivalent to the last case of Claim 5.19, already proved.

Finally, for Condition 6 for **LL**-frames, for any $x \in X$ we have

$$
\begin{aligned}
x \perp \omega \quad &\text{iff} \quad x \in \{\omega\}^\perp &&\text{iff} \quad x \in (\Gamma \omega)^\perp \quad (\text{using Lemma 5.2}) \\
&&&\text{iff} \quad x \in \Gamma \omega^* \quad (\text{using Lemma 5.2 again}) \\
&&&\text{iff} \quad x \in \Gamma x_{0^\perp} \quad (\text{using the fact } \omega = x_0 \text{ and } x_e^* = x_{e^\perp}) \\
&&&\text{iff} \quad x \in \Gamma x_{!1} \quad (\text{since } !1 = \mathbf{t} = 0^\perp) \\
&&&\text{iff} \quad x \in \Gamma(\mathord{\downarrow} x_1) \quad (\text{since } \mathord{\downarrow} x_a = x_{!a}, \text{ any } a) \\
&&&\text{iff} \quad x \in !\Gamma x_1 \quad (\text{given that } !\Gamma z = \Gamma(\mathord{\downarrow} z), \text{ any } z \in M) \\
&&&\text{iff} \quad x \in !X \quad (\text{since } \Gamma x_1 = \{u : 1 \in u\} = X) \\
&&&\text{iff} \quad x \in \{x : \exists y \, (x R_! y \text{ and } y \in X)\} \quad (\text{by definition of } !) \\
&&&\text{iff} \quad \exists y \, x R_! y
\end{aligned}
$$

This completes the proof that the canonical frame is an **LL**-frame in the sense of Definition 5.14.

For completeness, it remains to prove the following claim.

Lemma 5.20 (Canonical Interpretation Lemma for full **LL**). *The canonical interpretation satisfies the requisite recursive clauses in the definition of **LL**-models.*

Proof. The proof of the lemma rests on the truth of the next claim. $\qquad \square$

Claim 5.21. *For any a and any filter x,*

1. *$!a \in x$ iff $\exists y \, (x R_! y$ and $a \in y)$, and $!a \leq x$ iff $\forall y \, (a \in y \implies x R_!^\partial y)$;*

2. *$?a \in x$ iff $\forall y \, (a \leq y \implies x R_? y)$, and $?a \leq x$ iff $\exists y \, (a \leq y$ and $x R_?^\partial y)$.*

Proof. The claim is a special instance of Claims 4.9–4.12 proved in the course of the proof of the canonical interpretation Lemma 4.8 of [21], to which the reader is referred for details, though this instance can be easily proved by a reader who may wish to verify it. $\qquad \square$

Cases 1–2 of the above claim are clearly equivalent to the recursive clauses for the interpretation and co-interpretation of exponentials and therefore, by the above proofs, we can conclude with the completeness for full **LL**.

Theorem 5.22 (Completeness for full **LL**). *Linear Logic (with exponentials) is complete in the class of **LL**-frames (Definition 5.14).*

6 Conclusions

This article presented Kripke–Galois Frames and more specifically τ-frames (Definition 2.2) $\mathfrak{F} = (X, R, (R_\delta, R_\delta^\partial)_{\delta \in \tau})$, where the relations $R_\delta, R_\delta^\partial$ generate dual image operators $\bigcirc_\delta, \bigcirc_\delta^\partial$ (generalizing the Jónsson–Tarski image operators) on the families $\mathcal{G}_\lambda(X), \mathcal{G}_\rho(X)$ of stable, and respectively, co-stable subsets of X (where stability refers to the closure operator produced by composing the two maps of the Galois connection generated on subsets of X by the Galois relation R of the frame).

A τ-logic is the logic of some class of τ-frames, for a given similarity type τ. τ-logics include a number of familiar logical systems and we have examined in this article substructural systems ranging from the Full Lambek and Lambek–Grishin calculi, to the logics of De Morgan monoids and to Linear Logic (with, or without exponentials) and to non-distributive Relevance Logic.

Modal and temporal systems have been previously treated by this author in [22, 20], while the groundwork of the Kripke–Galois framework was first published in [21]. The present article is a revised and condensed version of the report [23]. The Kripke–Galois semantics approach is based on a Stone-type representation and duality result published several years ago [16] and the motivation has been to extend Dunn's theory of Generalized Galois Logics (GGLs, gaggles) [8, 6] to the case of an underlying non-distributive propositional calculus.

It is this author's opinion that the approach taken in [16, 22, 19, 20, 21, 23] and in this article overcomes the difficulties encountered in the approaches taken in [11, 27, 10, 7] as far as providing a meaningful semantic account of applied non-distributive logics is concerned, such as dynamic, temporal, epistemic, or more generally, modal logics over a non-distributive propositional basis.

Acknowledgements

This article constitutes a (very long) answer to some application related questions raised by participants of the *Third Workshop*, organized by Katalin Bimbo and held at the University of Alberta, Edmonton, Canada, May 16–17, 2016, where the groundwork for the Kripke–Galois semantic framework [21] was presented. My thanks go to Kata, for her kind invitation and to the participants of the workshop for their positive reception of the research work I presented.

References

[1] Gerard Allwein and J. Michael Dunn. Kripke models for linear logic. *Journal of Symbolic Logic*, 58(2):514–545, 1993.

[2] Alan Ross Anderson and Nuel D. Belnap. *Entailment: The Logic of Relevance and Necessity*, volume I. Princeton University Press, 1975.

[3] Alan Ross Anderson, Nuel D. Belnap, and J. Michael Dunn. *Entailment: The Logic of Relevance and Necessity*, volume II. Princeton University Press, 1992.

[4] Arnon Avron. The semantics and proof theory of linear logic. *Theoretical Computer Science*, 57:161–184, 1988.

[5] Katalin Bimbó. Functorial duality for ortholattices and De Morgan lattices. *Logica Universalis*, 1:311–333, 2007.

[6] Katalin Bimbó and J. Michael Dunn. *Generalized Galois Logics. Relational Semantics of Nonclassical Logical Calculi*, volume 188. CSLI Lecture Notes, CSLI, Stanford, CA, 2008.

[7] Andrew Craig, Maria Joao Gouveia, and Miroslav Haviar. TiRS graphs and TiRS frames: a new setting for duals of canonical extensions. *Algebra Universalis*, 74(1–2), 2015.

[8] J. Michael Dunn. Gaggle theory: An abstraction of galois coonections and resuduation with applications to negations and various logical operations. In *Logics in AI, Proceedings of European Workshop JELIA 1990, LNCS 478*, pages 31–51, 1990.

[9] J. Michael Dunn and Chunlai Zhou. Negation in the context of gaggle theory. *Studia Logica*, 80:235–264, 2005.

[10] Ivo Düntsch, Ewa Orlowska, Anna Maria Radzikowska, and Dimiter Vakarelov. Relational representation theorems for some lattice-based structures. *Journal of Relational Methods in Computer Science (JORMICS)*, 1:132–160, 2004.

[11] Mai Gehrke. Generalized Kripke frames. *Studia Logica*, 84(2):241–275, 2006.

[12] Mai Gehrke and John Harding. Bounded lattice expansions. *Journal of Algebra*, 238:345–371, 2001.

[13] Jean Yves Girard. Linear logic. *Theoretical Computer Science*, 50, 1987.

[14] Robert Goldblatt. Semantic analysis of orthologic. *Journal of Philosophical Logic*, 3:19–35, 1974.

[15] V. N. Grishin. On a generalization of the Ajdukiewicz–Lambek system. In A. I. Mikhailov, editor, *Studies in Nonclassical Logics and Formal Systems*, pages 315–334. Nauka, 1983.

[16] Chrysafis Hartonas. Duality for lattice-ordered algebras and for normal algebraizable logics. *Studia Logica*, 58:403–450, 1997.

[17] Chrysafis Hartonas. Pretopology semantics for bimodal intuitionistic linear logic. *Logic Journal of the IGPL*, 5(1):65–78, 1997.

[18] Chrysafis Hartonas. Reasoning with incomplete information in generalized Galois logics without distribution: The case of negation and modal operators. In Katalin Bimbó, editor, *J. Michael Dunn on Information Based Logics*, volume 8 of *Outstanding Contributions to Logic*, pages 303–336. Springer Nature, 2016.

[19] Chrysafis Hartonas. First-order frames for orthomodular quantum logic. *Journal of Applied Non-Classical Logics*, 26(1):69–80, 2016.

[20] Chrysafis Hartonas. Order-dual relational semantics for non-distributive propositional logics. *Logic Journal of the IGPL*, 2016. doi:10.1093/jigpal/jzw057.

[21] Chrysafis Hartonas. Order-dual relational semantics for non-distributive propositional logics: A general framework. *Journal of Philosophical Logic*, 2016. doi:10.1007/s10992-016-9417-7.

[22] Chrysafis Hartonas. Modal and temporal extensions of non-distributive propositional logics. *Logic Journal of the IGPL*, 24(2):156–185, 2016.

[23] Chrysafis Hartonas. Kripke-Galois semantics for substructural logics. Technical Report ResearchGate DOI: 10.13140/RG.2.2.19116.05760, University of Applied Sciences of Thessaly (TEI of Thessaly), September 2016.

[24] Chrysafis Hartonas and J. Michael Dunn. Stone duality for lattices. *Algebra Universalis*, 37:391–401, 1997.

[25] Gerd Hartung. A topological representation for lattices. *Algebra Universalis*, 29:273–299, 1992.

[26] Hiroakira Ono and Yuichi Komori. Logics without the contraction rule. *Journal of Symbolic Logic*, 50:169–201, 1985.

[27] Tomoyuki Suzuki. Bi-approximation semantics for substructural logic at work. In *Advances in Modal Logic vol 8*, pages 411–433, 2010.

[28] Alasdair Urquhart. A topological representation of lattices. *Algebra Universalis*, 8:45–58, 1978.

Received 5 September 2016

The Relevant Logic E and Some Close Neighbours: A Reinterpretation

Edwin Mares

School of History, Philosophy, Political Science & International Relations, Victoria University of Wellington, Wellington, New Zealand
<edwin.mares@vuw.ac.nz>

Shawn Standefer

School of Historical and Philosophical Studies, University of Melbourne, Melbourne, NSW, Australia
<shawn.standefer@unimelb.edu.au>

Abstract

This paper has two aims. First, it sets out an interpretation of the relevant logic E of relevant entailment based on the theory of situated inference. Second, it uses this interpretation, together with Anderson and Belnap's natural deduction system for E, to generalise E to a range of other systems of strict relevant implication. Routley–Meyer ternary relation semantics for these systems are produced and completeness theorems are proven.

Keywords: entailment, relevant logic, strict implication, situated inference, ternary relation semantics

1 Introduction

The logic E is supposed to be the logic of relevant entailment. E incorporates intuitions concerning both relevance and necessity. In the 1960s, Alan Anderson and Nuel Belnap constructed two central relevant logics. E and the logic of contingent relevant implication, R. They viewed the entailment connective of E as the strict version of the implication of R. To show that the two logics had this relationship,

Standefer's research was supported by the Australian Research Council, *Discovery Grant* DP150103801. Thanks to the audiences of the *Australasian Association for Logic Conference*, the *Pukeko Logic Workshop*, and the *Third Workshop* at the University of Alberta for feedback.

Robert Meyer constructed a modal version of R, $R^\square$, that adds to R a necessity operator and relevant versions of the axioms for S4 [13]. It was hoped that the conjunction, disjunction, negation, and strict implication fragment of $R^\square$ was the same as E. But Larisa Maksimova showed that these two logics are distinct [10].

In *Entailment*, volume 1, Anderson and Belnap wrote:

> we predict that if in fact it is found that $R^\square$ and E diverge, then we shall, with many a bitter tear, abandon E. [1, p. 351]

The logic E has not, however, been completely abandoned. It continues to be studied. *Entailment* volume 2 has both very interesting technical and historical information about E, including Belnap's elegant display logic proof theory for it [2, §62.5.3]. Mark Lance defends E over R as the central relevant logic [8] and Lance and Philip Kremer have developed a theory of linguistic commitment that had E as its logic [9]. Despite all of this, however, E seems to be ignored by most contemporary relevant logicians.

In this paper, we focus on E and some logics that closely resemble it. We give E an interpretation based on the theory of situated inference of [11] and generalise the intuitions behind this interpretation to develop a small class of *entailment logics*. These entailment logics are formulated first in terms of Fitch-style natural deduction systems which make clear both the components of relevance and modality incorporated into them. The logics are then formulated in terms of traditional axiom systems and these are shown complete with respect to classes of Routley–Meyer ternary relation models. The indices of these models are taken to be situations and the ternary relations in these models are interpreted in terms of the theory of situated inference.

The logics that we examine in this paper are negation-free. Although negation is easily added to the semantics of these systems using the Routley star operator, available treatments of it in the natural deduction system are not illuminating in the way that we desire. We promise to investigate the role and representation of negation in entailment logics in a future paper, but we do not do so here.

The plan of the paper is as follows. We begin by reviewing the theory of situated inference as it is applied to the logic R. We show how this theory can be used to understand the Routley–Meyer semantics for that logic. We then modify the theory to apply to the logic E. We examine E through its axiomatic formulation, natural deduction system, and Routley–Meyer semantics. The situated inference interpretation of E employs both situations and worlds. This interpretation is then generalised to treat a small class of other systems that incorporate principles from various modal logics. An E-like logic that is similar to the modal logic K, which we call E.K is explored, and so are its extensions E.KT, E.K4, and E.KT4 (which is just E itself). We end by exploring a suggestion of Alasdair Urquhart for an axiomatisation of an S5-ish entailment logic, E5.

2 Situated Inference

Routley–Meyer models for relevant logics are indexical models, that is, in them the truth or falsity of formulas is relativised to points. We call these points *situations*. A situation is a potential part or state of a possible world. Some situations actually obtain in some worlds, and so can be called possible situations and some do not and can be thought of as impossible situations. Impossible situations are important for the analysis of negation, which is not our main concern here and so we will not mention impossible situations further.

A situation contains or fails to contain particular pieces of information. For example, a situation that includes all the information available at a given time in a lecture room (in which no one is connected to the internet) may not contain information about the weather outside or about the current polls in the American presidential race. Whether a situation satisfies a formula in a model is given by an information condition rather than a truth condition. These information conditions abstract from the canonical ways in which information is made available in actual situations. For example, the way in which we usually tell that an object is not red is that it is of some colour that is incompatible with its being red. The information condition for negation is a more general representation of this sort of information condition. It states that $a \vDash \neg A$ if and only if a is incompatible with any situation b that satisfies A. This means that a contains the information that there is no situation in the same world as a that contains the information that A.

One feature of the informational interpretation of relevant logic is that the satisfaction conditions for the connectives need not be homomorphic. As we can see, we do not in every case set $a \vDash \neg A$ iff $a \nvDash A$. The requirement that satisfaction conditions be homomorphisms between the object language and semantic metalanguage does not hold for information. The way in which we find information structured in our environment need not mirror the structure of way that we express that information linguistically.

The main focus of this paper is the notion of entailment. We approach it through the closely related concept of relevant implication. As we have said, Anderson and Belnap think of entailment as modalised implication and implication as a contingent form of entailment. The Routley–Meyer satisfaction condition for implication is

$$a \vDash A \to B \quad \text{iff} \quad \forall x \forall y ((Raxy \wedge x \vDash A) \supset y \vDash B).$$

The information in one situation can be applied to the information in another situation. For example, if $C \rightarrow D$ is in a and C is in b, then combining this information we obtain D. One way of understanding this condition is to think of the information in a combined with the information in b in this way always results in information that is in c.[1]

The theory of situated inference explains why one would care about combining information in this way. Suppose that one is in a situation a and in a world w. She might hypothesise that another situation b also exists in w. Then she can combine the information in a and b to determine that other sort of situations are in w. For example, suppose that a person has available to him in situation a information that there are absolutely no ticks in New Zealand. Then, on the hypothesis that a particular woodland is in New Zealand, he has the licence to infer that there are no ticks in it. The theory of situated inference breaks this inference as being an inference from there being one situation in which the park is in New Zealand and (perhaps) another in which it contains no ticks. The theory of situated inference connects relevant implications in a direct way with ordinary inferences.

What the theory of situated inference tells us is that the ternary relation R relates a and b to a set of situations (call it Rab), such that given the information in a, on the hypothesis that b is in the same world as a, there is a situation c in Rab also in that world. We sometimes say that $Rabc$ says that some situation *like* c can be inferred from that application of a to b. The word 'like' here is not being used in any technical sense. It is just an abbreviation to say that c is one of the set of situations that contains all the information that can be inferred from the combination of the information in a with that in b.

The demodalised nature of relevant implication, as it is characterised by the logic R, is made explicit in the theory of situated inference by the reading of situated inferences as being inferences about situations and information all contained in a single world. As we have said $Rabc$ means that the hypothesis of a and b in the same world allows us to infer the existence of a situation like c in that world. The reading of inferences as being in a single world, together with a rather liberal notion of application, enables justifications of certain particular postulates of the Routley–Meyer semantics for R.

For example, consider the *permutation postulate* of the Routley–Meyer semantics for R: for any situations a, b, c, if $Rabc$ then $Rbac$. This postulate tells us that the

---

[1] A rather sophisticated reading of the notion of combination is in [3]. In that paper, combination is understood in terms of the application of one situation to another in the sense that functions are applied to arguments. The application reading of R works well for weaker relevant logics but can be used to interpret the logic R as well. It just takes quite a bit of effort to show how it fits with the Routley–Meyer semantics for R and so we do not use it here.

result of combining the information in a with that in b is the same as combining the information in b with that in a. This seems natural as the common notion of combination is symmetric. As we shall see later, however, in the semantics for E the permutation postulate fails, and it fails (on our reading) because E is a modal logic.

The permutation postulate makes valid the thesis of *assertion*: $A \rightarrow ((A \rightarrow B) \rightarrow B)$.

Derivation 1. The following is a proof of assertion in the Anderson–Belnap Fitch system for R:

1	A_1	hypothesis
2	$A \rightarrow B_2$	hypothesis
3	$A_{\{1\}}$	1, reiteration
4	$B_{\{1,2\}}$	2, 3, $\rightarrow$E
5	$(A \rightarrow B) \rightarrow B_{\{1\}}$	2–4, $\rightarrow$I
6	$A \rightarrow ((A \rightarrow B) \rightarrow B)_\emptyset$	1–5, $\rightarrow$I

The subscripts are to be understood as referring to situations. When we make a hypothesis, say, A_1, we are postulating the existence of a situation, say, a_1 that contains the information that A. When the subscript is the empty set, the formula is proven to hold in every *normal situation*. (We discuss normal situations in Section 4.) The expression $B_{\{1,2\}}$ is read as saying that an arbitrary situation c that is in the result of combining the information in a_1 with that in a_2. The use of permutation can be seen in this proof through the fact that it does not matter when a number is added to the subscript (in an application of the rule of implication elimination) nor when it is removed (in an application of implication introduction) which order the numbers are in. If we were to reject permutation, the subscript the minor premise in an implication elimination would have to be added at the end of the new subscript and likewise, when a hypothesis is discharged its number could only be removed from the end of the subscript. Having permutation allows us to commute the order the numbers in subscripts.

We can generalise the R relation to be an n-place relation for any positive integer n by taking products of R. We say that $Rabcd$ if and only if $\exists x(Rabx \wedge Rxcd)$ and more generally (for $n \geq 3$) $Ra_1 \ldots a_n c$ if and only if $\exists x(Ra_1 \ldots a_{n-1}x \wedge Rxa_n c)$. We read $Ra_1 \ldots a_n c$ as saying that the hypothesis of $a_1, \ldots, a_n$ all in the same world justifies the inference to there being a situation like c also in that world. The generalised R relation and its interpretation justifies a more general permutation postulate: if $Ra_1 \ldots a_m a_{m+1} \ldots a_n c$, then $Ra_1 \ldots a_{m+1} a_m \ldots a_n c$, for any m, $1 \leq m \leq n$. This generalised permutation postulate justifies the derivation of various theses such as the permutation of antecedents $((A \to (B \to C)) \to (B \to (A \to C)))$.

Before we leave R, let us look at a semantic postulate that is related to situated inference in a more complicated manner: the *contraction postulate*. The simple version of the contraction postulate says

$$Rabc \implies Rabbc.$$

The generalised version of contraction says that $Ra_1 \ldots a_m \ldots a_n c$ implies that $Ra_1 \ldots a_m a_m \ldots a_n c$. This generalisation follows from the simple version. We read contraction as saying that if we hypothesise that $a_1, \ldots, a_m, \ldots, a_n$ in a world to infer that there is a situation like c is also present in that world, in an inference we can really use the information in a_m twice as part of the inference the presence of a situation like c. We will return to the topic of contraction in our discussion of entailment logics weaker than E.

3   E

This paper is not about R, but about the logic of relevant entailment, E and some similar systems. The implication of E is usually understood as a form of strict relevant implication. One way of thinking about strict relevant implication is through combining relevant implication with modality. If we think of it that way, then it is natural to represent strict relevant implication in a modal extension of R. But we suggest that the notion of entailment be considered a unified notion that has the properties of being relevant and necessary.

The obvious difference between R and E is in their conditionals. The conditional of R is a contingent implication and that of E is entailment. We can compare the two logics in terms of axiomatisations of their conditional only fragments, $R_\to$ and $E_\Rightarrow$. Here are some axiom schemes that, together with the usual modus ponens rule, generate $E_\Rightarrow$:

1. $A \Rightarrow A$ (Identity);

2. $(A \Rightarrow B) \Rightarrow ((B \Rightarrow C) \Rightarrow (A \Rightarrow C))$ (Suffixing);

3. $(B \Rightarrow C) \Rightarrow ((A \Rightarrow B) \Rightarrow (A \Rightarrow C))$ (Prefixing);

4. $(A \Rightarrow (A \Rightarrow B)) \Rightarrow (A \Rightarrow B)$ (Contraction);

5. $((A \Rightarrow A) \Rightarrow B) \Rightarrow B$ (EntT).

The axiom EntT tells us that if any formula is entailed by a theorem it is true. It is, in effect, a form of the T-axiom from modal logic that tells us that any necessary formula is true. For contrast, consider a set of axioms for $R_\rightarrow$:

R1 $A \rightarrow A$ (Identity);

R2 $(A \rightarrow B) \rightarrow ((B \rightarrow C) \rightarrow (A \rightarrow C))$ (Suffixing);

R3 $(A \rightarrow (A \rightarrow B)) \rightarrow (A \rightarrow B)$ (Contraction);

R4 $(A \rightarrow (B \rightarrow C)) \rightarrow (B \rightarrow (A \rightarrow C))$ (Permutation of Antecedents).

The permutation axiom of $R_\rightarrow$ "demodalises" its implication. From $(A \rightarrow B) \rightarrow (A \rightarrow B)$, which is an instance of identity, it allows us infer that $A \rightarrow ((A \rightarrow B) \rightarrow B)$. The latter clearly makes $\rightarrow$ into a non-strict form of implication. If we read $\rightarrow$ as $\prec$, even in the sense of S5, this formula is not a logical truth.

The axiomatic basis for conjunction and disjunction are the same for both logics, if we replace $\Rightarrow$ with $\rightarrow$ throughout to obtain the axioms for positive R:

7. $A \Rightarrow (A \vee B)$; $B \Rightarrow (A \vee B)$;

8. $(A \wedge B) \Rightarrow A$; $(A \wedge B) \Rightarrow B$;

9. $((A \Rightarrow B) \wedge (A \Rightarrow C)) \Rightarrow (A \Rightarrow (B \wedge C))$;

10. $((A \Rightarrow C) \wedge (B \Rightarrow C)) \Rightarrow ((A \vee B) \Rightarrow C)$;

11. $(A \wedge (B \vee C)) \Rightarrow ((A \wedge B) \vee (A \wedge C))$.

The rules for positive R and positive E are just modus ponens and adjunction.

E also has the axiom,

$$(\blacksquare A \wedge \blacksquare B) \Rightarrow \blacksquare(A \wedge B) \text{ (Agg}\blacksquare)$$

The axiom Agg$\blacksquare$ (Aggregation for $\blacksquare$) is a translation of the usual aggregation thesis into the idiom of E. Here $\blacksquare A$ is defined as $(A \Rightarrow A) \Rightarrow A$.

We also add the Ackermann constant t, to facilitate the formulation of another, but more easily used, notion of necessity, $\square$. This notion of necessity is defined as follows:

$$\square A =_{df} t \Rightarrow A$$

The operator $\blacksquare$ is extremely difficult to use in proofs. Consider, for example, the axiom Agg$\blacksquare$. Written in primitive notation it is $((((A \Rightarrow A) \Rightarrow A)) \wedge ((B \Rightarrow B) \Rightarrow B)) \Rightarrow (((A \wedge B) \Rightarrow (A \wedge B)) \Rightarrow (A \wedge B))$. Using this formula to prove other modal theses can be quite difficult. Agg$\Box$, i.e., $(\Box A \wedge \Box B) \Rightarrow \Box(A \wedge B)$, however, is just $((t \Rightarrow A) \wedge (t \Rightarrow B)) \Rightarrow (t \Rightarrow (A \wedge B))$, and this is just an instance of axiom 9.

The axiom and rule for t are the following:

12. $(t \Rightarrow A) \Rightarrow A$ (Tt).

Rule Nt (necessitation for t)

$$\frac{\vdash A}{\vdash t \Rightarrow A}$$

$\vdash t$ follows from axiom 1, i.e., $t \Rightarrow t$, and axiom 12, $(t \Rightarrow t) \Rightarrow t$.

In the context of E, $\Box$ and $\blacksquare$ are equivalent. Here is a proof.

Lemma 3.1. *In E, it is a theorem that* $\Box A \Leftrightarrow \blacksquare A$.

Proof. First, the left-to-right direction of the biconditional:

1. $(t \Rightarrow A) \Rightarrow ((A \Rightarrow A) \Rightarrow (t \Rightarrow A))$ Suffixing
2. $((t \Rightarrow A) \Rightarrow A) \Rightarrow (((A \Rightarrow A) \Rightarrow (t \Rightarrow A)) \Rightarrow ((A \Rightarrow A) \Rightarrow A))$ Prefixing
3. $(t \Rightarrow A) \Rightarrow A$ Tt
4. $((A \Rightarrow A) \Rightarrow (t \Rightarrow A)) \Rightarrow ((A \Rightarrow A) \Rightarrow A)$ 2, 3, MP
5. $(t \Rightarrow A) \Rightarrow ((A \Rightarrow A) \Rightarrow A)$ 1, 4, Suffixing, MP
6. $\Box A \Rightarrow \blacksquare A$ 5, def $\Box$, def $\blacksquare$.

Now, the right-to-left direction:

1. $t \Rightarrow (A \Rightarrow A)$ Axiom 1 and Nt
2. $(t \Rightarrow (A \Rightarrow A)) \Rightarrow (((A \Rightarrow A) \Rightarrow A) \Rightarrow (t \Rightarrow A))$ Suffixing
3. $((A \Rightarrow A) \Rightarrow A) \Rightarrow (t \Rightarrow A)$ 1, 2, MP
4. $\blacksquare A \Rightarrow \Box A$ 3, def $\blacksquare$, def $\Box$.

Lemma 3.1 shows that Agg$\blacksquare$ is redundant in the logic with t. Moreover, the definition of $\blacksquare$ does not determine a modality with natural properties in some of the weaker systems we discuss later. These facts allow us to ignore $\blacksquare$ for the remainder of this paper.

Natural deduction proofs for E differ from those for R in the way in which subproofs are understood. In Anderson and Belnap's system [1, 2], only implicational formulas can be reiterated into subproofs. We modify that rule in order to produce proof systems for our other logics. We eliminate the reiteration rule altogether and change the implication eliminate proof to allow that the major premise be a previous step in a superior proof. We use $\Rightarrow$ for relevant entailment.

Derivation 2. The following is a derivation of the thesis of suffixing in the natural deduction system for E:

1	$A \Rightarrow B_1$	hypothesis
2	$B \Rightarrow C_2$	hypothesis
3	A_3	hypothesis
4	$B_{\{1,3\}}$	1, 3, $\Rightarrow$E
5	$C_{\{1,2,3\}}$	2, 4, $\Rightarrow$E
6	$A \Rightarrow C_{\{1,2\}}$	3–5, $\Rightarrow$I
7	$(B \Rightarrow C) \Rightarrow (A \Rightarrow C)_{\{1\}}$	2–6, $\Rightarrow$I
8	$(A \Rightarrow B) \Rightarrow ((B \Rightarrow C) \Rightarrow (A \Rightarrow C))_\emptyset$	1–7, $\Rightarrow$I

At step 4, the first hypothesis is used as the major premise of an implication elimination and the third hypothesis as its minor premise. We can think of scope lines as introducing new possible worlds at which situations indicated by the subscripts (as in proofs in R) hold. Thus, we can rewrite the above proof as:

1	$A \Rightarrow B_1$	w_1 hypothesis
2	$B \Rightarrow C_2$	w_2 hypothesis
3	A_3	w_3 hypothesis
4	$B_{\{1,3\}}$	w_3, 1, 3, $\Rightarrow$E
5	$C_{\{1,2,3\}}$	w_3 2, 4, $\Rightarrow$E
6	$A \Rightarrow C_{\{1,2\}}$	w_2 3–5, $\Rightarrow$I
7	$(B \Rightarrow C) \Rightarrow (A \Rightarrow C)_{\{1\}}$	w_1 2–6, $\Rightarrow$I
8	$(A \Rightarrow B) \Rightarrow ((B \Rightarrow C) \Rightarrow (A \Rightarrow C))_\emptyset$	0, 1–7, $\Rightarrow$I

The world parameters on the right indicate worlds that are hypothesised in each of the subproofs. (0 in the final line indicates that the formula is true at every normal situation. We will discuss the relationship between normal situations and worlds in Section 4.)

E has a ternary relation semantics like the semantics for R, but we wish to read it in a somewhat different way. We use E for the ternary relation in E-frames. The expression '$Eabc$' means that, for any worlds w_1, w_2, if a is in w_1, w_2 is *modally accessible* from w_1, and b is in w_2, then a situation like c is also in w_2. In order to

703

understand the accessibility relation E, we appeal to a second accessibility relation, in this case a binary modal accessibility relation. The status of worlds and the modal accessibility relation are best explained by appealing to the formal semantics for E. Let us move on, then, to this semantics.

4 Routley–Meyer Models for E

A positive E frame is a structure $(S, 0, E)$ such that S is a set (of situations), 0 is a non-empty subset of S, and $E \subseteq S^3$ such that all the following definitions and conditions hold. Where a, b, c, d are situations,

$$a \leq b =_{\text{df}} \exists x (x \in 0 \land Exab)$$

$$E^2 abcd =_{\text{df}} \exists x (Eabx \land Excd)$$

1. if $a \in 0$ and $a \leq b$, then $b \in 0$;

2. if $a \leq b$ and $Ebcd$ then $Eacd$;   if $c \leq d$ and $Eabc$ then $Eabd$;

3. there is a $b \in 0$ such that $Eaba$;

4. if $E^2 abcd$ then $\exists x (Eacx \land Ebxd)$;

5. if $Eabc$ then $E^2 abbc$.

Semantic postulates 3, 4, and 5 need some explanation. Below, we define a modal accessibility relation on *situations*, M, as $Mab =_{\text{df}} \exists x (x \in 0 \land Eaxb)$. Postulate 3 tells us that this accessibility relation on situations is reflexive. Postulate 4, however, has to do with the modal accessibility relation on *worlds*. $E^2 abcd$ says that if a is in a world w_1, b is in a world w_2, c is in w_3, then there is a situation like d in w_3 and w_2 is accessible from w_1 and w_3 is accessible from w_2. Since the modal accessibility relation on worlds is transitive, w_3 is accessible from w_1. $\exists x (Eacx \land Ebxd)$ tells us, in this instance, that there is a situation x such that if a is in w_1 and d is in w_3, then a situation like x is in w_3 and if b is in w_2 and x is in w_3, then a situation like d is also in w_3. The fact that the modal accessibility relation on worlds is transitive allows us to make sense of this postulate.

Semantic postulate 5 relies on reflexivity rather than transitivity. It tells us that if $Eabc$, then there if a is in w_1, b is in w_2, b is in w_3, then we can infer that there is a situation like c in w_3. This seems unintelligible, unless we read this as saying that if $Eabc$, then if a is in w_1, b is in w_2, b is in w_2, then we can infer that c is also in w_2. This makes so much sense as to seem obvious. What does the work here is identifying the worlds in which the first and second instance of b are located. We can do this if the modal accessibility relation is reflexive. To labour the point slightly, we read E^2abbc as saying that there is some situation x such that if a is in w_1 and b is in w_2, we can infer that there is a situation like x in w_2 and if x is in w_2 and b is in w_2 then there is a situation like c also in w_2, such that w_2 is accessible from w_1 and w_2 is accessible from itself.

At this point, we reflect on the nature of the set 0 of normal situations. In the natural deduction system, we can use $A_\emptyset$ at any stage, if A has been proved. In allowing this, we assume that every world contains at least one normal situation. A possible world (one that contains no contradictions) is *covered* by a normal situation. In full models for E, that contain mechanisms to deal with negation as well as the other connectives, normal situations are all bivalent. They make true the law of excluded middle. In this way, we can think of normal situations to some extent as surrogates for worlds in models. We do not, however, want to identify worlds with normal situations, at least as the latter are characterised in frames, since the way in which we understand the E relation in terms of worlds is not made explicit in frames.[2]

A positive E model is a quadruple $(S, 0, E, V)$ such that $(S, 0, E)$ is a positive E frame and V assigns sets of situations to propositional variables such that for any propositional variable p, $V(p)$ is closed upwards under $\leq$. Each value assignment V determines a satisfaction relation $\vDash_V$, between situations and formulas by means of the following inductive definition:

- $a \vDash_V p$ if and only if $a \in V(p)$;

- $a \vDash_V t$ if and only if $a \in 0$;

- $a \vDash_V A \wedge B$ if and only if $a \vDash_V A$ and $a \vDash_V B$;

- $a \vDash_V A \vee B$ if and only if $a \vDash_V A$ or $a \vDash_V B$;

- $a \vDash_V A \Rightarrow B$ if and only if $\forall b \forall c ((Eabc \wedge b \vDash_V A) \supset c \vDash_V B)$.

We write $\vDash$ instead of $\vDash_V$ where no confusion will result.

The following satisfaction condition for $\square$ can be derived:

$$a \vDash \square A \quad \text{iff} \quad \forall b \forall c ((Eabc \wedge b \in 0) \supset c \vDash A)$$

We can extract from this condition a definition of a modal accessibility relation:

$$Mab =_{\text{df}} \exists x (x \in 0 \wedge Eaxb)$$

---

[2] For contrast, see Urquhart's semantics discussed in Section 5.

Now we can state a Kripke-style satisfaction condition for necessity:

$$a \vDash \Box A \quad \text{iff} \quad \forall b(Mab \supset b \vDash A)$$

The relation M might not seem as if it is the right relation to represent necessity in E models. After all, this is a relation between situations, not worlds. But we have certain situations that can be treated as worlds. They might be mere proxies of "real" worlds or they might be the worlds themselves. This is a matter for metaphysicians to ponder. We set it aside here. These worlds are the members of the set 0. This is the set of normal situations — the situations at which all of the theorems of E are true under all interpretations.

The M relation, as defined above, has the properties of the accessibility relation in Kripke models for S4:

Proposition 4.1. *In any E-frame, M is transitive and reflexive.*

Proof. Suppose that Mab and Mbc. Then there is an $x \in 0$ and a $y \in 0$ such that $Eaxb$ and $Ebyc$. Therefore, E^2axyc. Thus, by semantic postulate 4, there is some situation z such that $Eayz$ and $Exzc$. Thus, Maz and $z \le c$. Thus, by semantic postulate 2, Mac. Generalising, M is transitive.

Reflexivity follow directly from semantic postulate 3 and the definition of M. □

In later sections of this paper, we will examine systems with weaker M relations.

5 Urquhart Semantics for E

In his PhD thesis [17] and in [16], Alasdair Urquhart gives a semantics for the implicational fragment of E. This semantics has a lot in common with the semantics for $R^\Box$. In particular, it is an extension of his semantics for the implicational fragment of R. The semantics uses a set of pieces of information and a semi-lattice join operator, $\cup$, between pieces of information. The information condition for relevant implication is

$$x \vDash A \rightarrow B \quad \text{iff} \quad \forall y(y \vDash A \supset x \cup y \vDash B).$$

Urquhart modifies this semantics to fit E by adding a set of worlds and a binary relation, N, on them. He also has formulas' being satisfied at a pair of a piece of information and a world. The condition for entailment becomes

$$(x, w_0) \vDash A \Rightarrow B \quad \text{iff} \quad \forall y \forall w_1((Nw_0w_1 \wedge (y, w_1) \vDash A) \supset (x \circ y, w_1) \vDash B).$$

The debt our interpretation owes to Urquhart's semantics is clearly extensive. The difference is that in Urquhart's semantics an R structure is present under the surface of the E structure.

In order to see what formal difference it makes to have an R structure underlying E models, let us consider using Urquhart's idea to model all of E. Consider a model $(S, 0, R, W, N, V)$ where $(S, 0, R)$ is a positive R frame, W is a non-empty set (of worlds), and N is a reflexive and transitive binary relation on W. Then we set

$$(a, w_0) \vDash A \Rightarrow B \quad \text{iff} \quad \forall b \forall c \forall w_1 ((N w_0 w_1 \wedge Rabc \wedge (b, w_1) \vDash A) \supset (c, w_1) \vDash B).$$

The conditions for the other connectives are the obvious ones.

We can prove that this model satisfies the formula that Maksimova constructed to show that NR is a proper extension of E — $((A \Rightarrow (B \Rightarrow C)) \wedge (B \Rightarrow (A \vee C))) \Rightarrow (B \Rightarrow C)$.

Proof. Suppose that $(a, w_0) \vDash (A \Rightarrow (B \Rightarrow C)) \wedge (B \Rightarrow (A \vee C))$. Also assume that $Rabc$, $N w_0 w_1$, and $(b, w_1) \vDash B$. We show that $(c, w_1) \vDash C$. Since $(a, w_0) \vDash B \Rightarrow (A \vee C)$ and $(b, w_1) \vDash B$, $(c, w_1) \vDash A \vee C$. $Rccc$, so $R^2 abcc$. Suppose that $(c, w_1) \vDash A$. **By the Pasch postulate**, $R^2 acbc$, and so there is some situation x such that $Racx$ and $Rxbc$. Since $(a, w_0) \vDash A \Rightarrow (B \Rightarrow C)$, $(x, w_1) \vDash B \Rightarrow C$. $Rxbc$, $N w_1 w_1 w_1$, and $(b, w_1) \vDash B$, so $(c, w_1) \vDash C$. Thus, if either $(c, w_1) \vDash A$ or $(c, w_1) \vDash C$, $(c, w_1) \vDash C$. Thus, $(a, w_0) \vDash B \Rightarrow C$. $\square$

The bolded step is not available, in general, for Routley–Meyer models for E. The Pasch Postulate — $\exists x (Rabx \wedge Rxcd) \supset \exists x (Racx \wedge Rxbd)$ — is in Routley–Meyer frames for R in order to make valid $(A \Rightarrow (B \Rightarrow C)) \Rightarrow (B \Rightarrow (A \Rightarrow C))$ (among other things), which would demodalise E. It would allow the inference from $(A \Rightarrow A) \Rightarrow (A \Rightarrow A)$, which is valid in E, to $A \Rightarrow ((A \Rightarrow A) \Rightarrow A)$, which is not.

We suggest, however, that Urquhart's semantics be used as a guide for the construction of entailment logics. It provides an intuitive treatment of modality in relevant logics. Although the semantics proves too much for E and for the generalisations of E that we examine below, it gives us upper bounds on the logics that we are to consider. It shows us what stronger forms of these logics (that incorporate elements from R) look like and our constructions remain weaker than these logics, but somewhat similar to them.

5.1 Note on $R^{\square}$

The system $R^{\square}$ — sometimes called "NR" — is a modal extension of R, formulated with a necessity operator and some axioms and rules taken from the modal logic S4.

The axioms are: $(\Box A \wedge \Box B) \rightarrow \Box(A \wedge B)$, $\Box(A \rightarrow B) \rightarrow (\Box A \rightarrow \Box B)$. And the additional rule is the rule of necessitation. The definition of a model for $R^\Box$ adds a second accessibility relation, N, for the necessity operator and the usual satisfaction condition for statements of the form $\Box A$ applies:

$$a \vDash \Box A \quad \text{iff} \quad \forall x(Nax \supset x \vDash A)$$

The difficulty in extending Urquhart's semantics to a semantics for all of E is replicated in the proof that $R^\Box$ is not a conservative extension of E. The underlying R frame in $R^\Box$ models creates the conditions for the proof of the Maksimova formula.

One difference between $R^\Box$ and E is that, viewed in terms of situated inference, the two logics represent different standards of information content. Consider the disjunction elimination rules for the two systems written in standard form. They look the same:

$$
\begin{array}{ll}
A \vee B_\alpha & \\
\quad \begin{array}{ll} A_k & \text{hypothesis} \\ \quad \vdots & \\ C_{\beta \cup \{k\}} & \end{array} & \\
\quad \begin{array}{ll} B_k & \text{hypothesis} \\ \quad \vdots & \\ C_{\beta \cup \{k\}} & \end{array} & \\
C_{\alpha \cup \beta} & \vee\text{E}
\end{array}
$$

where $k \notin \beta$. This similarity, however, is rather superficial. Given our situated informational interpretation of them, we can see a real difference here. We are licensed to make an inference from a disjunction by this rule in R when we have *contingent relevant implications* from both A and B. According to E, we can only make a similar inference when we have *entailments* from those two propositions. E places a stronger demand on what counts as the information available in a situation than does R. If we add an R semantic structure to an E frame, as happens in the Urquhart semantics and the semantics for $R^\Box$, then we undermine the E demand of stricter relations between the states of affairs of a situation and the further information that they carry. We can think of this distinction as a normative one. The two logics E and $R^\Box$ warrant different claims about what information is available in situations.

6 Two Notions of Necessity

In the semantics for E, there are really two notions of necessity. The first is the one that is incorporated into the entailment connective, $\Rightarrow$. Having $A \Rightarrow B$ true at a world (i.e., having the information that $A \Rightarrow B$ in some situation in that world) means that in any accessible world, if there is a situation that contains the information that A, there is also one that contains the information that B. We call this *closure necessity*, since it expresses closure conditions for worlds. The other sort of necessity is *fill necessity*. This sort of necessity is represented by $\square$.

Fill necessity can be understood both in terms of relationships between situations and relationships between worlds. As we said in Section 4, we place a modal accessibility relation between situations that acts in terms of necessity in the same way as accessibility relations in the standard worlds semantics do. A formula $\square A$ is satisfied by a situation a if and only if A is satisfied by all situations M-accessible to a. In terms of worlds, suppose that $\square A$ is true at a world w_1. Let's suppose that w_2 is accessible from w_1. As we said in Section 4, in each world there is at least one normal situation. Thus, there is some normal situation b in w_2 and there is at least one situation c in w_2 such that $Eabc$. Since $Eabc$, $c \vDash A$. Hence there is a situation in w_2 that contains the information that A, that is, A is true in w_2.

In E, both closure and fill necessity are formulated in terms of entailment. In $\mathrm{R}^{\square}$, they are both formulated in terms of $\square$. This turns out to be a very important difference between the two logics. As we have seen, they do not give us logically equivalent systems. They are also conceptually quite different. For E, and the associated logics that we will turn to presently, closure necessity is primary. In $\mathrm{R}^{\square}$, fill necessity is more important.

7 E.K

The foregoing analysis of necessity in E suggests that we look at logics in which the virtual accessibility relation between worlds has different properties. We call this relation virtual because it is present only in a very shadowy sense (in terms of the M relation between normal situations) in the formal semantics. As we have seen in Section 4, the modal accessibility relation of E is reflexive and transitive. It seems reasonable to look at systems in which the modal accessibility relation has different properties. The modal accessibility relation, however, is defined in terms of the ternary accessibility relation, which concerns entailment. Thus, we must adjust the ternary relation to modify the binary accessibility relation.

Our strategy is to formulate the properties of modality in terms of the way it is represented in the natural deduction system and then to modify this representation to incorporate different properties for modality. Then we axiomatise the resulting system and construct a Routley–Meyer semantics for it.

We begin with a logic we call E.K, to indicate that it is the basic system in much the same way that K is the basic normal modal logic.

The natural deduction rule is changed to allow applications of $\Rightarrow$E for cases in which the subproof in which the major premise is contained to be adjacent to the subproof in which the minor premise resides:

$$A \Rightarrow B_\alpha$$
$$\vdots$$
$$\vdots$$
$$A_\beta$$
$$B_{\alpha \cup \beta}$$

There is one exception to this. If the major premise has an empty-set subscript, then we allow $\Rightarrow$E to be applied to premises in the same subproof. We have to change the rule $\vee$E in a similar way:

$$A \Rightarrow C_\alpha$$
$$B \Rightarrow C_\alpha$$
$$\vdots$$
$$\vdots$$
$$A \vee B_\beta$$
$$C_{\alpha \cup \beta}$$

Again, we allow the exception that the entailment formulas can be in the same subproof as the disjunction if the subscript on the entailment formulas is the empty set. In addition, we add a rule to allow the closure of worlds under provable implications:

$$A \Rightarrow B_\emptyset$$
$$A_\alpha$$
$$\vdots$$
$$B_\alpha \qquad \text{Th}\Rightarrow$$

710

Otherwise, the rules are the same as for E.

The following is an axiomatisation of E.K:

Axioms

1. $A \Rightarrow A$

2. $A \Rightarrow (A \vee B); \quad B \Rightarrow (A \vee B)$

3. $(A \wedge B) \Rightarrow A; \quad (A \wedge B) \Rightarrow B$

4. $(A \wedge (B \vee C)) \Rightarrow ((A \wedge B) \vee (A \wedge C))$

5. $((A \Rightarrow B) \wedge (A \Rightarrow C)) \Rightarrow (A \Rightarrow (B \wedge C))$

6. $((A \Rightarrow C) \wedge (B \Rightarrow C)) \Rightarrow ((A \vee B) \Rightarrow C)$

7. t

Rules

$$\frac{\vdash A \Rightarrow B \quad \vdash A}{\vdash B} \text{ (MP)} \qquad \frac{\vdash A \quad \vdash A}{\vdash A \wedge B} \text{ (ADJ)} \qquad \frac{\vdash A}{\vdash \Box A} \text{ N}$$

$$\frac{\vdash B \Rightarrow C}{\vdash (A \Rightarrow B) \Rightarrow (A \Rightarrow C)} \text{ (PR)} \qquad \frac{\vdash A \Rightarrow B}{\vdash (B \Rightarrow C) \Rightarrow (A \Rightarrow C)} \text{ (SR)}$$

$$\frac{\vdash A^m \Rightarrow (A^{m+1} \Rightarrow \cdots (A^{n-1} \Rightarrow (B \Rightarrow C))\ldots) \quad \vdash A^p \Rightarrow (A^{p+1} \Rightarrow \cdots (A^n \Rightarrow B)\ldots)}{\vdash A^1 \Rightarrow (A^2 \Rightarrow \cdots (A^n \Rightarrow C)\ldots)} \text{ (RK)}$$

where $1 \leq m \leq n - 1$, $1 \leq p \leq n$, and at least one of $m = 1$ or $p = 1$.

The proof that all the axioms are provable in the natural deduction system and that the rules, with the exception of RK, are admissible in it is straightforward. To prove that the axiom system includes all the theorems provable in the natural deduction system, we show that in a given proof, if $C_{\{i_1,\ldots,i_n\}}$ is provable, then $A_{i_1} \Rightarrow (\cdots (A_{i_n} \Rightarrow C)\ldots)$ is provable in the axiom system, where $A_{i_1},\ldots,A_{i_n}$ are the i_1th, ..., i_nth hypotheses in the proof, respectively. Before we can prove this, we need to prove a crucial lemma.

Lemma 7.1. *If A_α is a step in a valid natural deduction proof in the system for E.K, then either $\alpha = \emptyset$ or the numbers in α are numerically consecutive and if α is non-empty, then α includes the numeral of the hypothesis of the subproof in which A_α occurs.*

Proof. By induction on the length of the proof of A_α. If α is empty, then the lemma follows. If A_α is a hypothesis, then it follows as well. The cases for conjunction introduction and elimination and disjunction introduction are straightforward, as is the case for implication introduction. Implication elimination and disjunction elimination are similar to one another. Suppose that we have a proof segment of the following form:

$$
\begin{array}{l}
A \Rightarrow B_\beta \\
\quad \vdots \\
A_\gamma \\
B_{\beta \cup \gamma} \qquad \Rightarrow\mathrm{E}
\end{array}
$$

By the inductive hypothesis, the numbers in β and γ are consecutive. By the entailment elimination rule the maximal number in γ is one higher than the maximal number in β. Let n be the maximal number in γ. Then $\gamma - \{n\}$ is either a subset of β or a proper superset. If it is a proper superset then $\gamma \cup \beta = \gamma$. If it is a subset, then $\gamma \cup \beta = \beta \cup \{n\}$. In either case, the lemma follows. As we said, the disjunction elimination case is similar. $\qquad \square$

The following proposition shows that E.K is at least as strong as the logic DJ, which in its class of theorems is the same as Ross Brady's logic of meaning containment, MC [4].

Proposition 7.2. $((A \Rightarrow B) \wedge (B \Rightarrow C)) \Rightarrow (A \Rightarrow C)$ *is a theorem of E.K.*

Proof. Let $(A \Rightarrow B) \wedge (B \Rightarrow C)$ be A^1 and A be A^2.

$$
\begin{array}{lll}
1. & ((A \Rightarrow B) \wedge (B \Rightarrow C)) \Rightarrow (B \Rightarrow C) & \text{axiom } 3 \\
2. & ((A \Rightarrow B) \wedge (B \Rightarrow C)) \Rightarrow (A \Rightarrow B) & \text{axiom } 3 \\
3. & ((A \Rightarrow B) \wedge (B \Rightarrow C)) \Rightarrow (A \Rightarrow C) & 1, 2, \text{RK}
\end{array}
$$

$\qquad \square$

Given the definition of $\Box A$ as $t \Rightarrow A$, Proposition 7.2 also shows that the following version of the K axiom is a theorem of E.K:

$$(\Box A \wedge (A \Rightarrow B)) \Rightarrow \Box B$$

Moreover, the following aggregation principle is an instance of axiom 5:

$$(\Box A \wedge \Box B) \Rightarrow \Box (A \wedge B)$$

Thus, E.K contains a good deal of what are relevant counterparts of the key theorems of the modal logic K.

For the following theorem, we use an abbreviation. Where $\{1, \ldots, n\}$ is a set of subscripts of hypotheses in a derivation, $A^1, \ldots, A^n$, respectively, $\{1, \ldots, n\} \Rightarrow C$ is the formula $A^1 \Rightarrow (A^n \Rightarrow C)$.

Theorem 7.3. *For any formula C, if $C_{\{i_1, \ldots, i_n\}}$ is a step in a valid natural deduction proof then $A_{i_1} \Rightarrow (\cdots (A_{i_n} \Rightarrow C) \ldots)$ is provable in the E.K axiom system.*

Proof. By induction on the length of the proof $C_{\{i_1, \ldots, i_n\}}$.

Base Case. Suppose that C_i is a hypothesis. By axiom 1, $C \Rightarrow C$ is a theorem of the axiom system.

The conjunction and negation cases are straightforward, as are the cases for the entailment and disjunction introduction rules. Thus, we prove only the cases for entailment and disjunction elimination.

Entailment Elimination. Suppose that $C_{\{i_1, \ldots, i_n\}}$ is proven from $A \Rightarrow C_\alpha$ and A_β. Then, by the rules of the natural deduction system, the maximal number in β is 1 greater than the maximal number in α. By the inductive hypothesis, $\vdash \alpha \Rightarrow (A \Rightarrow C)$ and $\vdash \beta \Rightarrow C$. By RK, then, $\vdash (\alpha \cup \beta) \Rightarrow C$.

Disjunction Elimination. Suppose that $C_{\{i_1, \ldots, i_n\}}$ is proven from $A \Rightarrow C_\alpha$ and $B \Rightarrow C_\alpha$ and $A \vee B_\beta$ by $\vee$E. Then $\alpha \cup \beta = \{i_1, \ldots, i_n\}$ and by the inductive hypothesis, $\vdash \alpha \Rightarrow (A \Rightarrow C)$, $\vdash \alpha \Rightarrow (B \Rightarrow C)$, and $\vdash \beta \Rightarrow (A \vee B)$. Thus, by axiom 5 and repeated applications of the prefixing rule, $\vdash \alpha \Rightarrow ((A \Rightarrow C) \wedge (B \Rightarrow C))$ and so, by Proposition 7.2 and repeated applications of the prefixing rule, $\vdash \alpha \Rightarrow ((A \vee B) \Rightarrow C)$. Thus, by $\vdash \beta \Rightarrow (A \vee B)$ and RK, we obtain $\vdash (\alpha \cup \beta) \Rightarrow C$. $\square$

We need the following lemmas for the completeness proof.

Lemma 7.4. *The following rule is derivable in E.K:*

$$\frac{\vdash A^1 \Rightarrow (\cdots (A^{n-1} \Rightarrow (A^n \Rightarrow C)) \ldots) \quad \vdash B \Rightarrow A^n}{\vdash A^1 \Rightarrow (\cdots (A^{n-1} \Rightarrow (B \Rightarrow C)) \ldots)}$$

Proof.

1. $\vdash A^1 \Rightarrow (\cdots (A^{n-1} \Rightarrow (A^n \Rightarrow C))\ldots)$ Premise

2. $\vdash B \Rightarrow A^n$ Premise

3. $\vdash (A^n \Rightarrow C) \Rightarrow (B \Rightarrow C)$ 2, PR

4. $\vdash (A^{n-1} \Rightarrow (A^n \Rightarrow C)) \Rightarrow (A^{n-1} \Rightarrow (B \Rightarrow C))$ 3, PR

5. $\ldots$

6. $\vdash (A^1 \Rightarrow (\cdots (A^{n-1} \Rightarrow (A^n \Rightarrow C))\ldots)) \Rightarrow$
$(A^1 \Rightarrow (\cdots (A^{n-1} \Rightarrow (B \Rightarrow C))\ldots))$

7. $\vdash A^1 \Rightarrow (\cdots (A^{n-1} \Rightarrow (B \Rightarrow C))\ldots)$ 1, 6, MP

$\square$

Lemma 7.5. *The following rule is derivable in E.K. Where $m \leq n$,*

$$\vdash (A^1 \Rightarrow (\cdots (A^n \Rightarrow (D \Rightarrow E))\ldots))$$
$$\vdash (B^1 \Rightarrow (\ldots (B^m \Rightarrow (C \Rightarrow D))\ldots))$$
$$\overline{\vdash (A^1 \Rightarrow (\cdots ((A^{n-m} \wedge B^1) \Rightarrow ((A^{(n-m)+1} \wedge B^2) \Rightarrow (\cdots ((A^n \wedge B^m) \Rightarrow (C \Rightarrow E))\ldots)))\ldots))}$$

Proof.

1. $\vdash (A^1 \Rightarrow (\cdots (A^n \Rightarrow (D \Rightarrow E))\ldots))$ Premise

2. $\vdash (B^1 \Rightarrow (\cdots (B^m \Rightarrow (C \Rightarrow D))\ldots))$ Premise

3. $\vdash (A^n \wedge B^m) \Rightarrow A^n$ Axiom 3

4. $\vdash (A^1 \Rightarrow (\cdots ((A^n \wedge B^m) \Rightarrow (D \Rightarrow E))\ldots))$ 1, 3, Lemma 7.4

5. $\ldots$

6. $\vdash (A^1 \Rightarrow (\cdots ((A^{n-m} \wedge B^1) \Rightarrow (\cdots ((A^n \wedge B^m) \Rightarrow$
$(D \Rightarrow E))\ldots))\ldots))$

7. $\vdash (B^1 \Rightarrow (\cdots ((A^n \wedge B^m) \Rightarrow (C \Rightarrow D))\ldots))$ 2, 3, Lemma 7.4

8. $\ldots$

9. $\vdash (A^{n-m} \wedge B^1) \Rightarrow (\cdots ((A^n \wedge B^m) \Rightarrow (C \Rightarrow D))\ldots)$

10. $\vdash (A^1 \Rightarrow (\cdots ((A^{n-m} \wedge B^1) \Rightarrow ((A^{(n-m)+1} \wedge B^2) \Rightarrow$
$(\cdots ((A^n \wedge B^m) \Rightarrow (C \Rightarrow E))\ldots)))\ldots))$ 6, 9 RK

$\square$

8 E.K Models

A positive E.K frame is a triple $(S, 0, E)$ just as for E frames, except that the semantic postulates are now the following:

1. $\leq$ is a partial order;

2. if $a \in 0$ and $a \leq b$, then $b \in 0$;

3. if $a \leq b$ and $Ebcd$ then $Eacd$;   if $c \leq d$ and $Eabc$ then $Eabd$;

4. if $Ea_1 \ldots a_n c$, then $\exists x \exists y (Ea_m \ldots a_{n-1} x \wedge Ea_p \ldots a_n y \wedge Exyc)$ (where $n - m \geq 2$ and $n - p \geq 1$).

Here we use an extension of the definition of E^2 given in Section 4. We define

$$E^{n+1} a_1 \ldots a_n a_{n+1} a_{n+2} c \quad \text{as} \quad \exists x (E^n a_1 \ldots a_{n+1} x \wedge Exa_{n+2} c).$$

For convenience, we drop the superscript from E^n and merely write $Ea_1 \ldots a_{n+1} c$.

A positive E.K model is a quadruple $(S, 0, E, V)$ such that $(S, 0, E)$ is a positive E.K frame and V assigns sets of situations to propositional variables such that for any propositional variable p, $V(p)$ is closed upwards under $\leq$. Each value assignment V determines a satisfaction relation $\vDash_V$, applying the same clauses as for E models. We write '$\vDash$' instead of '$\vDash_V$' where no confusion will result.

The meaning of entailment in E.K is, on one level, the same as it is for E: $A \Rightarrow B$ says that if, in an accessible world, there is a situation that contains the information that A, then there is a situation that contains B. Without reflexivity or transitivity, one use of E.K's implication could be to represent a form of doxastic entailment. An agent might be said to hold $A \rightarrow B$ in the sense of E.K if and only if she believes that B follows from A.

Excluding the rule RK, the axiomatic basis for E.K is the same as that of the minimal relevant logic B. The soundness of B over the Routley–Meyer semantics is well known [14, 15]. Thus it is sufficient to show that the class of E.K frames satisfy RK.

Lemma 8.1. *In any E.K model, if* $\vDash A^m \Rightarrow (\cdots (A^{n-1} \Rightarrow (B \Rightarrow C)) \ldots)$ *and* $\vDash A^p \Rightarrow (\cdots (A^n \Rightarrow B) \ldots)$, *then* $\vDash A^1 \Rightarrow (\cdots (A^n \Rightarrow C) \ldots)$.

Proof. Suppose that $\vDash A^m \Rightarrow (\cdots (A^{n-1} \Rightarrow (B \Rightarrow C)) \ldots)$ and $\vDash A^p \Rightarrow (\cdots (A^n \Rightarrow B) \ldots)$ and suppose that $Ea_1 \ldots a_n c$ and $a_i \vDash A_i$ for each i, $1 \leq i \leq n$. Let $m = 1$. The case in which $p = 1$ is similar. By semantic condition 4, there is a situation x such that $Ea_1 \ldots a_{n-1} x$ and a situation y such that $Ea_p \ldots a_n y$ and $Exyc$. By assumption and the information condition for implication, $x \vDash B \Rightarrow C$ and $y \vDash B$. So $c \vDash C$. $\qquad \square$

Thus, we can now state the following soundness theorem:

Theorem 8.2 (Soundness). *All the theorems of E.K are valid in the class of E.K frames.*

8.1 Completeness of E.K

In order to construct the canonical model, we define a form of logical consequence for a logic L:

$$\Gamma \vdash_L \Delta \text{ iff } \exists G_1, \ldots, G_m \in \Gamma \, \exists D_1, \ldots, D_n \in \Delta (\vdash_L (G_1 \wedge \ldots \wedge G_m) \Rightarrow (D_1 \vee \ldots \vee D_n)).$$

We use this consequence relation for a wide variety of purposes in what follows, first to define the notion of a theory.

Definition 8.3 (Theory). An L-theory Γ is a set of formulas such that if $\Gamma \vdash_L \{A\}$, then $A \in \Gamma$.

It is easy to show that if Γ is an L-theory, $A \in \Gamma$, and $B \in \Gamma$, then $A \wedge B \in \Gamma$. A theory Γ is said to be *prime* if and only if for all formulas $A \vee B \in \Gamma$, either $A \in \Gamma$ or $B \in \Gamma$. Γ is said to be *regular* if and only if $t \in \Gamma$.

We also use the consequence relation to define the notion of L-consistency: a pair of sets of formulas (Γ, Δ) is said to be *L-consistent* if and only if $\Gamma \nvdash_L \Delta$. The form of Lindenbaum extension theorem that is used for relevant logics employs L-consistency, rather than the more standard notion of negation consistency. This lemma was originally proven by Nuel Belnap and Dov Gabbay (see [5]).

Theorem 8.4. *If (Γ, Δ) is L-consistent, then there is a prime theory $\Gamma' \supseteq \Gamma$ such that (Γ', Δ) is L-consistent.*

Corollary 8.5. *A formula A is a theorem of L if and only if $A \in \Gamma$ for all regular prime L-theories Γ.*

Proof. If $\vdash_L A$, then $\vdash_L t \Rightarrow A$, by RN. If Γ is regular, then, by definition, $t \in \Gamma$, hence $\Gamma \vdash_L \{A\}$. Since Γ is a theory, $A \in \Gamma$. If $\nvdash_L A$ then $(L, \{A\})$ is L-consistent, where L is taken here to be the set of theorems of L. Thus, by the Lindenbaum theorem, there is a prime regular L-theory Γ such that $\Gamma \nvdash_L \{A\}$. □

In order to formulate our canonical model, we utilise a binary fusion operator on theories. Where a and b are L-theories for any of our logics L,

$$a \circ b =_{df} \{B \in Fml : \exists A((A \Rightarrow B) \in a \wedge A \in b)\}.$$

It is easy to prove that the fusion of two L-theories is an L-theory. Note, however, that the fusion of two prime theories may not be prime, but we can prove the following lemma.

Lemma 8.6. *(a) If a, b, and c are L-theories, a and c are prime, and $a \circ b \subseteq c$, then there is a prime L-theory $b' \supseteq b$ such that $a \circ b' \subseteq c$; (b) if a, b, and c are L theories, b and c are prime, and $a \circ b \subseteq c$, then there is a prime L-theory $a' \supseteq a$ such that $a' \circ b \subseteq c$; (c) where a, b, and c are L-theories, if $a \circ b \subseteq c$, then there are prime L-theories a', b', and c' such that $a' \circ b' \subseteq c'$.*

Proof. (a) Suppose that a, b, and c are L-theories, a and c are prime and $a \circ b \subseteq c$. Let X be the set of formulas A such that there is some $B \notin c$ and $A \Rightarrow B \in a$.

We show that (b, X) is L-consistent. Suppose that (b, X) is L-inconsistent. Then there are $B_1, \ldots, B_m \in b$ and $C_1, \ldots, C_n \in X$ such that $\vdash_L (B_1 \wedge \cdots \wedge B_m) \Rightarrow (C_1 \vee \cdots \vee C_n)$. By the definition of X, there are $A_1, \ldots, A_n$, not in c such that $C_1 \Rightarrow A_1 \in a, \ldots, C_n \Rightarrow A_n \in a$. By a simple logical derivation, $(C_1 \vee \cdots \vee C_n) \Rightarrow (A_1 \vee \cdots \vee A_n)$. Since c is prime, $(A_1 \vee \cdots \vee A_n) \notin c$. But if $\vdash_L (B_1 \wedge \cdots \wedge B_m) \Rightarrow (C_1 \vee \cdots \vee C_n)$, then $(C_1 \vee \ldots \vee C_n) \in b$. Hence, $a \circ b \not\subseteq c$. Thus, by reductio, (b, X) is L-consistent.

By Theorem 8.4, there is a prime theory b' extending b such that (b', X) is L-consistent. Hence $a \circ b' \subseteq c$.

(b) The proof is similar to that of (a).

(c) Suppose that $a \circ b \subseteq c$. Then, there is a prime L-theory c' extending c such that $a \circ b \subseteq c'$. Now we extend a to a prime L-theory a' such that $a' \circ b \subseteq c'$. We do so by noting that the set $X = \{A \Rightarrow B : A \in b \wedge B \notin c'\}$ is such that (a, X) is L-consistent (see the proof of (a) above). Then, by the Lindenbaum lemma, a can be extended to a prime L-theory a' such that $a' \circ b \subseteq c'$. By (a) above, there is a prime L-theory b' such that $a' \circ b' \subseteq c'$. $\qquad \square$

Lemma 8.7. *For every L-theory a, there is some regular prime L-theory o such that $o \circ a = a$.*

Proof. Let a be an L-theory. Then $Thm(L) \circ a = a$ and so $Thm(L) \circ a \subseteq a$. By Lemma 8.6(b), there is a prime L-theory o extending $Thm(L)$ such that $o \circ a \subseteq a$. Since o extends $Thm(L)$, o is regular. $\qquad \square$

We are now ready to construct the canonical model. The canonical model is a quadruple $\mathfrak{M}_L = (S, 0, E, V)$ such that

- S is the set of prime theories of L;

- 0 is the set of regular prime theories of L;

- $E \subseteq S^3$ is such that $Eabc$ if and only if $a \circ b \subseteq c$;

- V is a function from propositional variables to subsets of S such that $a \in V(p)$ if and only if $p \in a$.

Lemma 8.8. *For all L-theories, $a \leq b$ if and only if $a \subseteq b$.*

Proof. Suppose that $a \leq b$. Then there is some regular prime L-theory o such that $Eoab$. By the definition of E, $o \circ a \subseteq b$. Since $A \Rightarrow A \in o$, $a \subseteq b$.

Suppose now that $a \subseteq b$. By Lemma 8.7, there is a $o \in 0$ such that $o \circ a \subseteq a$, thus by the transitivity of subset, $o \circ a \subseteq b$. Therefore, $a \leq b$. $\qquad\square$

Lemma 8.9. *For $2 \leq n$, $Ea_1 \ldots a_n b$ if and only if $(\ldots (a_1 \circ a_2) \circ \cdots) \circ a_n \subseteq b$.*

Proof. By induction on n.
Base case: $n = 2$. Follows from the definition of E for the canonical model.
Inductive case: Suppose that, for all $b \in S$, $Ea_1 \ldots a_n b$ iff $(\ldots (a_1 \circ a_2) \circ \cdots) \subseteq b$. We show that for all $a_{n+1}, b \in S$, $Ea_1 \ldots a_{n+1} b$ iff $(\ldots (a_1 \circ a_2) \circ \cdots) \circ a_{n+1} \subseteq b$.

Suppose that $Ea_1 \ldots a_{n+1} b$. By definition, $Ea_1 \ldots a_n a_{n+1} b$ if and only if there is some $x \in S$, $Ea_1 \ldots a_n x$ and $Exa_{n+1}b$. By hypothesis, $(\ldots (a_1 \circ a_2) \circ \cdots) \circ a_n) \subseteq x$. Clearly, for all L-theories x, y, z, if $z \subseteq w$, then $z \circ y \subseteq w \circ y$. So, $(\ldots (a_1 \circ a_2) \circ \cdots) \circ a_n) \circ a_{n+1} \subseteq x \circ a_{n+1}$. Since $Exa_{n+1}b$, $x \circ a_{n+1} \subseteq b$. Thus, by the transitivity of subset, $(\ldots (a_1 \circ a_2) \circ \cdots) \circ a_{n+1} \subseteq b$.

Suppose now that $(\ldots (a_1 \circ a_2) \circ \cdots) \circ a_{n+1} \subseteq b$. $(\ldots (a_1 \circ a_2) \circ \cdots) \circ a_n$ is an L-theory. Thus, by Lemma 8.6(b), there is a prime L-theory x such that $(\ldots (a_1 \circ a_2) \circ \cdots) \circ a_n \subseteq x$ and $x \circ a_{n+1} \subseteq b$. By hypothesis, $Ea_1 \ldots a_n x$ and, by definition of E, $xa_{n+1}b$. Therefore, $Ea_1 \ldots a_{n+1} b$. $\qquad\square$

Lemma 8.10. *If $Ea_1 \ldots a_n b$, then for all p, $1 \leq p \leq n - 1$, $Ea_1 \ldots a_{n-1} x$ and $Ea_p \ldots a_n y$ and $Exyb$.*

Proof. Suppose that $Ea_1 \ldots a_n b$. Then, by Lemma 8.9, $(\ldots (a_1 \circ a_2) \circ \cdots) \circ a_n \subseteq b$. We show that $(\ldots (a_m \circ a_{m+1}) \circ \cdots) \circ a_{n-1}) \circ (\ldots (a_p \circ a_{p+1}) \circ \cdots) \circ a_n) \subseteq b$, where either m or p is 1. Case 1. $m = 1$. Suppose that $C \in (\ldots (a_1 \circ a_2) \circ \cdots) \circ a_{n-1}) \circ (\ldots (a_p \circ a_{p+1}) \circ \cdots) \circ a_n)$. We show that $C \in b$. Then, by the definition of fusion on theories, there is some formula B such that $B \Rightarrow C \in (\ldots (a_1 \circ a_2) \circ \cdots) \circ a_{n-1}$ and $B \in (\ldots (a_p \circ a_{p+1}) \circ \cdots) \circ a_n)$. Using the same reasoning, we can see that there are $A^2, \ldots, A^{n-1}$ such that for $2 \leq i \leq n - 1$, $A^i \in a_i$ and

$$A^2 \Rightarrow (\cdots (A^p \Rightarrow (\cdots (B \Rightarrow C) \ldots)) \ldots) \in a_1.$$

718

Similarly, there are formulas $D^{p+1}, \ldots, D^n$ such that for all j, $p+1 \leq j \leq n$, $D^j \in a_j$ and

$$D^p \Rightarrow (\cdots (D^n \Rightarrow B) \ldots) \in a_p.$$

Now, we know that E.K proves

$$(A^2 \Rightarrow (\cdots (A^p \Rightarrow (\cdots (B \Rightarrow C) \ldots)) \ldots)) \Rightarrow (A^2 \Rightarrow (\cdots (A^p \Rightarrow (\cdots (B \Rightarrow C) \ldots)) \ldots))$$

and

$$\vdash_{E.K} (D^p \Rightarrow (\cdots (D^n \Rightarrow B) \ldots)) \Rightarrow (D^p \Rightarrow (\cdots (D^n \Rightarrow B) \ldots)).$$

By Lemma 7.5, then, we can derive

$$\vdash_{E.K} (A^2 \Rightarrow (\ldots (A^p \Rightarrow (\ldots (B \Rightarrow C) \ldots)) \Rightarrow$$
$$((\ldots (A^p \wedge (D^p \Rightarrow (\ldots (D^n \Rightarrow B) \ldots)) \Rightarrow (D^p \Rightarrow (\ldots (D^n \Rightarrow B) \ldots)) \Rightarrow$$
$$((A^{p+1} \wedge D^{p+1}) \Rightarrow ((A^{n-1} \wedge D^{n-1}) \Rightarrow (D^n \Rightarrow C) \ldots).$$

Thus,

$$((\ldots (A^p \wedge D^p) \Rightarrow (\ldots (D^n \Rightarrow B) \ldots)) \Rightarrow (D^p \Rightarrow (\ldots (D^n \Rightarrow B) \ldots)) \Rightarrow$$
$$((A^{p+1} \wedge D^{p+1}) \Rightarrow ((A^{n-1} \wedge D^{n-1}) \Rightarrow (D^n \Rightarrow C) \ldots) \in a_1.$$

For all i, $2 \leq i < p$, and all j, $p+1 \leq j \leq n$, $A^i \in a_i$ and $(A^i \wedge D^j) \in a_j$. In addition, $(A^p \wedge D^p) \Rightarrow (\cdots (D^n \Rightarrow B) \ldots)) \in a_p$, so

$$C \in (\ldots (a_1 \circ a_2) \circ \cdots) \circ a_n).$$

Since $(\ldots (a_1 \circ a_2) \circ \cdots) \circ a_n \subseteq b$, $C \in b$, as required. Generalising, $(\ldots (a_1 \circ a_2) \circ \cdots) \circ a_{n-1}) \circ (\ldots (a_p \circ a_{p+1}) \circ \cdots) \circ a_n) \subseteq b$. Hence, by Lemma 8.6, there is a prime theory x extending $(\ldots (a_1 \circ a_2) \circ \cdots) \circ a_{n-1}$ and a prime theory y extending $(\ldots (a_p \circ a_{p+1}) \circ \cdots) \circ a_n$ and $x \circ y \subseteq b$, i.e., $Exyb$.

Case 2. $p = 1$. Similar to case 1. $\qquad \square$

9 E.KT and E.K4

We now look at two logics between E.K and E. These are E.KT and E.K4. In terms of their natural deduction systems, E.KT adds to E.K modified forms of the entailment and disjunction elimination rules. In E.KT, we can apply a major to a minor premise when the two are in the same subproof:

$$A \Rightarrow B_\alpha$$
$$A_\beta$$
$$B_{\alpha \cup \beta} \qquad \qquad \Rightarrow E$$

719

For E.K4, the $\Rightarrow$E rule is modified to allow, not premises in the same subproof, but a major premise that is in a subproof separated from the minor by one or more other subproofs.

$$
\begin{array}{l}
A \Rightarrow B_\alpha \\
\quad \ddots \\
\qquad A_\beta \\
\qquad B_{\alpha \cup \beta} \qquad \Rightarrow\text{E}
\end{array}
$$

For E.KT we replace the rule RK with the rule RKT:

$$
\frac{\vdash A^m \Rightarrow (A^{m+1} \Rightarrow \cdots (A^q \Rightarrow (B \Rightarrow C)) \ldots)}{\vdash A^p \Rightarrow (A^{p+1} \Rightarrow \cdots (A^n \Rightarrow B) \ldots)}{\vdash A^1 \Rightarrow (A^2 \Rightarrow \cdots (A^n \Rightarrow C) \ldots)}
$$

where either $p = 1$ or $q = 1$, $n - 1 \leq q \leq n$, and $1 \leq p \leq n$. For E.KT we also need to add the rule RTh:

$$
\frac{\vdash A^1 \Rightarrow (\cdots (A^n \Rightarrow (B \Rightarrow C)) \ldots)}{\vdash B}{\vdash A^1 \Rightarrow (\cdots (A^n \Rightarrow C) \ldots)}
$$

To obtain E.K4 we replace RK with the rule RK4. We begin with a finite sequence of formulas $\Sigma = \langle A_1, \ldots, A_n \rangle$. Let Γ and Δ be sequences, in which all of the formulas that occur in them occur in Σ and occur in the same order as in Σ. Moreover, for any A_i ($1 \leq i \leq n - 1$), the total number of times that it occurs in both Γ and Δ is at least the number of times that it occurs in $\langle A_1, \ldots, A_{n-1} \rangle$. (It follows from this that every formula that occurs in Σ occurs at least once in one of Γ or Δ.)

$$
\frac{\vdash \Gamma \Rightarrow (B \Rightarrow C)}{\vdash \Delta \Rightarrow (A^n \Rightarrow B)}{\vdash A^1 \Rightarrow (\cdots (A^n \Rightarrow C) \ldots)}
$$

Here $\langle D_1, \ldots, D_m \rangle \Rightarrow E$ is defined as $D_1 \Rightarrow (D_2 \Rightarrow (\cdots (D_m \Rightarrow E) \ldots))$.

Proposition 9.1. *Each of (i) $((A \Rightarrow B) \wedge A) \Rightarrow B$, (ii) $(A \Rightarrow (A \Rightarrow B)) \Rightarrow (A \Rightarrow B)$, and (iii) $((A \Rightarrow A) \Rightarrow B) \Rightarrow B$ are theorems of E.KT.*

Proof. (i)

$$
\begin{array}{lll}
1. & \vdash ((A \Rightarrow B) \wedge A) \Rightarrow (A \Rightarrow B) & \text{Axiom 3} \\
2. & \vdash ((A \Rightarrow B) \wedge A) \Rightarrow A & \text{Axiom 3} \\
3. & \vdash ((A \Rightarrow B) \wedge A) \Rightarrow B & 1, 2, \text{ RKT}
\end{array}
$$

(ii) Taking $A \Rightarrow (A \Rightarrow B)$ to be A^1 and A to be A^2, we get:

$$
\begin{array}{lll}
1. & \vdash (A \Rightarrow (A \Rightarrow B)) \Rightarrow (A \Rightarrow (A \Rightarrow B)) & \text{Axiom 1} \\
2. & \vdash A \Rightarrow A & \text{Axiom 1} \\
3. & \vdash (A \Rightarrow (A \Rightarrow B)) \Rightarrow (A \Rightarrow B) & 1, 2, \text{ RKT}
\end{array}
$$

(iii)

$$
\begin{array}{lll}
1. & \vdash ((A \Rightarrow A) \Rightarrow B) \Rightarrow ((A \Rightarrow A) \Rightarrow B) & \text{Axiom 1} \\
2. & \vdash A \Rightarrow A & \text{Axiom 1} \\
3. & \vdash ((A \Rightarrow A) \Rightarrow B) \Rightarrow B & 1, 2, \text{ RTh}
\end{array}
$$

$\square$

Proposition 9.2. *(a) $(A \Rightarrow B) \Rightarrow ((B \Rightarrow C) \Rightarrow (A \Rightarrow C))$ and (b) $(B \Rightarrow C) \Rightarrow ((A \Rightarrow B) \Rightarrow (A \Rightarrow C))$ are theorems of E.K4.*

Proof. (a) Let $A \Rightarrow B$ be A^1, $B \Rightarrow C$ be A^2, and A be A^3.

$$
\begin{array}{lll}
1. & (A \Rightarrow B) \Rightarrow (A \Rightarrow B) & \text{axiom 1} \\
2. & (B \Rightarrow C) \Rightarrow (B \Rightarrow C) & \text{axiom 1} \\
3. & (A \Rightarrow B) \Rightarrow ((B \Rightarrow C) \Rightarrow (A \Rightarrow C)) & 1, 2, \text{RK4}
\end{array}
$$

(b) Let $B \Rightarrow C$ be A^1, $A \Rightarrow B$ be A^2, and A be A^3.

$$
\begin{array}{lll}
1. & (B \Rightarrow C) \Rightarrow (B \Rightarrow C) & \text{axiom 1} \\
2. & (A \Rightarrow B) \Rightarrow (A \Rightarrow B) & \text{axiom 1} \\
3. & (B \Rightarrow C) \Rightarrow ((A \Rightarrow B) \Rightarrow (A \Rightarrow C)) & 1, 2, \text{RK4}
\end{array}
$$

$\square$

Proposition 9.2 shows that E.K4 is an extension of TW, which is Anderson and Belnap's system of ticket entailment, T, without the axiom of contraction. The entailment fragment, $TW_\Rightarrow$, is extraordinary because in any case in which an equivalence $A \Rightarrow B$ and $B \Rightarrow A$ is provable, then A and B are the same formula [12, 2]. We do not know at this point in time whether E.K4 is exactly the same logic (i.e., has the same theorems) as TW.

Propositions 9.1 and 9.2 together show that the logic E.KT4 which results from the axiom basis for E.KT together with the rule RK4 yields an extension of E. To show that it is E, it suffices to show that the rules RK, RKT, RTh, and RK4 are all derivable in E. This is easy (although somewhat tedious) to show, and so we can say that E.KT4 is equivalent to E.

Proposition 9.3. $\Box A \Rightarrow \Box\Box A$ *is a theorem of E.K4.*

Proof. Let $t \Rightarrow A$ be A^1 and t be A^2

$$
\begin{array}{lll}
1. & (t \Rightarrow A) \Rightarrow ((t \Rightarrow t) \Rightarrow (t \Rightarrow A)) & \text{Proposition } 9.2(b) \\
2. & t \Rightarrow (t \Rightarrow t) & \text{axiom 1 and RN} \\
3. & (t \Rightarrow A) \Rightarrow (t \Rightarrow (t \Rightarrow A)) & 1, 2, \text{ RK4} \\
4. & \Box A \Rightarrow \Box\Box A & 3, \text{ def. } \Box
\end{array}
$$

$\Box$

Proposition 9.3 shows that the fill necessity of E.K4 is very much like that of the classical modal logic K4.

10   E.KT and E.K4 Models

An E.KT frame is an E.K frame with two additional conditions. The first condition that we add is

(SRT)   If $Ea_1 \ldots a_n c$, then $\exists x \exists y (Ea_m \ldots a_n x \ \land \ Ea_p \ldots a_n y \ \land \ Exyc)$,

where at least one of m or p is 1. We do not require that either $n - m$ or that $n - p$ be at least 0, although we do require that $n \geq 2$. When $m = n$, then we read $Ea_m \ldots a_n x$ as $a_n \leq x$, and similarly for $p = n$.

Suppose that $Eabc$. By SRT we have $Eabx$ and $b \leq y$ and $Exyc$. By semantic condition 3 on E.K frames, we have $Exbc$ and so we have $Eabbc$. Thus, $Eabc$ implies $Eabbc$. This is the condition for contraction from the definition of an E frame.

We can also derive the condition $Eaaa$ for all situations a. Here is the proof. By semantic condition 1 on E.K frames, $\exists x(x \in 0 \land Exaa)$. By contraction, $Exaaa$, i.e., there is some y such that $Exay$ and $Eyaa$. By the definition of $\leq$, $a \leq y$ and so by semantic condition 3, $Eaaa$. The condition $Eaaa$ is called *complete reflexivity*.

Complete reflexivity allows us to prove simple instances of RKT such as:

$$\vdash A \Rightarrow (B \Rightarrow C)$$
$$\frac{\vdash A \Rightarrow B}{\vdash A \Rightarrow C}$$

The second condition we add is the following:

$$\text{(T)} \quad \exists x(x \in 0 \wedge Eaxa)$$

The condition T just says that M is reflexive for E.KT, as one would expect.

Lemma 10.1. *The rule RKT is valid in the class of E.KT frames.*

Proof. The only cases that are not covered by the soundness proof for E.K are instances of the rule in which $q = n$. We have already proven the case in which $q = n = 2$. Suppose that $A^1 \Rightarrow (A^{m+1} \Rightarrow \cdots (A^n \Rightarrow (B \Rightarrow C))\ldots)$ and $A^p \Rightarrow (A^{p+1} \Rightarrow \cdots (A^n \Rightarrow B)\ldots)$ are both valid in the class of E.KT frames. Now, consider an E.KT model and situations $a_1, \ldots, a_n$ and c such that $Ea_1 \ldots a_n c$ and $a_i \vDash A^i$ for all i, $1 \leq i \leq n$. By the assumption and SCT, $c \vDash B \Rightarrow C$ and $c \vDash B$. By $Eccc$ and the satisfaction condition, $c \vDash C$. Generalising, $\vDash A^1 \Rightarrow (\cdots (A^n \Rightarrow C)\ldots)$. $\square$

Theorem 10.2. *E.KT is sound over the class of E.KT frames.*

The proof of completeness for E.KT is very like the one for E.K. Lemma 8.10 has to be tweaked slightly, but the proof is essentially the same. Thus we merely state the completeness theorem:

Theorem 10.3. *E.KT is complete over the class of E.KT frames.*

The soundness and completeness theorems for E.KT show that there is an alternative axiomatisation of the logic that includes the axiomatic basis for E.K plus the two axiom schemes PMP and T.

The definition of an E.K4 frame is the same as for an E.K frame except that it includes the following condition. Where $\langle a_1, \ldots, a_n \rangle$ is a sequence of situations and $Ea_1 \ldots a_n c$, there are situations x and y such that $Ea_{i_1} \ldots a_{i_m} x$ and $Ea_{j_1} \ldots a_{j_p} a_n y$ and $Exyc$, where each of the a_is and a_js are in the original sequence and numbered in the same order as in the original sequence.

An E.K4 frame is an E.K frame with the addition of the condition SK4. Let $\sigma = \langle a_1, \ldots, a_{n-1} \rangle$ be a finite sequence of situations. Let γ and δ be sequences of situations taken from σ, such that in γ and δ every situation occurs in the same order as it occurs in σ and between γ and δ each situation occurs at least as many times as it occurs in σ.

$$(SK4) \quad \text{If } Ea_1 \ldots a_{n-1} a_n c, \text{ then } \exists x \exists y (E\gamma x \wedge E\delta a_n y \wedge Exyc).$$

Lemma 10.4. *In all E.K4 frames, for all situations a, b, c, d, if $Eabcd$ then there is some situation x such that $Eacx$ and $Ebxd$.*

Proof. Suppose that $Eabcd$. Then, there is some situation y such that $Eacy$ and some situation x such that Ebx and $Exyd$. By definition, Ebx is just $b \le x$, so by semantic condition 3 for E.K frames, $Ebxd$. $\square$

Lemma 10.4 shows that the condition used in E frames to prove the prefixing axiom is satisfied by E.K4 frames as well. This also shows that the modal accessibility relation M is transitive (see Proposition 4.1).

Lemma 10.5. *The rule RK4 is sound over the class of E.K4 frames.*

Proof. Let $A^1, \ldots, A^n$ be a sequence of formulas such that $\Gamma \Rightarrow (B \Rightarrow C)$ and $\Delta \Rightarrow B$ are valid in the class of E.K4 frames, where Γ is a subset of the sequence not including A^n and Δ is a subset of the sequence that includes A^n and $\Gamma \cup \Delta = \{A^1, \ldots, A^n\}$. Suppose that $Ea_1 \ldots a_n c$, where $a_i \vDash A^i$ for each i, $1 \le i \le n$. Let $S(\Gamma)$ be a subset of $\{a_1, \ldots, a_n\}$ such that for each $A^j \in \Gamma$, there is a situation $a_j \in S(\Gamma)$ such that $a_j \vDash A^j$ and similarly for $S(\Delta)$.

Let $\langle a_{j_1}, \ldots, a_{j_m} \rangle$ be the sequence of situations in $S(\Gamma)$ placed in the same order as they appear in $\langle a_1, \ldots, a_n \rangle$, and similarly let $\langle a_{k_1}, \ldots, a_{k_p} \rangle$ be the sequence of situations in $S(\Delta)$ placed in the same order as they appear in $\langle a_1, \ldots, a_n \rangle$. Then by the special semantic condition defining E.K4 frames, there are situations x and y such that $ES(\Gamma)x$, $ES(\Delta)y$ and $Exyc$. Therefore, $c \vDash C$. Generalising, $A^1 \Rightarrow (\cdots (A^n \Rightarrow C) \ldots)$ is valid on the class of E.K4 frames. $\square$

The completeness proof is a slightly more complicated version of the proof for E.K. We do not present it here.

An E.KT4 frame is an E.K frame that satisfies the conditions T and SKT4. The condition SKT4 is the following. Let $\sigma = \langle a_1, \ldots, a_n \rangle$ be a finite sequence of situations. Let γ and δ be sequences of situations taken from σ, such that in γ and δ every situation occurs in the same order as it occurs in σ, and between γ and δ each situation occurs the same number of times it occurs in σ.

$$(SKT4) \quad \text{If } Ea_1 \ldots a_n c, \text{ then } \exists x \exists y (E\gamma x \wedge E\delta a_n y \wedge Exyc)$$

We have shown that E.KT frames satisfy the contraction condition and that E.K4 frames satisfy the condition that $Eabcd$ implies $\exists x (Eacx \wedge Ebxd)$. Together with the conditions satisfied by every E.K frame and T, we are justified in stating the following theorem:

Theorem 10.6. *Every E.KT4 frame is an E frame.*

We think the converse is true as well, but we have no proof of this so we leave it open.

11 Symmetry

In order to axiomatise the logic that is characterised by symmetry of the modal accessibility relation, we suggest adding the Urquhart–Fine axiom:

$$\text{(UF)} \quad A \Rightarrow ((A \Rightarrow (B \Rightarrow C)) \Rightarrow (B \Rightarrow C)).$$

Urquhart used this axiom to distinguish between $E_\Rightarrow$ and the system $E5_\Rightarrow$, which is characterised by his semantics in which the modal accessibility relation is reflexive, transitive, and symmetric [16, 17]. Kit Fine proved that $E_\Rightarrow$ together with UF is $E5_\Rightarrow$, that is, that it is complete over Urquhart's semantics for it [6].

UF is a relative of the axiom of E sometimes called Restricted Assertion, $(A \Rightarrow B) \Rightarrow (((A \Rightarrow B) \Rightarrow C) \Rightarrow C)$, which is equivalent to E's Permutation axiom, $A \Rightarrow ((A \Rightarrow B) \Rightarrow C) \Rightarrow ((A \Rightarrow B) \Rightarrow (A \Rightarrow C))$. It is, in the presence of the transitivity axioms of Theorem 15, equivalent to a form of Permutation as well.

$$\text{(UF}') \quad (A \Rightarrow (B \Rightarrow (C \Rightarrow D))) \Rightarrow (B \Rightarrow (A \Rightarrow (C \Rightarrow D)))$$

UF follows from UF$'$ and an appropriate instance of axiom 1.

While Fine proved that $E5_\Rightarrow$ is complete for Urquhart's semantics where the accessibility relation is reflexive, transitive, and symmetric, E5 with negation and the conditional does not appear to validate the symmetry principle one would expect, namely, the B axiom: $A \Rightarrow \Box\neg\Box\neg A$. This leaves open the possibility that a different axiom is needed for completeness on the symmetric E.K frames, rather than the reflexive, transitive, symmetric frames. The lack of fit between the B axiom and symmetry is not peculiar to the entailment systems. The extension of $R^\Box$ with the B axiom is not characterised by the class of models in which the modal accessibility relation is transitive, reflexive and symmetric. Rather, a weaker postulate than symmetry is used [7].

UF does, however, capture a kind of symmetry. The axiom can be recovered in the Fitch system by adding another $\Rightarrow E$ rule.

$$A_\alpha$$

$$\ddots$$

$$A \Rightarrow (B \Rightarrow C)_\beta$$

$$B \Rightarrow C_{\alpha \cup \beta}$$

This permits one to use ⇒E when the antecedent is in a superior proof, provided the consequent is itself a conditional. If A is true at a situation a in w, Mww', and $A \Rightarrow (B \Rightarrow C)$ is true at a situation b in w' then we can infer that there is a situation c in w' in which $B \Rightarrow C$ is true, justified by the situation in w.

The strengthening of this rule that permits the consequent of the conditional to be a non-conditional, B, would permit the derivation of the R axiom of Assertion, $A \rightarrow ((A \rightarrow B) \rightarrow B)$, which is equivalent to the Permutation axiom whose proof was displayed in Derivation 1. This strengthening is unavailable to us, since it would move us to the non-modal logic R.

It seems that however we axiomatise E5, it should have UF as a theorem. This is perhaps easiest to see on the Urquhart semantics. Suppose that UF is not valid on a reflexive, transitive, symmetric frame, i.e., for some a, w, $a, w \nVdash A \Rightarrow ((A \Rightarrow (B \Rightarrow C)) \Rightarrow (B \Rightarrow C))$. Then there is a b, w' with Nww', such that $b, w' \vDash A$ and $a \circ b, w' \nVdash (A \Rightarrow (B \Rightarrow C)) \Rightarrow (B \Rightarrow C)$. There is, then, a c, w'' with $Nw'w''$ such that $c, w'' \vDash A \Rightarrow (B \Rightarrow C)$ but $a \circ b \circ c, w'' \nVdash B \Rightarrow C$. But, by transitivity and symmetry of N, $Nw''w$, so $c \circ a, w \vDash B \Rightarrow C$. Again by the symmetry of N, Nww'', so, by the properties of $\circ$, $a \circ b \circ c, w'' \vDash B \Rightarrow C$. We conclude that, contrary to the assumption, UF is valid. The preceding proof used the transitivity of N, which underlines the possibility that a different axiom is needed for symmetry in the absence of transitivity.

As we said in Section 5, we use Urquhart's semantics (and the systems characterised by it) as an upper bound of our E-based systems. E5 is the upper bound, although it should perhaps be called E.KT45. We will leave open the question of whether E.K5, E.K45, and E.KT5, obtained by adding UF to E.K, E.K4, and E.KT, respectively, are complete for the classes of symmetric, symmetric transitive, and symmetric reflexive frames, respectively.

Appendix I: The Natural Deduction System for E

Hypothesis: Any formula can be hypothesised with a new numeral as a subscript and introducing a new subproof.

Repetition: Any formula can be repeated within the same sub-proof.

Theorem: Any formula that has been previously proven or the constant t can be stated anywhere in any proof with the subscript $\emptyset$.

In ⇒E the premises can be in the same subproof or the minor premise may be in a (not necessarily immediate) subproof of the proof in which the major premise occurs. The same is true for ∨E.

$$A \Rightarrow B_\alpha$$

$$\ddots$$

$$A_\beta$$
$$\vdots$$
$$B_{\alpha \cup \beta} \qquad \Rightarrow\text{E}$$

$$A_i \qquad\qquad \text{hypothesis}$$
$$\vdots$$
$$B_\alpha$$
$$A \Rightarrow B_{\alpha - \{i\}} \qquad \Rightarrow\text{I}$$

$$\text{In} \Rightarrow\text{I}, \; i \in \alpha.$$

$$A_\alpha$$
$$B_\alpha$$
$$\vdots$$
$$A \wedge B_\alpha \qquad \wedge\text{I}$$

$$A \wedge B_\alpha$$
$$\vdots$$
$$A_\alpha \qquad \wedge\text{E}$$

$$A \wedge B_\alpha$$
$$\vdots$$
$$B_\alpha \qquad \wedge\text{E}$$

$$A_\alpha$$
$$\vdots$$
$$A \vee B_\alpha \qquad \vee\text{I}$$

$$B_\alpha$$
$$\vdots$$
$$A \vee B_\alpha \qquad \vee\text{I}$$

$$A \Rightarrow C_\beta$$
$$B \Rightarrow C_\beta$$
$$\ddots$$
$$A \vee B_\alpha$$
$$\vdots$$
$$C_{\alpha \cup \beta} \qquad \vee\text{E}$$

$$A_\emptyset$$
$$\vdots$$
$$t \Rightarrow A_\emptyset \qquad t \Rightarrow$$

$$A \wedge (B \vee C)_\alpha$$
$$\vdots$$
$$(A \wedge B) \vee (A \wedge C)_\alpha \qquad \text{Distribution}$$

Appendix II: Proof of Derivability of RK

Lemma 11.1. *The rule RK is admissible in the natural deduction system for E.K.*

Proof. Suppose that $A^m \Rightarrow (A^{m+1} \Rightarrow \cdots (A^{n-1} \Rightarrow (B \Rightarrow C))\ldots)$ and $A^p \Rightarrow (A^{p+1} \Rightarrow \cdots (A^n \Rightarrow B)\ldots)$ are provable in the natural deduction system. Let $m = 1$. The case in which $p = 1$ is similar. We then can construct a proof of C from $A^1, \ldots, A^n$ as follows:

A_1^1 · · · · · · · hypothesis

$A^1 \Rightarrow (A^2 \Rightarrow \cdots (A^{n-1} \Rightarrow (B \Rightarrow C))\ldots)_\emptyset$ · · · assumption

$A^2 \Rightarrow (A^3 \Rightarrow \cdots (A^{n-1} \Rightarrow (B \Rightarrow C))\ldots)_{\{1\}}$ · · · 1,2, $\Rightarrow$E

A_2^2 · · · · · · hypothesis

$A^3 \Rightarrow (A^4 \Rightarrow \cdots (A^{n-1} \Rightarrow (B \Rightarrow C))\ldots)_{\{1,2\}}$ · · · 3,4, $\Rightarrow$E

$\ddots$

A_p^p · · · · · · hypothesis

$A^{p+1} \Rightarrow (A^{p+2} \Rightarrow \cdots (A^{n-1} \Rightarrow (B \Rightarrow C))\ldots)_{\{1,\ldots,p\}}$ · · · $\ldots \Rightarrow$E

$A^p \Rightarrow (A^{p+1} \Rightarrow \cdots (A^n \Rightarrow B)\ldots)_\emptyset$ · · · assumption

$A^{p+1} \Rightarrow (A^{p+2} \Rightarrow \cdots (A^n \Rightarrow B)\ldots)_{\{p\}}$ · · · $\ldots \Rightarrow$E

A_{p+1}^{p+1} · · · · · · hypothesis

$A^{p+2} \Rightarrow (A^{p+3} \Rightarrow \cdots (A^{n-1} \Rightarrow (B \Rightarrow C))\ldots)_{\{1,\ldots,p\}}$ · · · $\ldots \Rightarrow$E

$A^{p+2} \Rightarrow (A^{p+3} \Rightarrow \cdots (A^n \Rightarrow B)\ldots)_{\{p\}}$ · · · $\ldots \Rightarrow$E

$\ddots$

A_{n-1}^{n-1} · · · · · · hypothesis

$B \Rightarrow C_{\{1,\ldots,n-1\}}$ · · · $\ldots \Rightarrow$E

$A^n \Rightarrow B_{\,1,\ldots,n-1\}}$ · · · $\ldots \Rightarrow$E

A_n^n · · · · · · hypothesis

$B_{\{1,\ldots,n\}}$ · · · $\ldots \Rightarrow$E

$C_{\{1,\ldots,n\}}$ · · · $\ldots \Rightarrow$E

$A^n \Rightarrow C_{\{1,\ldots,n-1\}}$ · · · $\ldots \Rightarrow$I

$\vdots$

$A^2 \Rightarrow (\cdots (A^n \Rightarrow C)\ldots)_{\{1\}}$ · · · $\ldots \Rightarrow$I

$A^1 \Rightarrow (A^2 \Rightarrow (\cdots (A^n \Rightarrow C)\ldots))_\emptyset$ · · · $\ldots \Rightarrow$I

$\square$

References

[1] Alan R. Anderson and Nuel D. Belnap. *Entailment: The Logic of Relevance and Necessity*, volume I. Princeton University Press, Princeton, 1975.

[2] Alan R. Anderson, Nuel D. Belnap, and J. Michael Dunn. *Entailment: The Logic of Relevance and Necessity*, volume II. Princeton University Press, Princeton, 1992.

[3] Jc Beall, Ross Brady, J. Michael Dunn, A. P. Hazen, Edwin Mares, Robert K. Meyer, Graham Priest, Greg Restall, David Ripley, John Slaney, and Richard Sylvan. On the ternary relation and conditionality. *Journal of Philosophical Logic*, 41:565–612, 2012.

[4] Ross Brady. *Universal Logic*. CSLI, Stanford, 2006.

[5] J. Michael Dunn. Relevance logic and entailment. In D.M. Gabbay and F. Guenthner, editors, *Handbook of Philosophical Logic*, volume III, pages 117–224. Kluwer, Dordrecht, 1984.

[6] Kit Fine. Completeness for the S5 analogue of E_I, (Abstract). *Journal of Symbolic Logic*, 41:559–560, 1976.

[7] André Fuhrmann. Models for relevant modal logics. *Studia Logica*, 49:501–514, 1990.

[8] Mark Lance. On the logic of contingent relevant implication: A conceptual incoherence in the intuitive interpretation of R. *Notre Dame Journal of Formal Logic*, 29:520–529, 1988.

[9] Mark Lance and Philip Kremer. The logical structure of linguistic commitment II: Systems of relevant entailment commitment. *Journal of Philosophical Logic*, 25:425–449, 1996.

[10] Larisa Maksimova. A semantics for the calculus E of entailment. *Bulletin of the Section of Logic*, 2:18–21, 1973.

[11] Edwin Mares. *Relevant Logic: A Philosophical Interpretation*. Cambridge University Press, Cambridge, 2004.

[12] Errol P. Martin and Robert K. Meyer. Solution to the P-W problem. *Journal of Symbolic Logic*, 47:869–886, 1982.

[13] Robert K. Meyer. Entailment and relevant implication. *Logique et analyse*, 11: 472–479, 1968.

[14] Richard Routley and Robert K. Meyer. Semantics for entailment III. *Journal of Philosophical Logic*, 1:192–208, 1972.

[15] Richard Routley, Robert K. Meyer, Ross Brady, and Val Plumwood. *Relevant Logics and their Rivals*. Ridgeview, Atascadero, 1983.

[16] Alasdair Urquhart. Semantics for relevance logics. *Journal of Symbolic Logic*, 37:159–169, 1972.

[17] Alasdair Urquhart. *Semantics of Entailment*. PhD thesis, University of Pittsburgh, 1973.

Received 2 November 2016

CHANNEL COMPOSITION AND TERNARY RELATION SEMANTICS

ANDREW TEDDER[*]

Department of Philosophy, University of Connecticut, Storrs, CT, U.S.A.
<andrew.tedder@uconn.edu>

Abstract

The focus of this paper is on the channel-theoretic interpretation given by Restall [19] as built on work by Barwise [2] in the framework of situation semantics. I characterise the notion of *serial composition* of channels, due to Barwise, and extend the ternary relation semantic framework to incorporate sets of points fitting the bill. It is shown that such an extension of the basic ternary relation semantic framework by such points is adequate for $\mathbf{B}_\wedge$, and so that restricting the class of $\mathbf{B}_\wedge$ models to those including composites is conservative over $\mathbf{B}_\wedge$. We close by noting directions for future research.

Keywords: channel theory, information based logic, relevant logic, situation semantics, ternary relation semantics

1 Ternary Relation Semantics and the Problem of Interpretation

The ternary relation semantic framework, as most famously developed in [23], though very powerful, has long presented difficulties in interpretation.[1] The proposal with

---

[*]I would like to thank Katalin Bimbó and J. Michael Dunn for organising the *Third Workshop* and these proceedings. In addition, I would like to thank members of the Melbourne Logic Seminar for comments on a draft of this paper. Particular mention is due to Rohan French, Greg Restall, Dave Ripley, Shawn Standefer, and an anonymous referee for helpful feedback and suggestions — remaining mistakes are purely my fault. For material support, thanks to the University of Connecticut, the Department of Philosophy, CLAS, and Global Affairs Office and the University of Melbourne SHAPS. An extended visit to Melbourne for research provided the impetus for this paper, and these organisations made that trip possible.

[1]Note the classic challenge in [8] that the semantics is merely a technical device with little of the intuitive grip had by truth-functional semantics or the Kripke semantic framework for intuitionist logic and modal logics.

which we are primarily concerned is that of *channel theory*, as developed by Barwise [2] and Restall [19], within the broader theoretical framework provided by *situation semantics* as developed in [3], [4], and elsewhere.[2]

The basic theoretical posit of situation semantics are *situations*. These can be understood, as following Barwise and Seligman [4], as *classification-systems* of a kind. A classification system is a collection of *types* and *tokens* and an assignment to each type, a set of tokens. In some slightly different terms, a situation is a collection of *objects* and *properties*, and an assignment to each property of some (possibly empty) set of objects. Formally, a situation can simply be modelled as a set of sentences over some vocabulary containing predicate-symbols (for the types) and name constants (for the tokens).[3]

Barwise [2] develops an account of how information flows *between* situations by considering an additional kind of entity, namely, an *information channel*. A channel, for Barwise, supports information flow from a situation (the *signal* of the channel) to a situation (its *target*), and just which kind of information flow it supports determines what conditional propositions are made true at that channel. He employs a ternary relation, evocatively notated $\beta \overset{\alpha}{\mapsto} \gamma$, to indicate that the channel α supports information flow from the signal β to the target γ. It was not long in the waiting for relevant logicians to recognise the similarity between this semantics and that employed in the ternary relation (Routley–Meyer) semantics for relevant logics.[4] Indeed, the formal match makes the interpretation of relevant semantics by channels quite natural.

As another brief point of motivation, while situation semantics has fallen out of fashion, it is, by my estimation, a very natural setting for a theory of inference, and due for a reappraisal in the broader philosophical community. Situated inference is general enough that it can provide a (somewhat) neutral background for debates about logic, and those contested principles of logic. That is, since something like the situation semantics (particularly, in the extended sense presaged by the *set-ups* of [24]) can be tweaked to provide semantics for many logics with a variety of consequence relations, non-classical and otherwise, it could provide a fairly natural setting for debates between proponents of these various logics. In any case, despite

---

[2]For a broad overview of the interaction between semantics for relevant logics and situation semantics/channel theory see [16].

[3]There is a great deal more to be said of the situation semantics and, for instance, the interpretation of *negation* on such a framework. However, our interests are, for the most part, restricted to the conditional, so we leave these other considerations to the side.

[4]Though this semantic framework often goes by the name "the Routley–Meyer semantics," a better name, given proper attention to the history, might be "the Maksimova–Urquhart–Routley(Sylvan)–Meyer–Fine semantics." What this name lacks in elegance, it makes up in correctness

the dearth of new work in situation semantics in the last decade or so, it is still a framework in which interested new work is desirable (or so I hold, having missed situation semantics in its heyday).

Now let me come back to the work at hand. There are a number of ways to proceed in fleshing out a channel theory in the ternary relation semantic framework. First, I shall focus on Greg Restall's approach as developed in [19], which is an interpretation of the ternary relation in terms quite similar to Barwise's theory. Of particular interest is Restall's treatment of *serial composition* (from here on just *composition*) and its relation to an operation of his related to the relevant connective *fusion*, which we shall come to define, as cashing out a notion of *application*. I shall provide some reasons to be dissatisfied with Restall's account, and shall instead go back to Barwise's approach, with a particular focus on his treatment of composition. The aim of this paper is to provide some more definition to the account by pulling Barwise's composition apart from Restall's application, and to set out an extension to the ternary relation semantics to provide a logic for channels and channel composition. I'll set out the basic semantic framework for the basic relevant logic **B** in the language of conjunction and implication, and show that this extension to the ternary relation semantics for this logic is conservative. Finally, I'll display some interesting differences between Restall's account and the account to be developed here as regards the idempotence and commutativity of composition, before closing with a problem about extending the approach to incorporate negation.

1.1 $B_\wedge$ Frames and Models

Our formal language includes a set of propositional atoms and the connectives $\wedge$ and $\rightarrow$ (both of arity 2).[5] In Section 5, we shall also discuss $\leftarrow$, but shall set out its semantics as we come to it. $A, B, C, \ldots$ are metavariables ranging over propositions.

A ternary relation *model* for $\mathbf{B}_\wedge$ is defined as follows:[6]

Definition 1.1. A ternary relation model for $\mathbf{B}_\wedge$ is a pair $\langle \mathcal{F}, \vDash \rangle$ of a frame $\mathcal{F}$ and valuation $\vDash$ meeting the following criteria:

$\mathcal{F} = \langle S, N, R, \sqsubseteq \rangle$

- $N \subseteq S$ and $N \neq \varnothing$

- $R \subseteq S^3$

---

[5]The situation-semantic story we shall be interested in here is given in a first order language, but for our purposes it is sufficient to stay at the level of propositional logic.

[6]We shall occasionally employ $\Rightarrow$ and $\&$ as metalanguage connectives which shall behave in accordance with material implication and classical conjunction, respectively. In addition, we shall occasionally use metalanguage quantifiers $\forall, \exists$ for brevity, and these always range over situations.

- $\exists x \in N(Rx\alpha\beta) \Leftrightarrow \alpha \sqsubseteq \beta$

- $\sqsubseteq$ is a partial order

- $\alpha' \sqsubseteq \alpha$, $\beta' \sqsubseteq \beta$, $\gamma \sqsubseteq \gamma'$, and $R\alpha\beta\gamma$ imply that $R\alpha'\beta'\gamma'$ Tonicity Conditions

- for any proposition A, $\alpha \sqsubseteq \beta \Rightarrow (\alpha \vDash A \Rightarrow \beta \vDash A)$ Heredity

The following conventions will be useful for a short expression of some features of the ternary relation.

- $R^2\alpha\beta\gamma\delta$ iff $\exists\epsilon(R\alpha\beta\epsilon \,\&\, R\epsilon\gamma\delta)$

- $R^2\alpha(\beta\gamma)\delta$ iff $\exists\epsilon(R\alpha\epsilon\delta \,\&\, R\beta\gamma\epsilon)$

- $R^3\alpha(\beta(\gamma\delta))\epsilon$ iff $\exists\zeta(R^2\alpha(\beta\zeta)\epsilon \,\&\, R\gamma\delta\zeta)$

Finally, to get a model $\mathcal{M} = \langle \mathcal{F}, \vDash \rangle$ of $\mathbf{B}_\wedge$, we define the valuation $\vDash$ as follows:

- $\alpha \vDash A \wedge B$ iff $\alpha \vDash A$ and $\alpha \vDash B$

- $\alpha \vDash A \rightarrow B$ iff   for all $\beta, \gamma \in S$, if both $R\alpha\beta\gamma$ and $\beta \vDash A$ then $\gamma \vDash B$

Given $\vDash$, we can define *theorem* and *model-validity* as usual.

Definition 1.2. A is valid on the $\mathbf{B}_\wedge$ model $\mathcal{M}$ ($\mathcal{M} \vDash A$) iff $x \vDash A$, for all $x \in N$.

Definition 1.3. A is a $\mathbf{B}_\wedge$ validity ($\vDash_{\mathbf{B}_\wedge} A$) iff for every model $\mathcal{M}$ of $\mathbf{B}_\wedge$, $\mathcal{M} \vDash A$.

1.2 $\mathbf{B}_\wedge$ — a Hilbert System

There are a handful of options regarding how to axiomatise $\mathbf{B}_\wedge$, but we use the following axioms and rules:[7]

A1 $A \rightarrow A$ I

A2 $(A \wedge B) \rightarrow A$ Simplification$_1$

A3 $(A \wedge B) \rightarrow B$ Simplification$_2$

A4 $((A \rightarrow B) \wedge (A \rightarrow C)) \rightarrow (A \rightarrow (B \wedge C))$ Lattice-$\wedge$

---

[7]See [24] for details. Upper case sans serif letters are used as names for some axioms and these refer to the names of the combinators of which these formulae are the principal type schemata. See [12], for more information about this convention.

R1 $A, B \vdash A \land B$ Adjunction

R2 $A \to B, A \vdash B$ Modus Ponens

R3 $A \to B, C \to D \vdash (B \to C) \to (A \to D)$ Affixing

Note here that our rules are *rules of proof*, in Smiley's sense (see Humberstone [13] for clarifications). That is, our statement of (R1) is intended to be understood as "when A and B are both theorems, then so is $A \land B$."

1.2.1 Axioms and Frame Conditions for Some Extensions of $\mathbf{B}_\land$

A11 $((A \to B) \land (B \to C)) \to (A \to C)$ Conjunctive Syllogism

A12 $(A \to B) \to ((C \to A) \to (C \to B))$ B

A13 $(A \to B) \to ((B \to C) \to (A \to C))$ B'

A14 $(A \to (A \to B)) \to (A \to B)$ W

The following conditions are associated with the above axioms.

S11 $R\alpha\beta\gamma \Rightarrow R^2\alpha(\alpha\beta)\gamma$

S12 $R^2\alpha\beta\gamma\delta \Rightarrow R^2\alpha(\beta\gamma)\delta$

S13 $R^2\alpha\beta\gamma\delta \Rightarrow R^2\beta(\alpha\gamma)\delta$

S14 $R\alpha\beta\gamma \Rightarrow R^2\alpha\beta\beta\gamma$

This *association* between the frame conditions and provable formulae amounts to a correspondence of the following sort: $\mathcal{M} \vDash$ A1i iff the frame of $\mathcal{M}$ obeys restriction S1i, for $1 \leq i \leq 4$. Proofs of these facts are sketched in [24, p. 313] and [22, pp. 203–204]. We shall have reason to refer to other pairs of provabilities and corresponding ternary relation conditions from time to time, but for the most part, our concern shall be with those above. The following logics are notable for our interests:

$\mathbf{DJ}_\land$ is $\mathbf{B}_\land$ plus A11.

$\mathbf{TW}_\land$ is $\mathbf{B}_\land$ plus A12 and A13.

$\mathbf{T}_\land$ is $\mathbf{TW}_\land$ plus A14.

$\mathbf{TW}$, $\mathbf{DJ}$ and their neighbours, including $\mathbf{B}$ itself, have long been of interest as potential homes for theories of naïve truth and sets.[8] While this paper won't involve

---

[8]The paradox to which we refer here is, of course, Curry's paradox and derivative paradoxes, like the validity curry [6]. For general information see [5].

any substantial comment on the paradoxes, it's worth a passing note that we are in the neighbourhood and that the channel theoretic interpretation seems a good fit for these very weak logics.

2   Information Channels

To begin with, some comments are in order to set out our focal notions of *application* and *composition*.

Some notion of application plays an essential role in many interpretations of relevant logics and particularly the ternary relation. For instance, Restall interprets $R\alpha\beta\gamma$ as "the conditional information given in α applied to β results in no more than γ" [19] and elsewhere as "applying the information in α to β gives information which is already in γ." [18] One can find similar intuitions and terminology at work in [25], [15], and parts of [7]. A natural way to make these intuitions concrete, following the lead of algebraic semantics for relevant logics, is to introduce a collection of points $\alpha \circ \beta$ into the semantics which, speaking loosely, are the results of *applying* α to β. Well-known problems with an operational semantics for relevant logics mean that we cannot, in general, assume that $\alpha \circ \beta$ is a unique point in the frame.[9] Thus, Restall [19] defines this as a set-forming operation:

Definition 2.1. $\alpha \circ \beta = \{\, \gamma \colon R\alpha\beta\gamma \,\}$.

In order to make sense of how points like this behave in the ternary relation semantics, we need to enforce at least the following condition:[10]

$$\alpha \circ \beta \vDash A \text{ only if for every } \gamma \in \alpha \circ \beta, \gamma \vDash A.$$

Even with this operation providing only a set of points, rather than a unique point, the application story is fairly natural, in abstract. The key fact here is that $(\alpha \vDash A \to B \ \& \ \beta \vDash A) \Rightarrow \alpha \circ \beta \vDash B$, as follows immediately from the definition of $\circ$. This fact provides the key intuition behind the ternary relation: when α is a channel from β to γ ($R\alpha\beta\gamma$), and $\beta \vDash A$ implies $\gamma \vDash B$, then $\alpha \vDash A \to B$.

On the other hand, we have *composition* $\alpha;\beta$ as channels. The key job we want composition to do is to enforce $(\alpha \vDash A \to C \ \& \ \beta \vDash C \to B) \Rightarrow \alpha;\beta \vDash A \to B$. Intuitively, when α is a channel supporting information flow from A-propositions to some propositions from which β supports information flow to B-propositions, there

---

[9]See [26] for details.

[10]In addition, we need some posits governing how $\circ$ interacts with $\sqsubseteq$ but these details are not necessary for the comments in this paper.

is a situation which is a channel cutting out the middle, as it were. As an example, suppose that a phone call allows for information to flow from my home in Connecticut to Edmonton, and an email allows for that (or some related) information to flow from Edmonton to my next-door neighbour in Connecticut. Then there is a channel resulting of the composition of the relevant bits of the phone-network connecting my house to Edmonton with the relevant bits of the internet and servers which support that email connection from Edmonton to my neighbour. It's the phone line *composed* with the email connection which allow for information to flow from me to my next door neighbour. Barwise [2] makes the following demands of composition, where 0 is a *logic channel* behaving essentially as a member of N as set out in Section 1 and $\beta \overset{\alpha}{\mapsto} \gamma$ is to be read as $R\alpha\beta\gamma$.[11]

- For any α, β, there is a unique $\alpha; \beta$.

- $\gamma \overset{\alpha;\beta}{\mapsto} \delta \Leftrightarrow \exists\epsilon(\gamma \overset{\alpha}{\mapsto} \epsilon \ \& \ \epsilon \overset{\beta}{\mapsto} \delta)$

- $0; \alpha = \alpha = \alpha; 0$[12]

- $\alpha; (\beta; \gamma) = (\alpha; \beta); \gamma$

He proceeds to show, given his very abstract framework, that composition has these features, and leaves open the questions of whether $\alpha = \alpha; \alpha$ and $\alpha; \beta = \beta; \alpha$, taking these as substantial questions to be filled in by fuller channel theories.

2.1 Restall and Application

Restall's move [19] is to identify $\alpha; \beta$ and $\alpha \circ \beta$. An equivalent statement of $R\alpha\beta\gamma$, given his account of $\circ$, is $\alpha \circ \beta \sqsubseteq \gamma$. This has some interesting features. One nice feature is that we can carry over intuitions about composition in order to explain some features of application, and hence the ternary relation. Barwise's desiderata for composition, when read in terms of $\circ$, are either built into the ternary relation semantics, or underwrite what many take to be plausible axioms and arguments, understood in information-theoretic terms. Consider Barwise's normality condition that $0; \alpha = \alpha = \alpha; 0$. For Restall's reading to capture this it must at least demand the following condition, where $\alpha \circ (\beta \circ \gamma) \sqsubseteq \delta$ holds just when there exists an ϵ s.t. $\alpha \circ \epsilon \sqsubseteq \delta$ and $\beta \circ \gamma \sqsubseteq \epsilon$, and $(\alpha \circ \beta) \circ \gamma \sqsubseteq \delta$ just in case some ϵ is s.t. $\alpha \circ \beta \sqsubseteq \epsilon$ and $\epsilon \circ \gamma \sqsubseteq \delta$.

---

[11]While, for the purposes of generality, we consider non-reduced models, that is, models with multiple normal points, in the remainder of Section 2, we follow Barwise and Restall in focusing on a distinguished normal point, 0.

[12]For this desideratum and some discussion, see [2, p. 19].

$$\alpha \circ \beta \sqsubseteq \gamma \Leftrightarrow (0 \circ \alpha) \circ \beta \sqsubseteq \gamma \Leftrightarrow (\alpha \circ 0) \circ \beta \sqsubseteq \gamma, \quad \text{i.e.,}$$
$$R\alpha\beta\gamma \Leftrightarrow R^2 0\alpha\beta\gamma \Leftrightarrow R^2 \alpha 0 \beta\gamma.$$

Note, that $R\alpha\beta\gamma \Leftrightarrow R^2 0\alpha\beta\gamma$ is immediate. For $R\alpha\beta\gamma \Rightarrow R^2 \alpha 0 \beta\gamma$, some more robust assumptions are required, but it is, perhaps, a plausible demand on this story.[13]

For Restall, satisfying the desired associativity property, $\alpha; (\beta; \gamma) = (\alpha; \beta); \gamma$, involves at least admitting:

$$(\alpha \circ \beta) \circ \gamma \sqsubseteq \delta \Rightarrow \alpha \circ (\beta \circ \gamma) \sqsubseteq \delta \qquad \text{i.e., } R^2 \alpha\beta\gamma\delta \Rightarrow R^2 \alpha(\beta\gamma)\delta,$$

which corresponds to (A12). In addition, in order to enforce the condition: $\alpha \vDash A \to B$ and $\beta \vDash B \to C$ imply $\alpha \circ \beta \vDash A \to C$, one needs the condition:

$$(\alpha \circ \beta) \circ \gamma \sqsubseteq \delta \Rightarrow \beta \circ (\alpha \circ \gamma) \sqsubseteq \delta \qquad \text{i.e., } R^2 \alpha\beta\gamma\delta \Rightarrow R^2 \beta(\alpha\gamma)\delta,$$

which corresponds to (A13). So, setting aside the concerns with our logic channels, the weakest logic which can be given a Restall-style channel account is around the strength of **TW**.[14]

There are a couple of potential problems here.

1. $\circ$ is not functional, so given some appropriate α, β, $\alpha \circ \beta$ is not unique.

2. This version renders the associativity of composition as a fairly substantial property, corresponding to the provability of the B' axiom $(A \to B) \to ((B \to C) \to (A \to C))$. On Restall's scheme, one also gets the B axiom $(A \to B) \to ((C \to A) \to (C \to B))$, and so rules out some very weak logics as underwriting a channel theory. However, its unclear how the provability of the axiom corresponding to this property of composition is to be justified on channel-theoretic terms. The frame conditions corresponding to B and B' involve at least blurring the distinction between channels and situations operated upon by channels.[15]

---

[13] A far too strong one, for instance, being $R\alpha\beta\gamma \Rightarrow R\beta\alpha\gamma$, which collapses the distinction between $\to$ and $\leftarrow$ (see the semantics of this arrow below), and corresponds to the axiom $A \to ((A \to B) \to B)$. Perhaps a more natural answer is just $R0\alpha\beta \Rightarrow R\alpha0\beta$, which is somewhat weaker.

[14] Adding disjunction and negation to **TW**$_\wedge$ involves adding the axioms A5–A9 of [24, p. 287] in addition to the contraposition axiom we consider in Section 6.

[15] While Restall takes this to be a feature, it is at least prima facie unclear if this is the case.

3. Surely composition and application are distinct operations with distinct notions, and even if it makes sense to identify them, such an argument would need more detail about how they operate separately.

4. The requirement that $R\alpha\beta\gamma \Rightarrow R^2\alpha 0\beta\gamma$ seems potentially problematic on the intended reading, in that it seems to demand that we accept something like $R\alpha 0\alpha$ which is quite implausible, when understanding 0 as some kind of logic *channel*, in line with Barwise. Why should it be that any channel given a logic channel as an input produces itself?

In the rest of this paper, we shall be interested to develop an account more in line with Barwise's initial proposal. This proposal has quite broad applications, applying naturally to classical, intuitionist, and some other logics, as Barwise showed, and as we shall show, to an important fragment of the basic relevant logic **B**.

3   Some Preliminaries

The ternary relation semantic framework set out in Section 1 by itself meets some of Barwise's desiderata for composition. In this section, we shall show that important consequences of the identity and associativity contraints are met.

The best way to proceed would be to define and fully work out the details of a composition operation or function on the ternary relation semantics of the type $S^2 \mapsto S$. This would involve defining an operation which interacts with R and $\sqsubseteq$ in ways which produce the desired behaviour. This project is difficult, and we start slow. The first demand is that for any pair of points in S there exists a point which behaves as their composite. That is, for any $\alpha \vDash A \to B$ and $\beta \vDash B \to C$, there is some $\alpha; \beta \vDash A \to C$.

First, however, we should set out just when a putative composite point is a channel between some points. The following condition fits the bill:

- $\exists\epsilon(R^2\beta(\alpha\gamma)\delta \Leftrightarrow R\epsilon\gamma\delta)$ Existence of Composites

We use $\alpha; \beta$ to name the ϵ for α, β in question. Then the above has the effect that $R^2\beta(\alpha\gamma)\delta \Leftrightarrow R(\alpha; \beta)\gamma\delta$, so long as $\alpha; \beta$ exists. On the intended reading, $R(\alpha; \beta)\gamma\delta$ tells us that there is some information one gets from applying α to γ which, when β is applied to it, results in δ. That is, there is a chain of channels along which we can reason where we take a signal γ for α, get its target, and then apply β to that target to get a situation which supports the target of β, namely δ.

That this fits the bill is easy to see. Suppose that $\alpha \vDash A \to B$ and $\beta \vDash B \to C$. Then since $\forall\gamma\forall\delta(R\alpha\gamma\delta \Rightarrow (\gamma \vDash A \Rightarrow \delta \vDash B))$ and $\forall\epsilon\forall\zeta(R\beta\epsilon\zeta \Rightarrow (\epsilon \vDash B \Rightarrow \zeta \vDash C))$, so there is an η s.t. $R\alpha^\frown\eta$ and $R\beta\eta\delta$ and $\gamma \vDash A$, then $\eta \vDash B$ and so $\delta \vDash C$ after all. So $R(\alpha;\beta)\gamma\delta \Rightarrow (\gamma \vDash A \Rightarrow \delta \vDash C)$ as desired.

We first present two preliminary results showing that if composite points are around in the frame, they behave in much the way that Barwise wanted. In lieu of $0;\alpha = \alpha = 0;\alpha$, we can show that something like $R\alpha\beta\gamma \Leftrightarrow R(0;\alpha)\beta\gamma \Leftrightarrow R(\alpha;0)\beta\gamma$, where we generalise to the non-reduced framework introduced in Section 1. This shows that when α is a channel from β to γ, then so is the composition of α with some normal point, which captures at least part of the spirit of Barwise's desideratum.

Theorem 3.1. *If $\mathcal{M}$ satisfies the existence of composites condition, then it also satisfies $R\alpha\beta\gamma \Leftrightarrow \exists x \in NR(\alpha;x)\beta\gamma \Leftrightarrow \exists y \in NR(y;\alpha)\beta\gamma$.*

Proof. Suppose that $R\alpha\beta\gamma$. Then we have that $R\alpha\beta\gamma$ and $\exists x \in N(Rx\gamma\gamma)$, and thus $\exists y\exists x \in N(R\alpha\beta y \ \& \ Rxy\gamma)$. That is, $\exists x \in NR^2x(\alpha\beta)\gamma$.

Suppose that $\exists x \in NR^2x(\alpha\beta)\gamma$. That is, $\exists x \in N\exists y(Rxy\gamma \ \& \ R\alpha\beta y)$. Since $x \in N$ and $Rxy\gamma$, we have that $y \sqsubseteq \gamma$, and so since $R\alpha\beta y$, it is the case that $R\alpha\beta\gamma$, by the tonicity conditions on R.

So, we have that $R\alpha\beta\gamma \Leftrightarrow \exists x \in NR^2x(\alpha\beta)\gamma$, that is $R\alpha\beta\gamma \Leftrightarrow \exists x \in NR(\alpha;x)\beta\gamma$.

Suppose that $R\alpha\beta\gamma$. Note that $\exists x \in NRx\beta\beta$ and hence $\exists y\exists x \in N(R\alpha y\gamma \ \& \ Rx\beta y)$, that is $\exists x \in NR^2\alpha(x\beta)\gamma$.

For the other direction, note that if $R^2\alpha(x\beta)\gamma$ then $\exists y$ s.t. $R\alpha y\gamma$ and $\beta \sqsubseteq y$, and so the tonicity conditions on R guarantee that $R\alpha\beta\gamma$. $\qquad\square$

In a similar vein, we can show that the following important consequence of associativity for composition holds in any ternary relation model:

Theorem 3.2. *If $\mathcal{M}$ satisfies the existence of composites condition, then it also satisfies $R(\alpha;(\beta;\gamma))\delta\epsilon \Leftrightarrow R((\alpha;\beta);\gamma)\delta\epsilon$.*

Proof. Suppose $R(\alpha;(\beta;\gamma))\delta\epsilon$, that is $R^2(\beta;\gamma)(\alpha\delta)\epsilon$. So $\exists\zeta(R(\beta;\gamma)\zeta\epsilon \ \& \ R\alpha\delta\zeta)$. So $\exists\zeta(R^2\gamma(\beta\zeta)\epsilon \ \& \ R\alpha\delta\zeta)$. Thus $\exists\zeta(\exists\lambda(R\beta\zeta\lambda \ \& \ R\gamma\lambda\epsilon) \ \& \ R\alpha\delta\zeta)$. Hence, since λ does not occur in $R\alpha\delta\zeta$, we have that $\exists\zeta\exists\lambda(R\beta\zeta\lambda \ \& \ R\gamma\lambda\epsilon \ \& \ R\alpha\delta\zeta)$. So $\exists\lambda(\exists\zeta(R\beta\zeta\lambda \ \& \ R\alpha\delta\zeta) \ \& \ R\gamma\lambda\epsilon)$. Thus $\exists\lambda(R^2\beta(\alpha\delta)\lambda \ \& \ R\gamma\lambda\epsilon)$. Hence $\exists\lambda(R(\alpha;\beta)\delta\lambda \ \& \ R\gamma\lambda\epsilon)$ and so $R^2\gamma((\alpha;\beta)\delta)\epsilon$ and so $R((\alpha;\beta);\gamma)\delta\epsilon$.

The other direction is similar (just do the above proof 'backwards', so to speak). $\qquad\square$

You can understand this proof as essentially giving us that both $R(\alpha;(\beta;\gamma))\delta\epsilon$ and $R((\alpha;\beta);\gamma)\delta\epsilon$ are equivalent to $R^3\gamma(\beta(\alpha\delta))\epsilon$ (from simply pulling the two existential quantifiers to the front as we did before). The series of equivalences can be nicely demonstrated linearly:

$$R^3\gamma(\beta(\alpha\delta))\epsilon \Leftrightarrow R^2(\beta;\gamma)(\alpha\delta)\epsilon \Leftrightarrow R(\alpha;(\beta;\gamma))\delta\epsilon \Leftrightarrow R^2\gamma((\alpha;\beta)\delta)\epsilon \Leftrightarrow R((\alpha;\beta);\gamma)\delta\epsilon.$$

Note that we only needed to appeal to the definition of R^2 and R^3, so nothing beyond the basic definition of ternary relation models is needed for the result. So, these facts hold even in **B**. So, our extension of the ternary relation framework by composite points will have at least these desired features.

4 Adequacy of $\mathbf{B}_\wedge$ for Channel Models

Let us call a $\mathbf{B}_\wedge$ ternary relation model a *channel model* just in case it includes composite points meeting our Existence of Composites condition from Section 3. Our goal is to prove that the proof theory we have presented for $\mathbf{B}_\wedge$ is adequate, i.e., sound and complete, for channel models. From this, we can obtain a conservative extension result that adding channel-composites to the $\mathbf{B}_\wedge$ ternary relation model structure does not alter the validities of that structure. First, we should make note of another salient adequacy fact.

Theorem 4.1. *The class of models in Section 1.1 is sound and complete with respect to $\mathbf{B}_\wedge$ as defined in Section 1.2.*

Proof. The proof can be found in [24], Chapter 4 using essentially the canonical model construction as set out below. $\qquad\square$

The composite points $\alpha; \beta$, for some α, β, will have to include all the arrow statements $A \to B$ s.t. α supports $A \to C$ and β supports $C \to B$. This feature is captured by simply incorporating the definition of $R(\alpha; \beta)\gamma\delta$ into something much like the usual valuation clause for $\to$:

$$\alpha; \beta \vDash A \to B \text{ iff } \forall\gamma\forall\delta(R^2\beta(\alpha\gamma)\delta \Rightarrow (\gamma \vDash A \Rightarrow \delta \vDash B)).$$

These are the points added to a ternary relation model for $\mathbf{B}_\wedge$ to result in a channel model. Essentially, all one needs to do to obtain a channel model from a $\mathbf{B}_\wedge$ model is to outfit the set of situations in that model with the appropriate composite points.

Theorem 4.2. *The class of channel models are sound with respect to $\mathbf{B}_\wedge$.*

Proof. This is obvious, as every channel model is a model of $\mathbf{B}_\wedge$. $\qquad\square$

Completeness is a bit more involved. Our strategy is to show that the canonical model of $\mathbf{B}_\wedge$ is a channel model, and to do this, we need only to show that each composite point is already in the canonical set S_c of $\mathbf{B}_\wedge$ situations, which we'll define shortly.

Definition 4.3. α is a $\mathbf{B}_\wedge$-theory just in case the following conditions hold:

- $A, B \in \alpha \Rightarrow A \wedge B \in \alpha$

- $(\vdash_{\mathbf{B}_\wedge} A \to B \,\&\, A \in \alpha) \Rightarrow B \in \alpha$

Furthermore, α is a *regular* $\mathbf{B}_\wedge$-theory just in case α is a $\mathbf{B}_\wedge$-theory and

- $\vdash_{\mathbf{B}_\wedge} A \Rightarrow A \in \alpha$.

Definition 4.4. The canonical frame of $\mathbf{B}_\wedge$ is $\mathcal{F}_c = \langle S_c, N_c, R_c, \rangle$, where

- S_c is the set of $\mathbf{B}_\wedge$ theories

- N_c is the set of regular $\mathbf{B}_\wedge$ theories

- $R_c \alpha \beta \gamma \Leftrightarrow ((A \to B \in \alpha \,\&\, A \in \beta) \Rightarrow B \in \gamma)$

To get the canonical model $\mathcal{M}_c = \langle \mathcal{F}_c, \vDash_c \rangle$, we need to add only the canonical valuation as follows:

$$\alpha \vDash_c A \Leftrightarrow A \in \alpha.$$

In this setting, we are interested to find a point $\alpha; \beta$ in $\mathcal{F}_c$ which obeys the following condition:

$$A \to B \in \alpha; \beta \Leftrightarrow \forall \gamma \forall \delta ((R^2 \beta(\alpha\gamma)\delta \,\&\, A \in \gamma) \Rightarrow B \in \delta).$$

The question of finding a point like this can be recast as one of whether one can take a set of conditional formulae meeting this condition and build a $\mathbf{B}_\wedge$-theory out of it. Importantly, the process of building a theory out of this set of conditionals must not involve adding any conditionals beyond those added to satisfy the above condition. If this were not to be the case, then one of these new conditionals $A \to B$ would not be such that $\alpha \vDash A \to C$ and $\beta \vDash C \to B$. If we can build such a set without any additional conditionals, then we'll have shown that $\alpha; \beta$ is in the canonical model after all and behaves as we want.

4.1 Construction of $\alpha; \beta$

We employ a standard theory construction. Let us begin with the following:

$$\alpha; \beta_0 = \{A \to B : \forall \gamma \forall \delta (R^2 \beta(\alpha\gamma)\delta \Rightarrow (A \in \gamma \Rightarrow B \in \delta))\}.$$

To make sure $\alpha; \beta$ is a theory in the language, we need only add conjunctions as follows:

$$A, B \in \alpha; \beta_n \Rightarrow A \wedge B \in \alpha; \beta_{n+1}$$

to get $\alpha; \beta$ as:

$$\alpha; \beta = \bigcup_{n < \omega} \alpha; \beta_n.$$

4.2 Verification of $\alpha; \beta$

Now, our concern is to verify that $\alpha; \beta$ is a $\mathbf{B}_\wedge$-theory after all. That it obeys the conjunction property clearly falls out of the construction (conjunctive formulae only get in when both conjuncts do). So, the remainder of the verification requires that we show that when $A \to B$ is a theorem of $\mathbf{B}_\wedge$, then $A \in \alpha; \beta \Rightarrow B \in \alpha; \beta$. First, we need some lemmata. The first is reported by Dezani-Ciancaglini et al. [9, p. 210]:

Lemma 4.5 (Bubbling). *Suppose* $\vdash_{\mathbf{B}_\wedge} \bigwedge_{i \in I} (A_i \to B_i) \to (A \to B)$ *for some proposi-tions indexed by* $I \subseteq \mathbb{N}$. *Then there is some non-empty finite* $J \subseteq I$ *s.t.* $\vdash_{\mathbf{B}_\wedge} A \to \bigwedge_{j \in J} A_j$ *and* $\vdash_{\mathbf{B}_\wedge} \bigwedge_{j \in J} B_j \to B$.

Proof. See Barendregt et al. [1, p. 933] and note that as we don't have anything like $\top$ matching their ω, so the initial non-identity clause, which would amount to $B \neq \top$ as in [9], is unnecessary for our purposes.[16] $\qquad \square$

It is easy to extend this lemma to an equivalence:

Lemma 4.6 (Double Bubbling). *Suppose that there exists a* $J \subseteq I$ *s.t.* $\vdash_{\mathbf{B}_\wedge} A \to \bigwedge_{j \in J} A_j$ *and* $\vdash_{\mathbf{B}_\wedge} \bigwedge_{j \in J} B_j \to B$. *Then* $\vdash_{\mathbf{B}_\wedge} \bigwedge_{i \in I} (A_i \to B_i) \to (A \to B)$.

Proof. The proof is completed in two stages. First is to show, in the ternary relation semantics for $\mathbf{B}_\wedge$, that if for every $x \in N_c$, $x \vDash A \to \bigwedge_{j \in J} A_j$ and $x \vDash \bigwedge_{j \in J} B_j \to B$, then for every $x \in N_c$, $x \vDash \bigwedge_{i \in J} (A_j \to B_j) \to (A \to B)$.

Suppose otherwise. That is, suppose that there exists α, β s.t. $\alpha \sqsubseteq \beta$ and $\alpha \vDash \bigwedge_{j \in J} (A_j \to B_j)$ and $\beta \nvDash A \to B$. Thus, there are γ, δ s.t. $R\beta\gamma\delta$ and $\gamma \vDash A$ and

---

[16]It is worth noting here that the proof given here bears a substantial resemblance to work done by Dunn and Meyer to provide ternary relation semantics for combinatory logic in [11]. Indeed, thanks are due to a referee for noting this point of resemblance.

$\delta \nvDash B$. Since $\gamma \sqsubseteq \gamma$ it follows that $\gamma \vDash \bigwedge_{j \in J} A_j$, and so $\gamma \vDash A_j$ for each $j \in J$. The tonicity properties of R guarantee that $R\alpha\gamma\delta$, and so $\delta \vDash B_j$ for each $j \in J$. Hence, since $\delta \sqsubseteq \delta$, we have that $\delta \vDash B$ after all. So, each $x \in N$ must satisfy $\bigwedge_{j \in J} (A_j \to B_j) \to (A \to B)$.

So if $\vdash_{\mathbf{B}} A \to \bigwedge_{j \in J} A_j$ and $\vdash_{\mathbf{B}} \bigwedge_{j \in J} B_j \to B$ then $\vdash_{\mathbf{B}} \bigwedge_{j \in J} (A_j \to B_j) \to (A \to B)$. By (R3) $\vdash_{\mathbf{B}} (\bigwedge_{i \in I} (A_i \to B_i) \to \bigwedge_{j \in J} (A_j \to B_j)) \to (\bigwedge_{i \in I} (A_i \to B_i) \to (A \to B))$ follows. However, when $J \subseteq I$, $\bigwedge_{i \in I} (A_i \to B_i) \to \bigwedge_{j \in J} (A_j \to B_j)$ is provable with either (A2) or (A3). Hence, if there exists $J \subseteq I$ s.t. $\vdash_{\mathbf{B}} \bigwedge_{j \in J} (A_j \to B_j) \to (A \to B)$ then $\vdash_{\mathbf{B}} \bigwedge_{i \in I} (A_i \to B_i) \to (A \to B)$. $\qquad\square$

Lemma 4.7. *If $\vdash_{\mathbf{B}_\wedge} C$ then there exists some $\bigwedge_{i \in I} (A_i \to B_i)$ s.t. $\vdash_{\mathbf{B}_\wedge} C \leftrightarrow \bigwedge_{i \in I} (A_i \to B_i)$, for formulae with indeces from some $I \subseteq \mathbb{N}$.*

Proof. Since $\vdash_{\mathbf{B}_\wedge} C$, we know that C must not be an atomic formula. Similarly, since a conjunction is provable in $\mathbf{B}_\wedge$ iff both conjuncts are provable, we can be sure that C does not have an atomic formula as a conjunct. So C must have some complex structure, and every non-atomic formulae in the language $\{\to, \wedge\}$ has the desired structure, hence every conjunction of non-atomic formulae in the language has the desired structure. $\qquad\square$

Theorem 4.8. $(\vdash_{\mathbf{B}_\wedge} A \to B \ \& \ A \in \alpha; \beta) \Rightarrow B \in \alpha; \beta$.

Proof. We proceed by structural induction on the proof of $A \to B$.
Base: $A \to B$ is an instance of a $\mathbf{B}_\wedge$ axiom.

(A1) That $A \in \alpha; \beta \Rightarrow A \in \alpha; \beta$ follows from the fact that $\alpha; \beta$ is a set.

(A2) Suppose that $A \wedge B \in \alpha; \beta$. Then at some stage n, $A \in \alpha; \beta_n$ and $B \in \alpha; \beta_n$. If $A \in \alpha; \beta_n$ then $A \in \alpha; \beta$. (A3) is similar.

(A4) Suppose that $(A \to B) \wedge (A \to C) \in \alpha; \beta$. Suppose that $A \to (B \wedge C) \notin \alpha; \beta$. So $\exists \gamma, \delta (R_c^2 \beta(\alpha\gamma)\delta \ \& \ A \in \gamma \ \& \ B \wedge C \notin \delta)$. Since $R_c^2 \beta(\alpha\gamma)\delta$ and $A \in \gamma$, it follows that $B \in \delta$ and $C \in \delta$. By supposition, δ is a $\mathbf{B}_\wedge$-theory, and so $B \wedge C \in \delta$ after all.

Induction Step: Consider the cases when $A \to B$ is a result of an application of a $\mathbf{B}_\wedge$ rule to some other $\mathbf{B}_\wedge$ theorems. In particular, suppose that each premise to the application of the rule is respected by $\alpha; \beta$, for induction, and we'll show that the consequence of the rule application is respected by $\alpha; \beta$.

744

(R2) Suppose that $\vdash_{\mathbf{B}_\wedge} C \to (A \to B)$ and $\vdash_{\mathbf{B}_\wedge} C$. From $\vdash_{\mathbf{B}_\wedge} C$, Lemma 4.7 guarantees that $\vdash_{\mathbf{B}_\wedge} C \leftrightarrow \bigwedge_{i \in I}(A_i \to B_i)$ for some collection of arrow formulae. Thus, the replacement property guarantees that both $\vdash_{\mathbf{B}_\wedge} \bigwedge_{i \in I}(A_i \to B_i) \to (A \to B)$ and $\vdash_{\mathbf{B}_\wedge} \bigwedge_{i \in I}(A_i \to B_i)$. Suppose that $A \in \alpha; \beta$. Since $\vdash_{\mathbf{B}_\wedge} \bigwedge_{i \in I}(A_i \to B_i) \to (A \to B)$, the Bubbling Lemma ensures that there exists a finite non-empty $J \subseteq I$ where $\vdash_{\mathbf{B}_\wedge} A \to \bigwedge_{j \in J} A_j$ and $\vdash_{\mathbf{B}_\wedge} \bigwedge_{j \in J} B_j \to B$. Under the supposition that both of these are respected by $\alpha; \beta$, given by the equivalence shown in Lemmas 4.5 and 4.6, we have that $\bigwedge_{j \in J} A_j \in \alpha; \beta$ and so each $A_j \in \alpha; \beta$. By IH, we have that $A_i \in \alpha; \beta \Rightarrow B_i \in \alpha; \beta$ for all $i \in I$, so for each $j \in J$, $B_j \in \alpha; \beta$ and so $\bigwedge_{j \in J} B_j \in \alpha; \beta$. Thus $B \in \alpha; \beta$.

(R3) Suppose that $\vdash_{\mathbf{B}_\wedge} A \to B$ and $\vdash_{\mathbf{B}_\wedge} C \to D$ and, in addition, that $B \to C \in \alpha; \beta$ and $A \to D \notin \alpha; \beta$. By this last supposition, we have $\exists \gamma \exists \delta (R^2 \beta(\alpha\gamma)\delta \,\&\, A \in \gamma \,\&\, D \notin \delta)$. It follows that $B \in \gamma$, since γ is a $\mathbf{B}_\wedge$-theory. Therefore, since $B \to C \in \alpha; \beta$, $C \in \delta$, and so $D \in \delta$ as δ is a $\mathbf{B}_\wedge$-theory as well.

$\square$

On our way to the conservative extension result, we note an additional result. Namely, that whenever both $\alpha, \beta \in N_c$, then $\alpha; \beta \in N_c$. This is a straightforward result of the construction, but one which guarantees that these composite points are not only around in the frame, but are sensitive to the points of which they are composites.

Theorem 4.9. *When α, β are regular $\mathbf{B}_\wedge$-theories, then so is $\alpha; \beta$.*

Proof. By supposition, we have that $\vdash_{\mathbf{B}_\wedge} A \Rightarrow A \in \alpha \cap \beta$. Suppose that $\vdash_{\mathbf{B}_\wedge} A$, to show that $A \in \alpha; \beta$. We have already shown that $\alpha; \beta$ is closed under the rules of the system, so it is sufficient for our purposes to show that if α, β are normal, then $\alpha; \beta$ must contain each *axiom* of the system, as the fact that $\alpha; \beta$ is a $\mathbf{B}_\wedge$ theory will ensure that anything provable from the axioms is in $\alpha; \beta$. Note that every axiom of $\mathbf{B}_\wedge$ has $\to$ as its main connective. So, we need only to consider the cases where the A in question is $B \to C$, and show that $B \to C \in \alpha; \beta$ when it is a theorem.

Suppose that $\vdash_{\mathbf{B}_\wedge} B \to C$, $\exists x(R_c\beta x\delta \,\&\, R_c\alpha\gamma x)$, $\alpha, \beta \in N_c$, and $B \in \gamma$. Since $\alpha \in N_c$, we have that $B \to C \in \alpha$ as $\vdash_{\mathbf{B}_\wedge} B \to C$. Since $R_c\alpha\gamma x$ and $B \in \gamma$, we have that $C \in x$. Now, as $\beta \in N_c$, we have that $C \to C \in \beta$, and so since $R_c\beta x\delta$, we have that $C \in \delta$. Hence, $B \to C \in \alpha; \beta$ given the construction of $\alpha; \beta$. $\square$

Corollary 4.10. $\exists \epsilon (R_c^2 \beta (\alpha \gamma) \delta \Leftrightarrow R_c \epsilon \gamma \delta)$.

Proof. We start by showing that $R_c^2 \beta (\alpha \gamma) \delta \Leftrightarrow R_c (\alpha; \beta) \gamma \delta$.

First, for the left-to-right direction suppose that $R_c^2 \beta (\alpha \gamma) \delta$ and let $\alpha; \beta$ be in accordance with the construction in Section 4.1. Theorems 4.8 and 4.9 show that the defined set is, indeed, a $\mathbf{B}_\wedge$ theory (which is regular if α, β are) — since S_c is the set of all $\mathbf{B}_\wedge$ theories, then, $\alpha; \beta \in S_c$. The construction ensures that $R_c (\alpha; \beta) \gamma \delta$, and so $R_c^2 \beta (\alpha \gamma) \delta \Rightarrow R_c (\alpha; \beta) \gamma \delta$.

Second, the right-to-left direction. Suppose that $R_c (\alpha; \beta) \gamma \delta$. We want to show that $\exists x (R_c \alpha \gamma x \,\&\, R_c \beta x \delta)$. Let $x = \{B : \exists A (A \to B \in \alpha \,\&\, A \in \gamma)\}$ — we can ensure that x is a theory using a construction very much like that given in Section 4.1. Note that we have immediately that $R_c \alpha \gamma x$. Suppose, then, that $A \to B \in \beta \,\&\, A \in x$. Then, there is a C s.t. $C \to A \in \alpha \,\&\, C \in \gamma$. The construction of $\alpha; \beta$ guarantees that $C \to B \in \alpha; \beta$ when $C \to A \in \alpha$ and $A \to B \in \beta$, so, since $R_c (\alpha; \beta) \gamma \delta$ and $C \in \gamma$, $B \in \delta$. Hence $R_c \beta x \delta$ after all, and thus $R_c^2 \beta (\alpha \gamma) \delta$. So $R_c (\alpha; \beta) \gamma \delta \Rightarrow R_c^2 \beta (\alpha \gamma) \delta$.

Thus, $R_c^2 \beta (\alpha \gamma) \delta \Leftrightarrow R_c (\alpha; \beta) \gamma \delta$, and so $\exists \epsilon (R_c^2 \beta (\alpha \gamma) \delta \Leftrightarrow R_c \epsilon \gamma \delta)$. $\qquad \square$

Note that $R_c^2 \beta (\alpha \gamma) \delta \Rightarrow \exists \epsilon R_c \epsilon \gamma \delta$ follows from Corollary 4.11. This is, perhaps, the most natural statement that for any α, β standing in the correct relation, there is a composite point in the canonical model.

Corollary 4.11. *The logic characterised by the class of channel models conservatively extends that characterised by the class of $\mathbf{B}_\wedge$ models.*

Proof. Given Corollary 4.10, for any α, β in the $\mathbf{B}_\wedge$ canonical model, there exists a $\mathbf{B}_\wedge$ theory (which is regular whenever α, β are). That is, the canonical model satisfies the existence of composites property. So, the canonical model of $\mathbf{B}_\wedge$ is a channel model, and hence the class of channel models is complete for $\mathbf{B}_\wedge$. Hence, the class of channel models admits no validities not already admitted by the class of $\mathbf{B}_\wedge$ models, and so the extension of the $\mathbf{B}_\wedge$ model structures by additional composite points is conservative over $\mathbf{B}_\wedge$. $\qquad \square$

The facts proven in Section 3 also provide an argument for the associativity and left/right normality of our composition 'operation'. Theorem 4.8 implies that Theorems 3.1 and 3.2 also hold in the canonical model (as they hold in any $\mathbf{B}_\wedge$-model). It is a key question whether or not this construction continues to work in logics extending $\mathbf{B}_\wedge$.

This proof strategy, at very least, does not extend any further, due to its reliance on the Bubbling Lemma. This lemma fails in $\mathbf{B}_{\wedge\vee}$, more commonly referred to as $\mathbf{B}^+$, because it includes all instances the following theorem for $\vee$:

$$\vdash_{\mathbf{B^+}} ((A \rightarrow C) \wedge (B \rightarrow C)) \rightarrow ((A \vee B) \rightarrow C),$$

some instances of which are counterexamples to the Bubbling Lemma.[17] So, we cannot extend this proof to cover any of the extensions of $\mathbf{B_\wedge}$ including the usual disjunction. However, we can say something interesting about other potential extensions of $\mathbf{B_\wedge}$ and how composition as we have defined it operates in those logics. First, we can say something about the question whether ; is commutative and idempotent under the intended interpretation, and how our answers stack up against Restall's. Second, there is a more serious problem into which this approach runs as soon as we consider logics with negations obeying a certain contraposition property.

5 Idempotence and Commutativity of Composition

Whether $\alpha;\alpha = \alpha$ and $\alpha;\beta = \beta;\alpha$ are interesting questions, and ones which Barwise leaves open. Our concern here is to consider what results enforcing these conditions has on our approach, as opposed to Restall's. Unsurprisingly, we get quite different answers, and answers which provide some indication of the split in the approaches. We'll display some of the key differences, and reflect on which approach is more natural for the channel theoretic interpretation.

$\alpha;\alpha = \alpha$ looks to be some kind of contraction principle. Understood as Restall does, the most salient consequence of this is, in Restall's notation, $\alpha \circ \alpha \sqsubseteq \alpha$, or more commonly $R\alpha\alpha\alpha$, which corresponds with the axiom $((A \rightarrow B) \wedge A) \rightarrow B$, which we call WI. As has been known since at least [17], this theorem is bad news for the usual naïve theories, as it provides for a straightforward Curry paradox. This is an interesting result, and, understood as Restall wants, there is a nice story one can give for why this principle ought to fail along channel-theoretic terms. However, understood in our terms, the important related upshot of this idempotence principle is $R\alpha\beta\gamma \Leftrightarrow R(\alpha;\alpha)\beta\gamma \Leftrightarrow R^2\alpha(\alpha\beta)\gamma$. Under interpretation, this is a somewhat different kind of 'reuse of resources' than one has in something like WI. With our interpretation, what this tells us is that exploiting a situation α *qua*-channel to get from β to γ is in no way different from exploiting α twice. This is to be contrasted with what contraction allows, namely, that one can exploit a situation *qua*-signal twice. In other words, that one can use the same proposition as a *premise* as many times as one likes, with no change in the validity of the argument.

Consider not $R\alpha\alpha\alpha$ and WI, which is contraction 'mixed' with identity, but rather the pure contraction axiom W: $(A \rightarrow (A \rightarrow B)) \rightarrow (A \rightarrow B)$. The frame condition corresponding to this axiom is $R\alpha\beta\gamma \Rightarrow R^2\alpha\beta\beta\gamma$. Using Restall's notation, this

---

[17]This can be seen by noting that, in general, $\nvdash_{\mathbf{B^+}} (A \vee B) \rightarrow (A \wedge B)$.

comes to $\alpha \circ \beta \sqsubseteq \gamma \Rightarrow (\alpha \circ \beta) \circ \beta \sqsubseteq \gamma$, which can naturally be read in terms of allowing the exploitation of an antecedent-supporting situation multiple times when it can be exploited once. Our way of cashing out composition provides for a kind of 'contraction' of the channel in use which is quite distinct from employing some premise information twice. Note that $R\alpha\beta\gamma \Rightarrow R^2\alpha(\alpha\beta)\gamma$ is a consequence of our precisification of the idempotence clause. This frame condition corresponds to (A11):

$$((A \to B) \wedge (B \to C)) \to (A \to C).$$

A natural reading of this formula, and its associated ternary relation condition, is that whenever some situation supports two constraints $A \to B$ and $B \to C$, then it must also 'act as its own composition' and support $A \to C$. This is, indeed, what $R\alpha\beta\gamma \Rightarrow R(\alpha;\alpha)\beta\gamma$ most naturally gives us. That is, that when α is a channel from β to γ, so is $\alpha;\alpha$. There may be reasons to accept this principle, perhaps resulting in a theory not too dissimilar from that of [14], but it is at least not obvious.

Consider a channel understood in purely physical terms, as a part of the world which connects some site of information to another site. Then employing a channel, as it were by applying it to some signal, is one instance of application of that channel. If the same channel supports $A \to B$ and $B \to C$, it is at least questionable that one can pass across both conditionals with only one application of the channel.[18]

We leave off the interpretation for now simply to point out two interesting features of this approach. These have to do with the interaction between $\to$ and the $\leftarrow$ which is available in this logical setting where we don't have $R\alpha\beta\gamma \Rightarrow R\beta\alpha\gamma$.[19] Briefly, the valuation clause for $\leftarrow$ is as follows:

$$\alpha \vDash B \leftarrow A \text{ iff } \forall\beta\forall\gamma(R\beta\alpha\gamma \Rightarrow (\beta \vDash A \Rightarrow \gamma \vDash B)).$$

When we enforce idempotence for our composition, $R\alpha\beta\gamma \Rightarrow R(\alpha;\alpha)\beta\gamma$, an immediate consequence is a contraction principle for $\leftarrow$, namely:

$$((B \leftarrow A) \leftarrow A) \to (B \leftarrow A).$$

This is noteworthy because of its connection to the commutativity of composition. An immediate consequence of the commutativity of ; in our sense is:

$$R(\alpha;\beta)\gamma\delta \Leftrightarrow R(\beta;\alpha)\gamma\delta \qquad (\text{i.e., } R^2\beta(\alpha\gamma)\delta \Leftrightarrow R^2\alpha(\beta\gamma)\delta)$$

---

[18]Thanks go to Dave Ripley for pushing me on this point.

[19]Details on this connective and its relation to $\to$ are available in many places, and [20] provides a nice overview.

which enforces:

$$((C \leftarrow B) \leftarrow A) \rightarrow ((C \leftarrow A) \leftarrow B),$$

which is a kind of permutation for $\leftarrow$. However, there seems to be no obvious connection to the frame conditions which enforce prefixing or suffixing for $\leftarrow$, i.e., $R^2\alpha(\beta\gamma)\delta \Rightarrow R^2(\alpha\beta)\gamma\delta$ and $R^2\alpha(\beta\gamma)\delta \Rightarrow R^2(\alpha\gamma)\beta\delta$, respectively. This is an avenue for some future work.

6 The Axiom Form of Contraposition

Something difficult happens here when we consider a common contraposition axiom, namely:

A15 $(A \rightarrow B) \rightarrow (\neg B \rightarrow \neg A)$.

Negation is generally interested in the ternary relation semantics for relevant logics by means of an operation * on worlds, the Routley–Routley star (see [24]). For the small point we make here, it is enough to note that the valuation condition on negation in this semantics is $\alpha \vDash \neg A$ iff $\alpha^* \nvDash A$, and that the above axiom corresponds to the following frame condition:

S15 $R\alpha\beta\gamma \Rightarrow R\alpha\gamma^*\beta^*$

Note, given what we have up to this point, $\alpha;\beta$ does not provide what we want if we are to consider logics including this axiom. Suppose that $\alpha \vDash A \rightarrow B$ and $\beta \vDash B \rightarrow C$, so that $\alpha \vDash \neg B \rightarrow \neg A$ and $\beta \vDash \neg C \rightarrow \neg B$. We have $\alpha;\beta \vDash A \rightarrow C$, but not necessarily $\alpha;\beta \vDash \neg C \rightarrow \neg A$.

Fact 6.1. If $\alpha;\beta \vDash A \rightarrow B$, then $\beta;\alpha \vDash \neg B \rightarrow \neg A$.

Proof. Suppose that $\alpha;\beta \vDash A \rightarrow B$ and $\beta;\alpha \nvDash \neg B \rightarrow \neg A$. Unpacking the latter, we have $\exists\gamma\exists\delta(R^2\alpha(\beta\gamma)\delta \,\&\, \gamma \vDash \neg B \,\&\, \delta \nvDash \neg A)$. Since $\gamma \vDash \neg B$, we get that $\gamma^* \nvDash B$, and since $\delta \nvDash \neg A$, we have $\delta^* \vDash A$. Since $R^2\alpha(\beta\gamma)\delta$ we know there is an ϵ s.t. $R\alpha\epsilon\delta \,\&\, R\beta\gamma\epsilon$. Therefore, $R\alpha\delta^*\epsilon^* \,\&\, R\beta\epsilon^*\gamma^*$, and thus $R^2\beta(\alpha\delta^*)\gamma^*$. Since $\alpha;\beta \vDash A \rightarrow B$ and $\delta^* \vDash A$, we get that $\gamma^* \vDash B$, contrary to hypothesis. $\square$

If we impose merely $R\alpha\beta\gamma \Rightarrow R\alpha\gamma^*\beta^*$, what we get by assuming $R^2\beta(\alpha\gamma)\delta$ is not $R^2\beta(\alpha\delta^*)\gamma^*$, as we'd want, but rather $R^2\alpha(\beta\delta^*)\gamma^*$. So, we don't have:

$$R(\alpha;\beta)\gamma\delta \Rightarrow R(\alpha;\beta)\delta^*\gamma^*,$$

749

but rather,

$$R(\alpha; \beta)\gamma\delta \Rightarrow R(\beta; \alpha)\delta^*\gamma^*.$$

One way to enforce the result we want is to enforce that $R(\alpha; \beta)\gamma\delta \Rightarrow R(\beta; \alpha)\gamma\delta$, but this is a fairly hefty assumption, and hard to justify. At the very least, I don't see any intuitive reason to think it holds.

For the general case, we need only to enforce:

$$R^2\alpha(\beta\gamma)\delta \Rightarrow R^2\alpha(\beta\delta^*)\gamma^*$$

Now, we needn't necessarily enforce the above commutativity principle to get this. The question is, what does this add to the frames of logics containing **DW**?

This raises a more general question, namely, what should negation look like in channel theory in the first place? While the ternary relation of the relevant semantics is fairly natural in the channel-setting, the Routley–Routley star is another matter entirely.[20]

7 Concluding Remarks

We have proved the main result, that one can supplement the ternary relation semantics for $\mathbf{B}_\wedge$ with points behaving as the composites of other points in the frame. This is a good first step to the larger project of fully laying out how Barwise's initial approach to channel theory helps to interpret the ternary relation semantics, but, as we have seen, there are some difficulties ahead. Some, like extending the result to full $\mathbf{B}^+$, seem plausible, but clearly require some other proof tactic. From there, one might hope that it is not a difficult manner to extend the picture upwards into other positive logics gotten by extending $\mathbf{B}^+$ by various additional axioms and rules. Incorporating negation into the picture, on the other hand, seems to require some more foundational work before the formalism can get off the ground.

In addition to these forward looking comments, as mentioned at the start of Section 2, the work we have done here, though it is a step to giving us some insight into channel composition in relevant semantics, doesn't yet settle the behaviour of this operation in any other than purely extensional terms. Future work is, certainly, needed to expand the picture we've given to one which provides a more robust insight into just what kind of critter this composition operation really is, and not just that we can find composites when we want them, at least as far as $\mathbf{B}_\wedge$ is concerned.

---

[20] As are Dunn's 'perp' in [10] and Restall's compatibility in [21].

References

[1] Henk Barendregt, Mario Coppo, and Mariangiola Dezani-Ciancaglini. A filter lambda model and the completeness of type assignment. *Journal of Symbolic Logic*, 48(4):931–940, 1983.

[2] Jon Barwise. Constraints, channels, and the flow of information. In P. Aczel, D. Israel, Y. Katagiri, and S. Peters, editors, *Situation Theory and Its Applications, 3*, pages 3–27. CSLI Publications, Stanford, CA, 1993.

[3] Jon Barwise and John Perry. *Situations and Attitudes*. MIT Press, Cambridge, MA, 1983.

[4] Jon Barwise and Jeremy Seligman. *Information Flow: The Logic of Distributed Systems*. Number 44 in Cambridge Tracts in Theoretical Computer Science. Cambridge University Press, 1997.

[5] Jc Beall. Curry's paradox. In E. N. Zalta, editor, *Stanford Encyclopedia of Philosophy*. Spring 2013 edition, 2013.

[6] Jc Beall and Julien Murzi. Two flavors of Curry's paradox. *Journal of Philosophy*, 110(3):143–165, 2013.

[7] Jc Beall, Ross Brady, J. Michael Dunn, Allen Hazen, Edwin Mares, Robert K. Meyer, Graham Priest, Greg Restall, David Ripley, John Slaney, and Richard Sylvan. On the ternary relation and conditionality. *Journal of Philosophical Logic*, 41(3):595–612, 2012.

[8] B. Jack Copeland. On when a semantics is not a semantics: Some reasons for disliking the Routley–Meyer semantics for relevance logic. *Journal of Philosophical Logic*, 8(1):399–413, 1979.

[9] Mariangiola Dezani-Ciancaglini, Robert K. Meyer, and Yoko Motohama. The semantics of entailment omega. *Notre Dame Journal of Formal Logic*, 43(3): 129–145, 2002.

[10] J. Michael Dunn. Star and perp: Two treatments of negation. *Philosophical Perspectives*, 7:331–357, 1993.

[11] J. Michael Dunn and Robert K. Meyer. Combinators and structurally free logic. *Logic Journal of the IGPL*, 5(4):505–537, 1997.

[12] J. Michael Dunn and Greg Restall. Relevance logic. In D. Gabbay and F. Guenthner, editors, *Handbook of Philosophical Logic*, volume 6, pages 1–136. Kluwer, 2 edition, 2002.

[13] Lloyd Humberstone. Smiley's distinction between rules of inference and rules of proof. In T. J. Smiley, J. Lear, and A. Oliver, editors, *The Force of Argument: Essays in Honour of Timothy Smiley*, pages 107–126. Routledge, 2010.

[14] Edwin Mares. Relevant logic and the theory of information. *Synthese*, 109(3): 345–360, 1996.

[15] Edwin Mares. *Relevant Logic: A Philosophical Interpretation*. Cambridge University Press, 2004.

[16] Edwin Mares, Jeremy Seligman, and Greg Restall. Situations, constraints, and channels. In J. van Benthem and A. ter Meulen, editors, *Handbook of Logic and Language*, pages 329–344. Elsevier, 2011.

[17] Robert K. Meyer, J. Michael Dunn, and Richard Routley. Curry's paradox. *Analysis*, 39:124–128, 1979.

[18] Greg Restall. A useful substructural logic. *Bulletin of the Interest Group in Pure and Applied Logics*, 2:137–148, 1994.

[19] Greg Restall. Information flow and relevant logics. In J. Seligman and D. Westerståhl, editors, *Logic, Language and Computation: The 1994 Moraga Proceedings*, pages 463–477. CSLI Publications, Stanford, CA, 1995.

[20] Greg Restall. *An Introduction to Substructural Logics*. Routledge, 2000.

[21] Greg Restall. Defining double negation. *Logic Journal of the IGPL*, 8(6):853–860, 2000.

[22] Richard Routley and Robert K. Meyer. The semantics of entailment: III. *Journal of Philosophical Logic*, 1(2):192–208, 1972.

[23] Richard Routley and Robert K. Meyer. The semantics of entailment. In H. Leblanc, editor, *Truth, Syntax, and Modality. Proceedings of the Temple University Conference on Alternative Semantics*, pages 199–243. North-Holland, Amsterdam, Netherlands, 1973.

[24] Richard Routley, Val Plumwood, Robert K. Meyer, and Ross T. Brady. *Relevant Logics and Their Rivals: The Basic Philosophical and Semantical Theory*. Ridgeview, 1982.

[25] John Slaney and Robert K. Meyer. Logic for two: The semantics of distributive substructural logics. Technical report, Research School of Information Sciences and Engineering and Centre for Information Science Research Australian National University, January 1997.

[26] Alasdair Urquhart. Semantics for relevant logics. *The Journal of Symbolic Logic*, 37(1):159–169, 1972.

Received 6 September 2016

Some Concerns Regarding Ternary-relation Semantics and Truth-theoretic Semantics in General

Ross T. Brady*

La Trobe University, Melbourne, VIC, Australia
<ross.brady@latrobe.edu.au>

Abstract

This paper deals with a collection of concerns that, over a period of time, led the author away from the Routley–Meyer semantics, and towards proof-theoretic approaches to relevant logics, and indeed to the weak relevant logic MC of meaning containment.

Keywords: completeness deception, meaning discrepancies, proof-theoretic methods, Routley–Meyer semantics, truth-theoretic semantics

1 Introduction

It is hoped that the concerns dealt with in this paper will help to round out the philosophical discussion around ternary relation semantics, which was the subject of the Third Workshop, held at the University of Alberta, Edmonton, Canada, during May 16–17, 2016, for which this paper was written.

Whilst these concerns form a rather motley collection, the main point to be made is that proof theory and semantics have distinct interpretations, with special reference to disjunction and existential quantification. (See Section 3.1 below for the details.) One should especially note that the author supports the proof-theoretic interpretation over the semantic one, this contributing to the concern about semantics in general. (See Section 3.2 for this point.) Though the author has given some

---

*I wish to thank the audiences of the *Third Workshop*, held at the University of Alberta, Edmonton, May 16–17, 2016, and the *Australasian Association for Logic* conference, held at La Trobe University, Melbourne, during June and July, 2016, to which this paper was presented. A special thanks goes to Kit Fine, Alasdair Urquhart, Arnon Avron, Mike Dunn, David Ripley and Lloyd Humberstone for their discussion and comments.

earlier presentations of these two distinct interpretations, this is the first time such material would appear in published form.

Another major point is the importance of decidability in establishing a two-valued meta-logic, whilst completeness is of lesser value than the current literature espouses. (Sections 3.3 and 3.4 deal with these respective points.) This again covers new ground.

2 Some Concerns about the Routley–Meyer Semantics

2.1 The Complexity of Fine's Semantics for Quantified Relevant Logics

Logics need to include quantifiers and lead into (non-logical) applications in order for logic to be applied and thus be a worthwhile study. The central value of logic is in its application to mathematics, computer science, science in general, and to familiar everyday arguments. However, sentential matters largely define the logic one is using, since the main differentiation between logics occurs at the sentential level. The addition of quantifiers is more clear-cut and its other premises are determined largely by the (non-logical) concepts to which one is applying the logic. So, sentential logic is still a worthwhile study in itself, but one still should be able to indicate how such a logic would be extended into applications, which would pass through quantification.

At the sentential level, the ternary-relation semantics of Routley and Meyer, culminating in their book [43], does have a somewhat bearable complexity, with its string of semantic postulates capturing the appropriate properties of the ternary relation R, and Priest in [39] sets out corresponding tree methods that can by-and-large be used to determine validity or invalidity of sentential formulae. Bear in mind that some strong relevant logics such as R and T are undecidable at the sentential level. This result being due to Urquhart in [48], but nevertheless some weaker relevant logics such as RW, TW, DW and DJ are decidable (see Fine [26] and Brady [10, 11] and [15]). The undecidability for these strong logics does mean that there is a certain inner complexity in them. (For axiomatizations of the above logics, see Section 2.3 below.)

However, when we pass to Fine's variable-domain semantics for quantified relevant logics in [28], there is a step up in complexity mainly due to the variability of domains from world to world. This complexity is so much so that there is a general lack of corresponding use in the literature to determine the validity or otherwise of quantified formulae. Further, there is also a general lack of research work into Fine's quantified semantics, with Mares' addition of the identity relation and his and

Goldblatt's alternative semantics being the main contributions the author is aware of. (See Mares [31] and Mares and Goldblatt [29].) Also, Priest in [40] sets out a tree method for the quantified relevant logic BQ and its familiar relevant extensions, and with identity, but for constant-domain semantics only. Without knowledge of completeness for a quantified logic such as B with respect to such a semantics, this tree method also works by-and-large, as it does for the strong sentential logics such as R in his [39].

Furthermore, one also has the task of interpreting Fine's semantics, leaving one with the two questions of what a semantics for a logic ought to look like and what the logical concepts of the connectives and quantifiers are. (More on this in Sections 3.1 and 4.1.) Recall too that Routley in [42] put forward the proposal for a constant-domain quantified semantics, but Fine showed that the logic RQ is incomplete with respect to this type of semantics. As far as the author is aware, this incompleteness is still an open question for weaker relevant logics such as DW or DJ, though this will seem doubtful when Section 3.4 is taken into account. It was at this point that the author started to become disillusioned with the ternary semantics in general, especially as I had spent a lot of time trying to make this constant-domain semantics work.[1] Here, I was having difficulty in establishing witnesses for existential quantification within a domain that is constant across worlds. The author is now of the view that in applying quantified inferential logic, the domain of quantification should be constant. It is understandable that for quantified modal logics that possible worlds might have differing domains from world to world, but this is not clear for practical non-modalized examples such as Peano arithmetic. Indeed, logical applications generally have fixed domains of objects, such as natural numbers or sets, and one should not have to vary such a domain when replacing classical logic by a supposedly superior logic. Further, in any proof theory with quantification, except maybe in a modal logic, it is understood that each quantifier is applied to the same domain, although sub-domains can be determined by a restricting predicate. (There will be some more on this point later in Section 4.1.)

The author's general concern with complexity is as follows. Put oneself in the mind of a reasoner conducting a simple inference step and ask the question: what is the rationale or justification for the inference? Here, we are assuming that any complexity that does occur results from the transitivity of a sequence of inferences. One can understand if the reasoner says that he or she is preserving truth in that the truth of the consequent or conclusion follows, given the truth of the premises and/or antecedent. One can also understand if the reasoner says that he or she

---

[1]Indeed, this paper will sketch out the subsequent journey undertaken by the author over many years, although not in strict historical order.

is preserving meaning through a meaning analysis of premises and/or antecedents. The author cannot think of any other criteria that could be going on in the mind of such a reasoner, except perhaps a combination of both, though many systems have been created and their semantics has been studied over the years. In the case of a combination however, one can reduce it to single inference steps involving only one of the two preservations. Further, such criteria need to be simple as such a reasoner is not going to embrace much complexity in making and justifying a single inference step. The logic governing the step would be clean and clear, based on well-understood concepts. (There is some complexity to follow, but the reason for it will be explained in each circumstance.)

2.2 The Lack of a Single Logical Concept Captured by the Routley–Meyer Semantics

In a group paper Beall et al. [3], it was argued that the inferential concept captured by the ternary relation R of the Routley–Meyer semantics was that of conditionality. This is a broad concept intended by the above group of authors to accommodate the various logics which can be given such a semantics and this is very inclusive, ranging from B to R and beyond to classical logic, with very few logics missing out. However, this does not tell us what the specific inferential concepts pertaining to these logics are.

Len Goddard verbally made the point at the time when the Routley–Meyer semantics was introduced in the early 1970s that the semantic postulates are just in one-one correspondence with the axioms. Essentially, his idea was that one could more-or-less determine the semantic postulate for a given axiom and vice versa. What he thought was needed here was a semantics that characterised a particular logic through a semantic rendition of a particular concept of inference.

The problem for relevant logics is that there are far too many of them and, as such, there is a lack of definition in the concept of relevance. If we take relevance as meaning relatedness, which is its immediately intuitive concept, this is, by itself, not a suitable concept upon which to base a logic as it is too vague. Relevance, as determined in its sharper form by the variable-sharing property (if $A \to B$ is a theorem then A and B share a sentential variable), has been taken as a necessary condition for a good logic, but not a sufficient one, leaving a plethora of systems to consider. The strong relevant logics such as R, satisfying this property, are based on technical criteria such as the neatness in the presentation of their natural deduction systems rather than on a specific logical concept. (That natural deduction systems are more complex for logics weaker than R can be seen in Brady [6].) This lack of concept makes application difficult, as there needs to be some logical concept prior to the non-logical applied concepts so that the application can be completed by embracing both logical and non-logical concepts.

Historically, this difficulty has borne out in a number of ways. Meyer tried to prove that the Disjunctive Syllogism rule, $\sim A, A \to B \Rightarrow B$, was admissible for relevant arithmetic based on the logic R, but ultimately a counter-example was found (see Meyer and Friedman [36]). This is an example of what was seen as an important admissible rule of the logic R, not extending to arithmetic based on R. Furthermore and more clearly, Meyer showed that the irrelevant implication, $x = y \to .p \leftrightarrow p$, is derivable from the Extensionality Axiom in the form: $x = y \to .x \in \{ x : p \} \leftrightarrow y \in \{ x : p \}$, also based on R. This example shows up the difficulty of maintaining relevance in an application, given that the variable-sharing property holds for this form of Extensionality Axiom when applied to free variables. (See Meyer [35] and [33] on relevant arithmetic and Brady [12] for Meyer's example from set theory.)

So, one does need some further specification to fix upon a particular logic, which we will consider within the following section.

2.3 The Lack of Facility to Drop the Distribution Axiom in the Routley–Meyer Semantics

As an answer to Section 2.2 and as stated in Section 2.1, there are essentially two key semantic concepts relating to logic, that of truth and meaning. These two concepts can be used to provide an understanding of inferential connectives and rules. Truth-preservation clearly applies to rules $A \Rightarrow B$, these being meta-theoretic in nature and based on the notion of a deductive argument. (See Brady [23] for discussion of this, and also the relationship between this (classical) deduction and relevant deduction.) The material implication $\supset$ of classical logic is essentially truth-preservation, expressed as a connective, as can be seen from its truth-table. Furthermore, the relationship between the classical connective $A \supset B$ and the rule $A \Rightarrow B$ can be seen from their deductive equivalence, assuming that the Law of Excluded Middle (LEM), $A \vee \sim A$, and the Disjunctive Syllogism (DS), $\sim A, A \vee C \Rightarrow C$, both hold for the antecedent A. (We base the deductive equivalence on the basic system B^{d}, which is the logic B with the addition of the meta-rule: if $A \Rightarrow B$ then $C \vee A \Rightarrow C \vee B$.)

The inferential connective associated with meaning is an entailment, representing the containment of the meaning of the consequent in that of the antecedent. Such a logic, called MC, based on the connective $\to$ representing meaning containment was introduced by Brady, after some tweaking which dropped the distribution axiom from an earlier version DJ^{d}, set out in Brady [12] and [18]. The logic DJ^{d}, was initially determined using 'set-theoretic containment' properties which are in evidence in its content semantics (see below in Section 4.1 for such content semantics). This was then modified to 'intensional set-theoretic containment' to form the distribution-less MC, as set out in Brady and Meinander [5], where MC was introduced. The axiomatisation of MC and also its quantificational extension MCQ are below.

Indeed, there is a strong case made in Brady and Meinander [5] for dropping the distribution in axiom-form from such a logic, as distribution does not follow from the standard meanings of conjunction and disjunction. Although the problem with distribution was initially pointed out in a review of Brady [18] by Restall in [41], it was Schroeder-Heister in [45] who made it clear to the author that the introduction and elimination rules for conjunction and disjunction sufficed to uniquely specify these two concepts, i.e., without the addition of distribution. (See also Schroeder-Heister [44].)

What Schroeder-Heister showed was the following. Let & and $\&'$ satisfy the introduction and elimination rules:

$$\frac{A \quad B}{A \& B} \qquad \frac{A \quad B}{A \&' B} \qquad \frac{A \& B}{A} \qquad \frac{A \& B}{B} \qquad \frac{A \&' B}{A} \qquad \frac{A \&' B}{B}$$

Then: $\dfrac{A \& B}{A \&' B}$ and $\dfrac{A \&' B}{A \& B}$

Thus, $A \& B$ and $A \&' B$ are equivalent and can be substituted in all contexts of the logic. The same sort of argument applies to $A \vee B$ and $A \vee' B$, where the introduction and elimination rules that apply to $\vee$ also apply to $\vee'$. Thus, we also have the following equivalence:

$$\frac{A \vee B}{A \vee' B} \qquad \frac{A \vee' B}{A \vee B}$$

So, the introduction and elimination rules uniquely specify standard conjunction and disjunction concepts. Anderson and Belnap in [2] indicated, for their natural deduction systems for strong relevant logics such as R, that distribution requires a separate rule, i.e., $\&\vee$, over and above the introduction and elimination rules for conjunction and disjunction. The separation of this rule is also required for logics weaker than R, as can be seen in Brady [6].

The logic MC is set out as follows:
Primitives: $\sim$, $\&$, $\vee$, $\rightarrow$.
Axioms:

1. $A \rightarrow A$

2. $A \& B \rightarrow A$

3. $A \,\&\, B \to B$

4. $(A \to B) \,\&\, (A \to C) \to . A \to B \,\&\, C$

5. $A \to A \lor B^2$

6. $B \to A \lor B$

7. $(A \to C) \,\&\, (B \to C) \to . A \lor B \to C$

8. $A \to {\sim}B \to . B \to {\sim}A$

9. ${\sim}{\sim}A \to A$

10. $(A \to B) \,\&\, (B \to C) \to . A \to C$ \qquad (conjunctive syllogism)

Rules:

1. $A, A \to B \Rightarrow B$

2. $A, B \Rightarrow A \,\&\, B$

3. $A \to B, C \to D \Rightarrow B \to C \to . A \to D$

Meta-rule:

1. If $A, B \Rightarrow C$, then $D \lor A, D \lor B \Rightarrow D \lor C$

This two-premise meta-rule is deductively equivalent to the one-premise meta-rule 'if $A \Rightarrow B$ then $C \lor A \Rightarrow C \lor B$', together with the distribution rule '$A \,\&\, (B \lor C) \Rightarrow$

---

[2]Over the years, there has been some discussion as to whether $A \to A \lor B$ and $B \to A \lor B$ should be included as axioms in a logic based on meaning containment. Indeed, Kit Fine raised this issue in Edmonton and we exchanged a series of e-mails on this topic. The objections to these axioms, in accordance with Analytic Implication, first introduced by Parry (see his account in [38]) and more recently taken up by Fine (see his re-publication in [27]), are based on the respective lack of B in A and of A in B in these axioms. However, I contend that logic is a representation of the meanings of the connectives and quantifiers and other non-logical concepts, expressed in a logical language. This language is only a vehicle for transmitting the logical meanings and one must thus look into the meaning of $A \lor B$ as an 'either ... or' rather than how it is expressed syntactically. A (alternatively B) clearly adds to the meaning of 'either A or B' by creating the witness, establishing that the meaning of $A \lor B$ is contained within A (and within B). Further, Lloyd Humberstone pointed out in discussion that $A \leftrightarrow A \,\&\, (A \lor B)$ holds in MC (as originally pointed out by Dunn) and if the right hand equivalent is substituted for the first A in $A \to A \lor B$ then it becomes $A \,\&\, (A \lor B) \to A \lor B$, which is then an instance of $A \,\&\, B \to B$, which is not under contention. This then has the effect of linking $A \to A \lor B$ with its De Morgan dual $A \,\&\, B \to A$.

$(A\&B)\vee(A\&C)$'. Thus, distribution in rule-form still holds in MC, with the inclusion of MR1.

For the general purposes of this paper, we extend the logic MC to MCQ with the following quantificational additions:

Primitives: $\forall$, $\exists$.

$a, b, c, \ldots$ range over free variables. $x, y, z, \ldots$ range over bound variables. Terms $s, t, u, \ldots$ can be individual constants (when introduced) or free variables.

Quantificational Axioms:

  1. $\forall x A \to At/x$, for any term t.

  2. $\forall x(A \to B) \to . A \to \forall x B$

  3. $At/x \to \exists x A$, for any term t.

  4. $\forall x(A \to B) \to . \exists x A \to B$

Quantificational Rule:

  1. $Aa/x \Rightarrow \forall x A$, where a is not free in A.

Meta-Rule:

1. If $A, Ba/x \Rightarrow Ca/x$, then $A, \exists x B \Rightarrow \exists x C$, where QR1 is not used to generalize on any free variables occurring in the A nor in the Ba/x of the rule $A, Ba/x \Rightarrow Ca/x$. This restriction on QR1 also applies to the rule $A, B \Rightarrow C$ of MR1 for the sentential component.

Note that the existential distribution rule, $A \,\&\, \exists x B \Rightarrow \exists x(A \,\&\, B)$, follows from R2 and QMR1. However, as with intuitionist logic, the universal distribution rule, $\forall x(A \vee B) \Rightarrow A \vee \forall x B$, fails, as the universal quantifier $\forall$ and the disjunction $\vee$ are essentially the same for intuitionist logic as for MCQ, since they are both constructively interpreted concepts.

Largely, the sentential distribution axiom is an add-on, except where it falls into place in extensional contexts such as classical logics. Routley–Meyer semantics, set in classical meta-logic, has no means at its disposal of dropping distribution, without some further complexity, as can essentially be seen from Dunn and Allwein [1] on linear logic.

Thus, the Routley–Meyer semantics cannot be used as it stands for the logic MC of meaning containment. For proof theory however, the lack of distribution generally simplifies the rules and makes results easier to prove. For the Fitch-style natural deduction systems, which can be seen in Anderson and Belnap [2] and in Brady [6], there is an additional distribution rule &∨, over and above the introduction and elimination rules for conjunction and disjunction. This rule would be removed for a distribution-less system. For Gentzen systems, the lack of distribution usually allows one to have only one structural connective instead of two, as can be seen from Dunn's Gentzenization of R+ in Anderson and Belnap [2], which has two structural connectives. Compare this with Brady [13] and [14], where there is only one structural connective required in the final Gentzen systems for a range of distribution-less contraction-less logics, with the added bonus of decidability at the quantificational level. Thus, whilst the lack of distribution would add significant complexity to the Routley–Meyer semantics when attempted, such a lack would considerably simplify the standard proof theories.

So, this lack of distribution essentially puts paid to the standard ternary semantics and we will need to consider other structures to provide such a semantics and to enable invalidity of formulae and other results to be shown. Here, the structures need not be the same as these two functions do differ, unlike the Routley–Meyer semantics and other semantics that attempt to play both the role of semantics and the role of being a technical vehicle for the proof of a range of results. (There is further discussion on this point in Section 4.3.)

3   Some Concerns about Truth-theoretic Semantics in General

3.1   The Discrepancy in the Meanings of Disjunction and Existential Quantification

The most telling differentiation between proof theory and semantics is the discrepancy in the meanings of disjunction and existential quantification. Both disjunction and existential quantification are characterised by the expression 'at least one of ...', the disjunction applying to two sentences and the existential quantification applying to a predicate expression. The issue is whether this means that there is a witness disjunct and a witness existential instantiation or whether there need not be such witnesses. Note that the priming property, 'if $A \vee B$ then either A or B' requires a disjunctive witness and the existential property, 'if $\exists x A$ then Aa/x, for some a' requires an existential witness.

In standard semantics, which are all based on formula-induction, there must be such witnesses, as seen by the Henkin-style completeness proofs where a canonical model is built up so as to ensure that all such witnesses are in place. This is achieved by constructing the canonical model using an enumeration of formulae and, as each formula is selected, a decision is made on whether to admit it to the constructed model or not, so as to ensure each required witness is present. In his completeness proof for the predicate calculus in [30], Henkin added existential witnesses in the construction of maximally consistent sets based on an initial consistent set. Since the LEM holds in classical logic, we can alternatively replace maximal consistency (ensuring negation-completeness) with disjunctive closure, i.e., requiring that each disjunction be witnessed by one of its disjuncts, yielding negation-completeness in particular. Indeed, disjunctive closure occurs in the completeness proof for the Routley–Meyer semantics, and this can be used for classical logic as well with some additional semantic postulates. (See Chapter 4 of Routley, Meyer, Plumwood and Brady [43].)

Fine's semantics in [27] makes this point clearer. There, Fine uses theories to capture conjunction and implication, whilst prime theories are used to capture disjunction and negation. (Fine used the term 'saturated theories'.) Thus, Fine separated theories and prime theories in his semantics, instead of bundling them all into worlds, which are all prime, as in the Routley–Meyer semantics. This emphasises the fact that it is negation and disjunction that minimally require priming. The reason negation is affected here is largely because of De Morgan's Laws which relate disjunction with conjunction through negation.

In proof theory however, there is no requirement for such witnesses. Consider the $\vee E$ and $\exists E$ rules of Fitch-style natural deduction below, where a general argument to a common conclusion suffices, whether for each disjunct or for any possible instantiation.

$\vee S$: If $A \vee B_a$, $A \to C_b$ and $B \to C_b$, then $C_{a \cup b}$.

$\exists E$: If $\exists x A_a$ and $\forall x (A \to B)_b$, then $B_{a \cup b}$.

Both these rules can have a restriction on the index sets a and b, in accordance with the particular logic involved. (See Brady [6] for details of this.) This is appropriate as hypotheses can be disjunctive or existential, where no specification of a witness would be required. An assumption of $A \vee B$ need not spell out which disjunct applies and an assumption of $\exists x A$ need not spell out which instantiation of x in A applies. This is quite appropriate as $A \vee B$ or $\exists x A$ can be assumptions or premises of an argument, where it may not be part of such an assumption to name or imply a particular disjunct or a particular existential instantiation. So, there is a discrepancy in the meaning of disjunction and existential quantification between the semantics and this proof theory (and other proof theories follow suit). And, it does seem overly restrictive to insist on a witness disjunct or existential instantiation that may not exist in many cases.

It depends too on one's view of logic here, i.e., whether logic is about arguments from premises to conclusions or about worlds specified by formula-induction based on atoms.

3.2   What is Logic About?

To attempt to answer this last question, we need to briefly examine the classical account and its influence on worlds semantics. There, logic is about propositions which are either true or false, but not both. This essentially locks in classical two-valued logic, with its truth-tables determining the truth-values of the connectives and hence its subsequent analyticities. It relies on a body of truths, initially taken from the real world. Falsity is just a fall-back position for sentences that are not true, since there is no other value they can take. Worlds are built up from atomic propositions by using a formula-induction process, which then extends to the universal and existential quantifiers. The objects of the domain of quantification initially consist of the existing things of the real world. As Quine said "To be is to be the value of a bound variable." In such a case, logic is abstracted from the real world. However, this domain can be extended in various ways by using truth-bearers. It is this world semantics that is taken to be the semantics of the classical predicate logic, and extended to possible worlds by Kripke for modal logics, using binary relations between these worlds. Then, Routley and Meyer, using ternary relations, extended this style of semantics to include relevant implications and entailments, but importantly the worlds can be impossible as well as possible. (Initially, such worlds were called set-ups by Routley to distinguish them from possible worlds.) And, people like David Lewis have attempted to give some reality to possible worlds whilst Meinong and Priest attempted to give some sort of reality to impossible worlds. Nevertheless, such semantics is still based on worlds of a sort, which are determined using formula-induction, based on atoms, but incorporating binary or ternary relations between worlds, largely to capture modal and inferential concepts, whilst maintaining the true-false dichotomy. One should note that for the Routley–Meyer semantics, the ∗-function is used, under the Australian Plan, to maintain this true-false dichotomy, despite its negation being non-classical.

The main alternative is to take logic to be about the proof of conclusions from premises. This broadens logic to include arguments and concepts that do not fit the worlds picture, i.e., as considered in Section 3.1 above, they may include witness-less disjunctions and existentials. In the case of classical worlds, before one can even proceed with a piece of reasoning, every question must have a yes-or-no answer in order to establish the two values, truth and falsity. In most practical reasoning, not every question is answered, though allowances are made for this in the Routley–Meyer semantics through the use of their ∗-function but not in Kripke semantics. Nevertheless, Routley–Meyer semantics still requires each disjunction to have a witness, and an extension of the semantics to include quantifiers over a constant domain would still require every existential quantification to have an instantiation.

Further, a proof account does attempt to capture the logical and non-logical concepts being dealt with in its axioms, premises and rules. However, a concept may not be completely captured in the logical system that underspecifies it, giving rise to negation-incompleteness. On the other hand, concepts may sometimes be overspecified in a logical system, giving rise to contradictions, but here we generally try to eliminate them by removing a conceptual clash or tightening up the concepts so that they are not overspecified. (Brady [24] and [25] has some recent discussion on this point.) This attempt to capture concepts axiomatically is, we believe, at the heart of logic and the worlds approach is too restrictive as not all logical concepts precisely fit the worlds picture. Indeed, most logical reasoning proceeds without all questions being answered and without all witnesses being determined in advance.

3.3   What is the Meta-logic?

What is not generally realized is that a proof-theoretic view of logic requires one to review the meta-logic. Firstly, the meta-logic should be the same logic as that used for the object language, specialised to that logic which applies to formal systems. Following on from Section 3.2, in the classical account, the meta-logic would be classical as it is pre-determined by the nature of propositions, regardless of whether they apply in the object theory or its meta-theory. This carries over to the Kripke semantics for classical modal logics as well. For the Routley–Meyer semantics, even though negation can be non-classical, the use of the Routley ∗ enables the meta-theory to be two-valued in that each valuation in the semantics can only take the values T and F. So, for such semantics as these, classical meta-logic is universally used. Part of the reason for this too is that formalized logic is taken to be an object of mathematical study and that mathematics uses classical logic, having done so for at least a century. Whilst this is hard to shift, one should consider logic seriously, for its own sake, and hence apply it in accordance with its own principles to meta-theory and to mathematics generally, despite the fact that more work is needed in this process.

We now consider the case where logic is about proof. Here, decidability is important for a classical meta-logic as any undecided formula would clearly amount to a proof-gap, which would then become a truth-value-gap for the meta-logic. This would make the meta-logic three-valued with respect to proof, given the formal system is not contradictory regarding proofs. Such a contradiction would require a formula to be both provable and not provable, and we would assume this to be not so, on the grounds that a formal system is a conceivable concept, with 'not provable' at least lying within the classical fall-back position of 'non-proof'.

A decidable logical system would then have a classical two-valued meta-logic, as each formula could be either proved or its non-proof be established as not provable. However, a problem here is that some strong relevant logics such as R, E and T are undecidable even at the sentential level (see Urquhart [48]). Nevertheless, the logic DJ^d, which is of some related interest, is decidable at the sentential level (see Fine [26] and Brady [15] and [16]). And, the logic MC is decidable using normalised natural deduction, as sketched in Section 4.3, whilst its quantified logic MCQ has good prospects for being decidable. However, this latter result is still work to be completed.

3.4 The Deception of Completeness

Semantic completeness is deceptive in that it fails for many applied logics and holds mainly for pure logics and, as argued in Section 2.1 above, logics need to be applied to be worthwhile. Soundness, however, is not a problem and so we do focus on completeness. If one considers the classical semantics for predicate logic, for example, completeness is proved in Henkin [30]. In the application to Peano arithmetic, assuming consistency, completeness will fail, due to Gödel's first theorem. In order to make such a Henkin-style completeness proof work for standard truth-theoretic semantics for a quantified logic, one would need an infinite supply of existential witnesses, which may not be available if one is focussed on a specific domain as one often is for applications, where a standard model is invoked. Consider Peano arithmetic with its standard domain of natural numbers, where one cannot add an infinite number of witnesses that may be needed over and above the natural numbers. That is, one cannot guarantee that a given domain is adequate for all the witnesses needed in a completeness proof. Even for disjunctive witnesses, one may need to add disjuncts that may be too specific for the concepts that are being captured in some other applications. A case here would be a concept based on an unwitnessed disjunctive property. So, logics need to be applied and completeness can quite often fail for such applications.

To clarify this, the difference between pure and applied logics is that standard models are used in applications that have domains specific to the application. Pure logics, on the other hand, use all models that are appropriate for the generality of the connectives and quantifiers. This then provides sufficient generality for completeness to be proved using disjunctive and existential witnesses. Further, as we saw in Section 2.1, there are difficulties even for the pure quantified relevant logics to be complete with respect to constant-domain semantics, due to this need for existential witnesses, and constant-domain semantics is appropriate for applications generally. This would then reduce the main completeness results to the sentential level for such logics, which reduces their usage even further. As seen from Section 3.3 above, decidability is important for all logics, so that they can have a classical meta-logic and indeed this is more important than completeness that can easily fail above the sentential level.

In operating applied systems such as arithmetic, one needs to go back to the mathematical-style of proof, which consists of a Hilbert-style axiomatization, with some use of natural deduction to make deductions easier and more perspicuous. Semantical methods, such as truth-trees, are not of much value here. So, the use of completeness is largely limited to pure logics and of major value for sentential logics at that.

Completeness, together with soundness, enables one to say that proof theory and semantics are different representations of the logic involved. However, logic is about capturing concepts, and proof theory and semantics do differ conceptually in their interpretation of disjunction and existential quantification, as argued above in Section 3.1. Further, we favoured proof theory in Section 3.2, as truth-theoretic semantics does not capture the connectives and quantifiers precisely, and proof-theoretic semantics, as studied for example by Schroeder-Heister in [44] and [45], could offer a better capture of these concepts.

4 Concluding Directions

In conclusion, we examine two alternatives to standard semantics, followed by discussion of some proof-theoretic systems.

4.1 Content Semantics

Content semantics, as set out in Brady [12] and [18], offers a good capture of logical concepts. The contents used are logical contents that are best understood as analytic closures. One considers a sentence, engages in repeated meaning analysis of concepts from the sentence until such a process closes, and the set of sentences thus obtained is the analytic closure of that initial sentence. This analytic closure of the sentence is then its logical content, which is a deductively closed set. We can then quite reasonably use this to capture an entailment $A \rightarrow B$, based on meaning containment by simply taking its content as that of the set-theoretic containment statement of the content of the consequent B in the content of the antecedent A. However, more recently after dropping distribution in Brady and Meinander [5], we have had to modify this set-theoretic containment to reflect the absence of distribution, as argued in Section 2.3. Now, we use the term '*intensional* set-theoretic containment' instead, in order to represent set-theoretic containment but narrowed down to reflect the intensional meanings of conjunction, disjunction and the two quantifiers, in particular.

We now consider the contents of the connectives and quantifiers. The content of a conjunction of two sentences is the closure of the set-theoretic union of the two contents. The reason the closure is needed is that the two sentences may interact in producing conclusions that are not provable from either of the two sentences individually. The content of a disjunction is simply the set-theoretic intersection of their respective contents, whilst the content of an entailment is as described above. Negations can be dealt with using the dual concept of range, related to contents via De Morgan properties, but below we use the simpler $*$-function on contents. (Note that in Brady [18], it was seen that this $*$-function relates ranges and contents through its definition, but in the final analysis the ranges can be dropped in favour of the $*$.) The contents of the two quantifiers are similar unions and intersections to those of conjunction and disjunction, but are set unions and intersections where the set is controlled by the individual predicates used to generate them. We also have to take bound variables into account. With apologies, the quantificational extension of the content semantics is omitted which, though understandable in its interpretation, does add quite some complexity, due to the predicates and the bound variables. However, as discussed below and also in Section 4.3, being a semantics of meaning rather than truth, it is not so clear cut in its determinations and hence not so able to act as a vehicle for wide-ranging technical results.

The *content semantics* for the logic MC is set out as follows, as in Brady [18], but taking into account the tweaking of the logic occurring in Brady and Meinander [5] and the adopting of Mares' treatment of the closed union as a content of the set-theoretic union of two contents in his [32].[3]

A *content model structure (c.m.s)* consists of the following 4 concepts: $T, C, {}^*, c$, where C is a set of sets (called contents), $T \neq \emptyset$, $T \subseteq C$ (the non-empty set of all true contents), $*$ is a 1-place function on C (the $*$-function on contents), and c is a 1-place function from containment sentences, $c_1 \supseteq c_2$, and also unions $c_1 \cup c_2$, concerning contents c_1 and c_2 of C, to members of C, subject to the semantic postulates p1–p15, below. The concepts $\cap, \cup, =$ and $\supseteq$, are taken from the background set theory, $\cup$ and $\cap$ being a 2-place functions on C (the union and intersection of contents,

---

[3]Mares in his [32, on p. 202], expresses this union as $c(c_1 \cup c_2)$, where c is the same content operator as used for $c_1 \supseteq c_2$ but applied to the set-theoretic union $c_1 \cup c_2$. This is preferable to the author's treatment in [12] and [18], as it reinforces the use of standard set-theoretic concepts within the content semantics.

respectively), $=$ being a 2-place relation on C (identity), and $\supseteq$ being a 2-place relation on C (content containment).

The *semantic postulates* are:

p1. $c(c_1 \cup c_2) \supseteq c_1$, $\quad c(c_1 \cup c_2) \supseteq c_2$

p2. If $c_1 \supseteq c_2$ and $c_1 \supseteq c_3$, then $c_1 \supseteq c(c_2 \cup c_3)$.

p3. $c_1 \supseteq c_1 \cap c_2$, $\quad c_2 \supseteq c_1 \cap c_2$

p4. If $c_1 \supseteq c_3$ and $c_2 \supseteq c_3$, then $c_1 \cap c_2 \supseteq c_3$.

p5. $c_1^{**} = c_1$

p6. If $c_1 \supseteq c_2$, then $c_2^* \supseteq c_1^*$.

p7. If $c_1 \supseteq c_2$ and $c_1 \in T$, then $c_2 \in T$.

p8. If $c_1 \in T$ and $c_2 \in T$, then $c(c_1 \cup c_2) \in T$.

p9. If $c_1 \cap c_2 \in T$, then $c_1 \in T$ or $c_2 \in T$.

p10. $c(c(c_1 \supseteq c_2) \cup c(c_2 \supseteq c_3)) \supseteq c(c_1 \supseteq c_3)$

p11. $c(c(c_1 \supseteq c_2) \cup c(c_1 \supseteq c_3)) \supseteq c(c_1 \supseteq c(c_2 \cup c_3))$

p12. $c(c(c_1 \supseteq c_3) \cup c(c_2 \supseteq c_3)) \supseteq c(c_1 \cap c_2 \supseteq c_3)$

p13. $c(c_1 \supseteq c_2) \supseteq c(c_2^* \supseteq c_1^*)$

p14. $c(c_1 \supseteq c_2) \in T$ iff $c_1 \supseteq c_2$.

p15. If $c_1 \supseteq c_2$, then $c(c_3 \supseteq c_1) \supseteq c(c_3 \supseteq c_2)$ and $c(c_2 \supseteq c_3) \supseteq c(c_1 \supseteq c_3)$.

An *interpretation I on a c.m.s.* is an assignment, to each sentential variable, of an element of C. An interpretation I is extended to all formulae, inductively as follows:

(i) $I(\sim A) = I(A)^*$

(ii) $I(A \,\&\, B) = c(I(A) \cup I(B))$

(iii) $I(A \vee B) = I(A) \cap I(B)$

(iv) $I(A \rightarrow B) = c(I(A) \supseteq I(B))$

A formula A is true under an interpretation I on a c.m.s. M iff $I(A) \in T$.

A formula A is valid in a c.m.s. M iff A is true under all interpretations I on M.

A formula A is valid in the content semantics iff A is valid in all c.m.s.

Soundness (if A is a theorem of MC then A is valid in the content semantics) follows readily and completeness (if A is valid in the content semantics then A is a theorem of MC) follows by the usual Lindenbaum method for algebraic-style semantics, but here there is a slight difference. In constructing the canonical models, instead of taking equivalence classes of formulae as the contents, we put the content $[A]$ of A as $\{C \colon A \to C \in T'\}$, where T' is constructed as a prime extension of the set of theorems which does not include a non-theorem B. This essentially means that these canonical contents are closed under entailment, i.e., they are *analytic closures* of the sentence (or sentences) involved, since the set T of theorems is already prime, due to the logic MC being metacomplete (see Section 4.2). Since entailments here are understood as meaning containments, closure under entailment is closure under meaning containment and hence closure under the analysis of the meanings of words.

Thus, this semantics captures the meaning of the logical concepts in that it has transparency of concepts, shown by using the real set theoretic concepts of union, intersection, identity and containment in setting up the semantics. With the use of ranges to capture negation, this semantics is a "real" semantics, this being quite different from the use of semantic primitives with postulates, which have completely general interpretations restricted only by the postulates themselves, as occurs in the algebraic-style of content semantics in Brady [8] and [9]. Unlike truth-theoretic semantics, this semantics requires some understanding of content containment to work the semantics in showing invalidity (as was pointed out by Restall in discussion). As such, it would reject the axiom-form of distribution, for example. Further, this content semantics represents the logics MC and MCQ alone, unlike the author's earlier contents semantics of [8] and [9], which were quite wide-ranging, generally applying to logics in the range from BB right through to classical logic.

4.2 Metavaluations

Metavaluations combine features of proof theory and semantics to yield a technique that can produce results that would be hard or impossible for a truth-theoretic semantics to emulate. In particular, it can be used to prove the simple consistency of Peano arithmetic using finitary methods, for a quantified version of the logic MC (see Brady [21]). Though metavaluations are set up using truth-functions, it is essentially a proof theory, using formula-induction without worlds to capture that part of proof that behaves in a formula-inductive fashion. This inductive part of proof focuses on that part of a formula that sits between maximal entailment sub-formulae and the whole formula in a formula tree, i.e., the technique does not enable one to access any sub-formula inside a maximal entailment sub-formula. However, there is an exception for negated entailments where, for the so-called M2-metavaluations (see below), the metavaluation is essentially expressed in terms of its antecedent and its negated consequent.

Meyer introduced metavaluations in his [34] where he showed that the metavaluation technique can be very generally applied to positive logics, both sentential and quantified. However, the technique only works for certain logics once negation is added, as shown by Slaney in [46] and [47]. Once soundness and completeness is derived, with or without negation, such a logic is called a metacomplete logic. Slaney in [47] introduced two types of metavaluation, M1 and M2, depending respectively on whether there are no negated entailment theorems in the logic or whether a negated entailment is a theorem if and only if its antecedent and its negated consequent are both theorems. Indeed, the corresponding M2-logics have $A, {\sim}B \Rightarrow {\sim}(A \to B)$ as a derived rule, with the converse as an admissible rule, where for corresponding M1-logics this derived rule is absent from the logic. MC and MCQ are M1-metacomplete logics, and as such contain no negated entailment theorems. Thus, they are entailment-focussed logics.

Slaney's metavaluations v and v^* are as follows, with my symbolism and a slightly simplified layout:

(i) $v(p) = F$; $v^*(p) = T$, for sentential variables p.

(ii) $v(A \,\&\, B) = T$ iff $v(A) = T$ and $v(B) = T$;
$v^*(A \,\&\, B) = T$ iff $v^*(A) = T$ and $v^*(B) = T$.

(iii) $v(A \vee B) = T$ iff $v(A) = T$ or $v(B) = T$;
$v^*(A \vee B) = T$ iff $v^*(A) = T$ or $v^*(B) = T$.

(iv) $v({\sim}A) = T$ iff $v^*(A) = F$; $v^*({\sim}A) = T$ iff $v(A) = F$.

(v) $v(A \to B) = T$ iff $\vdash A \to B$, if $v(A) = T$ then $v(B) = T$, and if $v^*(A) = T$ then $v^*(B) = T$.
$v^*(A \to B) = T$, for M1-logics.
$v^*(A \to B) = T$ iff, if $v(A) = T$ then $v^*(B) = T$, for M2-logics.

We add the quantificational metavaluations, as follows:

(vi) $v(\forall x A) = T$ iff $v(At/x) = T$, for all terms t.
$v^*(\forall x A) = T$ iff $v^*(At/x) = T$, for all terms t.

(vii) $v(\exists x A) = T$ iff $v(At/x) = T$, for some term t.

$v^*(\exists x A) = T$ iff $v^*(At/x) = T$, for some term t.

The following key properties are then derivable (see Meyer [34] and Slaney [46] and [47]):

Completeness: If $v(A) = T$ then $\vdash A$, for all formulae A, and hence if $v^*(A) = F$ then $\vdash \sim A$.

Consistency: If $v(A) = T$ then $v^*(A) = T$.

Soundness: If $\vdash A$ then $v(A) = T$, and hence if $\vdash \sim A$ then $v^*(A) = F$.

Metacompleteness: $\vdash A$ iff $v(A) = T$, and hence $\vdash \sim A$ iff $v^*(A) = F$.

Priming Property: If $\vdash A \vee B$ then $\vdash A$ or $\vdash B$.

Negated Entailment Property: Not-$\vdash \sim(A \to B)$ (for M1-logics);
$\vdash \sim(A \to B)$ iff $\vdash A$ and $\vdash \sim B$ (for M2-logics).

Existential Property: If $\vdash \exists x A$ then $\vdash At/x$, for some term t.

The metavaluational technique can reject some non-theorems of the stronger metacomplete logics for which the technique applies. In particular, it can be used to reject the LEM, $A \vee \sim A$, and the Modus Ponens Axiom, $A \,\&\, (A \to B) \to B$ in the logic MC. Still to be researched is the good prospect of further metavaluations affecting negated entailments in different ways, which would reject other non-theorems in particular metacomplete logics. (This possibility is flagged in Slaney [46].) As stated above, Peano arithmetic can be shown to be simply consistent using finitary methods using metavalutions and this proof relies heavily on specific properties of metavaluations that cannot be duplicated using standard truth-theoretic semantics. (Here, in accordance with finitary methodology, mathematical induction is incorporated into the formulation of the metavaluations for the quantifiers, so as to ensure that all universal formulae can be proved through use of mathematical induction.) The specific properties of metacomplete logics can also be used to establish the simple consistency of naive set theory, the proof of which uses a single transfinite sequence of metavaluations (see Brady [22] for details).

4.3 Proof-Theoretic Methods

More familiar proof-theoretic methods include cut-free Gentzen systems and normalized natural deduction systems. Both of these can take advantage of the lack of distribution to make simplifications.

Of the three cut-free Gentzen systems set out in Brady [18, pp. 93–140], the best one for our purposes would be the left-handed cut-free Gentzen system for the logic DJ, which is MC + $A \,\&\, (B \vee C) \to (A \,\&\, B) \vee (A \,\&\, C)$ − MR1, i.e., essentially adding back the distribution axiom. Such a Gentzen system just consists of structures, to which an initial axiom together with rules apply. As it stands, the system has four structural connectives, and one would expect to delete the extensional and the corresponding k-intensional ones, with the removal of distribution as required for MC. That would leave us with what are called the i-intensional and j-intensional structural connectives, the i-intensional one ':' being interpreted as cotenability $\oplus$, defined as $A \oplus B =_{df} \sim(A \to \sim B)$, and the j-intensional one ';' being interpreted as fusion '$\circ$', axiomatically introduced by the two-way rule, $A \circ B \to C \Leftrightarrow A \to .B \to C$. The j-intensional connective can be inverted around the standard i-intensional connective, much as Belnap did in his [4] paper on Display Logics, with each derivable structure having an i-intensional connective as its main structural connective, which serves in lieu of a turnstile. To illustrate, these inversion rules, for structures α, β and γ, are as follows (see Brady [18, p. 133]):

$$(\text{Ii}) \quad \alpha : (\beta : \gamma) \,/\, (\alpha \,;\beta) : \gamma \qquad\qquad (\text{Ij}) \quad \alpha : (\beta \,;\gamma) \,/\, (\gamma : \alpha) : \beta$$

Further, as stated in Brady [16, pp. 350–351], quantifiers can be added to this Gentzen system in a fairly standard way.

However, there is a problem with proving decidability of DJ with this system in that the rule, (CSij) $(\alpha \,;\beta) : (\alpha \,;\delta) \,/\, \alpha : (\beta : \delta)$, representing conjunctive syllogism, is a form of contraction for the structure α, with the structures β and δ as parametric. Though decidability of DJ has been proved in Fine [26], and in Brady [15] and [16] by a semantic method, it remains to be proved in this setting. It should be noted that this semantic method as it stands, cannot be used once the distribution axiom is removed from DJ to form the logic MC.

In any case, Gentzen systems are rather stylized and are good if suitable systems are available for the logics of interest. This leads us to our preferred proof-theoretic method of representing the logics MC and MCQ or indeed other similar logics, that is, by normalized natural deduction systems. This is because they would capture reasoning most closely, roughly as it would occur in practice in closely reasoned contexts and, due to normalisation, in a way that proceeds straight to the point of the conclusion without detouring in and out of connectives and quantifiers. In Brady [19], there is such a system for DW, which is DJ $- (A \to B) \& (B \to C) \to .A \to C$. This would need further work to make it suitable for MC in not only removing the distribution axiom but also adding conjunctive syllogism. Another version of normalized natural deduction, called 'Free Semantics', appears in Brady [20], where, in the process of establishing a semantics based on natural deduction, a normalized version of that is given for the logic LDW, which is DW without the distribution axiom. Here, we would need to add to the restrictions on the $T \to E$ rule to embrace conjunctive syllogism, i.e., we add case (iii) to $T \to E$ (i) and (ii) below. So, in Brady [20], the logic MC is covered as well as LDW, giving a normalised natural deduction system for MC in the process of establishing such a "free semantics."

We set up the following natural deduction system MMC for the logic MC. The system MMC is a modified natural deduction system that is set up as a preliminary system so that normalization can then take place. MMC is somewhat simpler, but we will subsequently indicate what is needed for this normalisation process. MMC is taken from Brady [19], which contains a normalized natural deduction system for the logic DW, but we make a slight simplification to remove distribution and we also extend it to include conjunctive syllogism, yielding a system for MC. Reference to this normalised natural deduction system is made in Brady [20], but there the principle focus was on tableau and reductio systems for the logic LDW rather than that for MC.

We now present the rules of MMC, which is a Fitch-style natural deduction system, set out in the manner of Anderson and Belnap [2], but with modifications to help pin down the structure of it to suit the logic in hand. As part of this process, we use signed formulae TA and FA instead of a formula A, and structure them inductively as follows.

(1) If S is a sign T or F, and A is a formula then SA is a structure.

(2) If α and β are structures then (α, β) is a structure.

Each whole structure has a single index set, which is of one of the two types: $\emptyset$ or a complete set of natural numbers $\{\, j, \ldots, k \,\}$, which is a finite set of natural numbers in order with no numerical gaps. Structures are to be understood disjunctively and threads of proof within a subproof are obtained by following the signed formulae through in a particular position within a structure, just like a subproof within the subproof. (These threads of proof are defined in detail on [19, pp. 40–42], except, to remove distribution, one needs to remove the concept of a thread of proof extending another thread of proof, but the linkage between the previous thread of proof and the its continuation after the removal of these extended threads of proof remains. See [20, p. 522] for details of this.) That is, threads act as mini-subproofs, but without introducing a new index in the process as occurs in Anderson and Belnap [2].

Hyp. A signed formula of the form TH may be introduced as the hypothesis of a new subproof, with a subscript $\{\, k \,\}$, where k is the depth of this new subproof in the main proof. (Depth is defined as on p. 70 of Anderson and Belnap [2], but is called 'rank'. Also, see Brady [7] for 'depth relevance'.) Any hypothesis thus introduced must subsequently be eliminated by an application of the rule $T \to I$ below.

$T \to I$. From a subproof with conclusion TB_a on a hypothesis $TA_{\{\, k \,\}}$, infer $TA \to B_{a-\{\, k \,\}}$ in its immediate superproof, where $a = \{\, j, \ldots, k \,\}$ and either:

(i) $a - \{\, k \,\} = \emptyset$ with $j = k = 1$, or

(ii) $a - \{k\} = \{j, \ldots, k-1\}$ with $k \geq 2, 1 \leq j \leq k - 1$.

The conclusion and hypothesis need not be distinct in (i). In both cases, $TA \rightarrow B$ can also occur inside a structure, within a thread of proof.

$T \rightarrow E$. From TA_a and $TA \rightarrow B_b$, infer $TB_{a \cup b}$. (Direct version) From FB_a and $TA \rightarrow B_b$, infer $FA_{a \cup b}$. (Contraposed version) Whilst TA_a (or FB_a) and its conclusion $TB_{a \cup b}$ (or $FA_{a \cup b}$) are located in a proof P, either $TA \rightarrow B_b$ is in the main proof or it is located in P's immediate superproof, in accordance with the proviso below. $T \rightarrow E$ carries the proviso that either:

(i) $b = \emptyset$, in which case $a \cup b = a$, or

(ii) $a = \{k\}, k \geq 2, b = \{j, \ldots, k-1\}, 1 \leq j \leq k-1$, in which case $a \cup b = \{j, \ldots, k\}$, or

(iii) $a = \{j, \ldots, k\}, k \geq 2, b = \{j, \ldots, k-1\}, 1 \leq j \leq k-1$, and hence $a \cup b = \{j, \ldots, k\}$.

We say that $T \rightarrow E$ is applied to a proof containing $TA \rightarrow B$ into a proof containing TA (or FB) and TB (or FA). Such applications of $T \rightarrow E$ (ii) must be made en bloc (into a proof) to all the signed formulae of a (whole) structure, thereby, maintaining its common index set. However, $T \rightarrow E$ (iii) can be subsequently applied singly, i.e., to a single thread of proof, with the following exception.

For $T \rightarrow E$ (iii) to be applied for the first time into a thread of proof, it must also be applied to its adjacent thread(s) of proof. (See $F \& E$ and $T \vee E$ below for adjacent threads of proof. Also, $F \& E$ and $T \vee E$ can initiate further adjunct pairs to which $T \rightarrow E$ (iii) would also be applied in this case.) Subsequent applications of $T \rightarrow E$ (iii) into these threads of proof can then be made singly.

$T \sim I$. From FA_a, infer $T \sim A_a$.

$T \sim E$. From $T \sim A_a$, infer FA_a.

$F \sim I$. From TA_a, infer $F \sim A_a$.

$F \sim E$. From $F \sim A_a$, infer TA_a.

$T \& I$. From TA_a and TB_a, infer $TA \& B_a$. (all applied within the same thread)

$T \& E$. From $TA \& B_a$, infer TA_a.   From $TA \& B_a$, infer TB_a.

$F \& I$. From FA_a, infer $FA \& B_a$.   From FB_a, infer $FA \& B_a$.

$F \& E$. From $FA \& B_a$, infer (FA_a, FB_a). (introducing an adjacent pair of threads)

$T \vee I$. From TA_a, infer $TA \vee B_a$.   From TB_a, infer $TA \vee B_a$.

$T \vee E$. From $TA \vee B_a$, infer (TA_a, TB_a). (introducing an adjacent pair of threads)

$F \vee I$. From FA_a and FB_a, infer $FA \vee B_a$. (all applied within the same thread)

$F \vee E$. From $FA \vee B_a$, infer FA_a.   From $FA \vee B_a$, infer FB_a.

$, E$. From SA_a, SA_a, infer SA_a. (eliminating an adjacent pair of threads)

A formula A is a theorem of MMC iff $TA_\emptyset$ is provable in the main proof (with a null index set).

To convert MMC into the normalized natural deduction system NMC, we need to be able to contrapose an entire subproof in the process, and to do this we distinguish T- and F-subproofs, introduce two new rules, $F \to I$ and $F \to E$, which are contraposed versions of the corresponding $T \to E$ and $T \to I$ rules, interchange the signs T and F, and also interchange each thread with a corresponding strand. These strands are introduced in a similar way to that of threads but through $T \& E$ and $F \vee E$ and eliminated through $T \& I$ and $F \vee I$. Thus, these strands are conjunctively separated, in a similar way to the separation of threads by disjunction. (See Brady [19] for further details about how all this is done.)

As in Brady [19], any formula instance B occurring in a normalized proof of a formula A is a subformula of A, and the index set a of such a formula instance B lies in a subproof whose depth, $\max(a)$, is equal to the depth of B in A. (We take $\max(\emptyset)$ to be 0 and the depth of the main proof to be 0, to make this identity work.) Thus, using these properties, decidability can be shown for MC, as it is for DJ in Brady [17], for DW in Brady [19] and for LDW in Brady [20]. However, the quantificational logic MCQ still needs a normalised natural deduction system, and this still needs to be researched.

4.4   In Conclusion

In reference to the differing interpretations of proof theory and semantics, as presented in Section 3.1, we conclude the discussion by considering two related pairs of concepts that need separating: truth and meaning, on one hand, and semantics and technical systems used for proving results, on the other.

The content semantics shows us that meaning does not need to be expressed using truth-conditions, and this real semantics expressing the meanings of the logical words is not necessarily ideal for the proof of results, as it may not have the sharpness required to produce good technical results.

Truth and meaning drive different inferences: deductive inference and meaning containment. Deductive inference preserves truth, but is represented by a rule '$\Rightarrow$', as it concerns deduction within the formal system as a whole, whilst meaning containment is represented by a connective '$\rightarrow$', relating two sentences. (See Brady [23] for discussion on this.) It is instructive in this regard to examine negated inferences. Truth-preservation can be easily falsified when the antecedent or premise is true and the consequent or conclusion is false. It is not so easy to falsify meaning containment and there are different positions one can take on this. This can be clearly seen by considering the M1- and M2-metavaluations, where the corresponding M1-logics have no negated entailment theorems whilst for the corresponding M2-logics $\sim(A \rightarrow B)$ is a theorem iff A and $\sim B$ are theorems. And, as predicted in Slaney [46], there are likely to be other metavaluations where negated entailment theorems have different properties again. This illustrates the difficulty in pinning down meaning containment.

The true-false dichotomy provides a sharpness which is needed for proofs of technical results and so truth-theoretic semantics is ideal for such purposes. On the other hand, meaning is not so clear-cut and thus real semantics is less suitable for these technical purposes. This adds an interesting twist to the direction of this paper, giving some value to the truth-theoretic approach to semantics.

References

[1] Gerard Allwein and J. Michael Dunn. Kripke models for linear logic. *Journal of Symbolic Logic*, 58:514–545, 1993.

[2] Alan R. Anderson and Nuel D. Belnap. *Entailment: The Logic of Relevance and Necessity*, volume I. Princeton University Press, Princeton, NJ, 1975.

[3] Jc Beall, Ross Brady, J. Michael Dunn, A. P. Hazen, Edwin Mares, Robert K. Meyer, Graham Priest, Greg Restall, David Ripley, John Slaney, and Richard Sylvan. On the ternary relation and conditionality. *Journal of Philosophical Logic*, 41:595–612, 2012.

[4] Nuel D. Belnap. Display logic. *Journal of Philosophical Logic*, 11:375–417, 1982.

[5] R. T. Brady and A. Meinander. Distribution in the logic of meaning containment and in quantum mechanics. In K. Tanaka, F. Berto, E. Mares, and F. Paoli, editors, *Paraconsistency: Logic and Applications*, pages 223–255. Springer, Dordrecht, 2013.

[6] Ross T. Brady. Natural deduction systems for some quantified relevant logics. *Logique et Analyse*, 27:355–377, 1984.

[7] Ross T. Brady. Depth relevance of some paraconsistent logics. *Studia Logica*, 43:63–73, 1984.

[8] Ross T. Brady. A content semantics for quantified relevant logics – I. *Studia Logica*, 47:111–127, 1988.

[9] Ross T. Brady. A content semantics for quantified relevant logics – II. *Studia Logica*, 48:243–257, 1989.

[10] Ross T. Brady. The gentzenization and decidability of RW. *Journal of Philosophical Logic*, 19:35–73, 1990.

[11] Ross T. Brady. The gentzenization and decidability of some contraction-less relevant logics. *Journal of Philosophical Logic*, 20:97–117, 1991.

[12] Ross T. Brady. Relevant implication and the case for a weaker logic. *Journal of Philosophical Logic*, 25:151–183, 1996.

[13] Ross T. Brady. Gentzenizations of relevant logics without distribution – I. *Journal of Symbolic Logic*, 61:353–378, 1996.

[14] Ross T. Brady. Gentzenizations of relevant logics without distribution – I. *Journal of Symbolic Logic*, 61:379–401, 1996.

[15] Ross T. Brady. Semantic decision procedures for some relevant logics. *Australasian Journal of Logic*, 1:4–27, 2003.

[16] Ross T. Brady, editor. *Relevant Logics and Their Rivals, vol. II. A Continuation of the Work of R. Sylvan, R. Meyer, V. Plumwood and R. Brady.* Number 59 in Western Philosophy Series. Ashgate, Burlington, VT, 2003.

[17] Ross T. Brady. Normalized natural deduction systems for some relevant logics II: Other logics and deductions. Talks at conferences of the Australasian Association for Logic, 2003 and 2004.

[18] Ross T. Brady. *Universal Logic.* CSLI Publications, Stanford, CA, 2006.

[19] Ross T. Brady. Normalized natural deduction systems for some relevant logics I: The logic DW. *Journal of Symbolic Logic*, 71(1):35–66, 2006.

[20] Ross T. Brady. Free semantics. *Journal of Philosophical Logic*, 39:511–529, 2010.

[21] Ross T. Brady. The consistency of arithmetic, based on a logic of meaning containment. *Logique et Analyse*, 55:353–383, 2012.

[22] Ross T. Brady. The simple consistency of naive set theory using metavaluations. *Journal of Philosophical Logic*, 43:261–281, 2014.

[23] Ross T. Brady. Logic – The big picture. In J.-Y. Beziau, M. Chakraborty, and S. Dutta, editors, *New Directions in Paraconsistent Logic*, pages 353–373. Springer, Dordrecht, 2015.

[24] Ross T. Brady. The use of definitions and their logical representation in paradox derivation. Talk at the 5th Universal Logic Conference, Istanbul, Turkey, 2015.

[25] Ross T. Brady. On the law of excluded middle. Talk at the Logic Seminar, University of Melbourne, 2015.

[26] Kit Fine. Models for entailment. *Journal of Philosophical Logic*, 3:347–372, 1974.

[27] Kit Fine. Analytic implication. *Notre Dame Journal of Formal Logic*, pages 169–179, 1986.

[28] Kit Fine. Semantics for quantified relevance logic. *Journal of Philosophical Logic*, 17:27–59, 1988.

[29] Robert Goldblatt and Edwin D. Mares. An alternative semantics for quantified relevant logic. *Journal of Symbolic Logic*, 71(1):163–187, 2006.

[30] Leon Henkin. The completeness of the first-order functional calculus. *Journal of Symbolic Logic*, 14:159–166, 1949.

[31] Edwin D. Mares. Semantics for relevance logic with identity. *Studia Logica*, 51: 1–20, 1992.

[32] Edwin D. Mares. *Relevant Logic. A Philosophical Interpretation*. Cambridge University Press, Cambridge, UK, 2004.

[33] Robert K. Meyer. Arithmetic formulated relevantly. Typescript, A.N.U.

[34] Robert K. Meyer. Metacompleteness. *Notre Dame Journal of Formal Logic*, 17:501–516, 1976.

[35] Robert K. Meyer. The consistency of arithmetic. Typescript, A.N.U.

[36] Robert K. Meyer and Harvey Friedman. Can we implement relevant arithmetic? Technical Report TR–ARP–12/88, Automated Reasoning Project, R.S.S.S., A.N.U., 1988. 11 pp.

[37] Jean Norman and Richard Sylvan, editors. *Directions in Relevant Logic*. Number 1 in Reason and Argument. Kluwer, Dordrecht, 1989.

[38] W. T. Parry. Analytic implication, its history, motivation and possible varieties. In J. Norman and R. Sylvan, editors, *Directions in Relevant Logic*, volume 1 of *Reason and Argument*, pages 101–117. Kluwer, Dordrecht, 1989.

[39] G. Priest. *An Introduction to Non-Classical Logic*. Cambridge University Press, Cambridge (UK), 1st edition, 2001.

[40] G. Priest. *An Introduction to Non-Classical Logic. From If to Is*. Cambridge University Press, Cambridge (UK), 2nd edition, 2008.

[41] G. Restall. Review of *Universal Logic*. *Bulletin of Symbolic Logic*, 13:544–547, 2007.

[42] R. Routley. Problems and solutions in the semantics of quantified relevant logics. In A. I. Arruda, R. Chuaqui, and N. C. A. da Costa, editors, *Proceedings of the fourth Latin American Symposium on Mathematical Logic*, pages 305–340, Amsterdam, 1980. North-Holland.

[43] Richard Routley, Robert K. Meyer, Val Plumwood, and Ross T. Brady. *Relevant Logics and Their Rivals*, volume 1. Ridgeview Publishing Company, Atascadero, CA, 1982.

[44] P. Schroeder-Heister. Validity concepts in proof-theoretic semantics. *Synthese*, 148:525–571, 2006.

[45] P. Schroeder-Heister. Proof theoretic semantics. 2007.

[46] J. K. Slaney. A metacompleteness theorem for contraction-free relevant logics. *Studia Logica*, 43:159–168, 1984.

[47] J. K. Slaney. Reduced models for relevant logics without WI. *Notre Dame Journal of Formal Logic*, 28:395–407, 1987.

[48] Alasdair Urquhart. The undecidability of entailment and relevant implication. *Journal of Symbolic Logic*, 49:1059–1073, 1984.

 Received 13 July 2016

A Preservationist Perspective on Relevance Logic

Bryson Brown

Department of Philosophy, University of Lethbridge, Lethbridge, AB, Canada
<brown@uleth.ca>

Abstract

A common way to tame an apparent inconsistency is to re-interpret the conflicting commitments. While $\{\alpha, \neg\alpha\}$ is obviously inconsistent, if α is ambiguous, the inconsistency may be merely apparent. This essay presents two distinct but closely related ways of formalizing this obvious point, one already proposed and the other a variant. Both are closely related to the well-known paraconsistent logics LP and FDE. The paper concludes by suggesting ways to extend the application of formal ambiguity to interpret Routley's "star worlds" and suggest an alternative approach to relevant conditionals.

Keywords: aggregation, ambiguity, ambiguity measures, paraconsistency, preservationism, property tables, relevance

1   Introduction: Inconsistency or Ambiguity?

Preservationism is a distinctive approach to non-classical logic. Rather than propose heterodox treatments of truth values, preservationists have sought new properties of sentences and ensembles of sentences which are worth preserving, and developed logics that preserve them. The first such value to be investigated is called the *level* of a premise set, defined as the minimum number of cells in a *consistent covering* of a premise set, that is, a set of sets whose union is the premise set and each member of which is consistent. Level preservation weakens logical *aggregation*, the way in which premises are "gathered together" to arrive at consequences that follow from a set of premises but from no individual premise in the set alone. In particular, repeated application of the classical rule ∧-intro produces single sentences that are consistent only so long the finite collection of sentences "conjoined" is collectively consistent. Consequently, ∧-intro allows us to "gather together" any finite set of inconsistent but non-contradictory premises to produce a contradiction. However, this violates level-preservation, since the maximum level of any such finite set is the cardinality of the set, while any set including a contradiction has no level, not even the countably infinite level. (As an example of the latter, take the level of the set of all atoms together with their negations.[1] See [7].)

One result that has had an important influence on preservationist thinking goes back to Dana Scott. In [8], Dana Scott proves that any reflexive, monotonic and transitive (RMT) relation on a set can be given a 1/0 semantics. That is, such relations are determined by a set of "allowed" 1/0 valuations on the members of the set. A preservationist take on Scott's result grows out of a simple observation: it follows from Scott's result that there is some property of items in the domain which is preserved by any RMT relation. This observation leaves it open to philosophers and logicians to explore just what property it is that a given such relation preserves, and to find new consequence relations by seeking new properties of our premises and our conclusions that are *worth* preserving.

The motive for seeking such new properties is particularly clear when we consider classically inconsistent premise sets and classically trivial conclusion sets (such as sets all of whose members cannot be *consistently denied*). Such sets are classically trivial, because they lack the only properties classical logic seeks to preserve (from left to right in the case of premise sets, and right to left in the case of conclusion sets). But we can be a little more precise about how we express this. Logic in general, and classical logic in particular, preserves consistency in a very conservative way. For example, the closure of a premise set Γ under the classical $\vdash$ is the set of all sentences that are classically consistent extensions of every consistent extension of Γ. Inversely, the "closure" of a conclusion set Δ under the classical $\vdash$ is the set of all sentences that are consistently deniable extensions of every consistently deniable extension of Δ.

The preservationist project has been summed up in a pair of slogans.

Hippocrates: Don't make things worse. (P. K. Schotch)

Making do: Find something you like about your premises, and preserve it.

(R. E. Jennings)

The two slogans fit together nicely — following the second prescription is a good way of obeying the first. There are many properties we might value in a set of premises Γ, some of which aren't, or aren't always, preserved when we close Γ under the classical $\vdash$. Preserving such properties ensures we don't make things worse when we *reason* with those premises.

---

[1]To provide a label for the (non-)level of sets including contradictions, Schotch and Jennings simply defined ∞ as the "level" of such sets.

For instance, closure under ⊢ includes closure under conjunction. However, one valuable property of a set of sentences, considered as reasonable grounds of action, is having only members with a probability higher than some threshold of "minimal acceptability." Closure of a consistent set under conjunction is consistency preserving, but probability degrading, as Henry Kyburg famously observed in [5].

Keeping these broader observations in mind, we begin our voyage here by focusing on *ambiguity measures*, which measure properties of sets of sentences in a classical propositional language, whose *preservation* gives us the consequence relations of LP, $K3$ and FDE. Afterwards, we turn to some more speculative points about preservation and relevance.

2 Ambiguity and Ambiguity Measures

Ambiguity is a wonderful device for rhetoricians and other tricksters, but it turns out that we can also perform some useful logical tricks with the help of ambiguity. Beginning with sentential logic, we can simply treat different instances of some atom(s) as having different truth values. Semantically, this allows us to freely assign different truth values to different instances of those atoms, treating some as true and others as false. I call this unconstrained form of ambiguity "chaotic," because it treats instances of the ambiguous atoms as if they were just independent atoms. Each different instance of such atoms within a wff (i.e., a formula) and even in different instances of the very same wff can have different values. The results, as far as which instances of wffs including these atoms are assigned the value 1 and which are assigned the value 0 goes, can be "captured" by appealing to a single pair of new independent atoms, one assigned the value 1 and the other assigned the value 0, with different instances (including those appearing in different inscriptions of the very same wff) semantically "disambiguated" by assigning, in whatever way we like, 0 to some and 1 to others.

However, this extreme form of ambiguity is described here only to provide a kind of limit case against which more constrained applications of ambiguity and ambiguity measures can be contrasted. From a logical perspective, it is much more interesting to exploit ambiguity in *systematic* ways. For example, we can exploit the hypothetical ambiguity of some atoms with the aim of assigning (we might say *imposing*) either the value 1 or the value 0 to the wffs that include those ambiguous atoms.

Consider the first approach. Instances of an ambiguous atom ϕ as a stand-alone wff will be assigned the value 1, while instances of ϕ in the wff $\ulcorner \neg \phi \urcorner$ are assigned the value 0. More generally, instances of ϕ in wffs of the form $\ulcorner \neg \cdots \neg \phi \urcorner$ would be assigned 1 if the number of '¬'s is even, and 0 if the number of '¬'s is odd, while instances in $\ulcorner(\phi \vee \psi)\urcorner$ and $\ulcorner(\phi \wedge \psi)\urcorner$ would be assigned the value 1. In the first of these last two cases, this is enough to ensure that the wff as a whole gets the value 1, but (of course) in the second the value of the wff as a whole still hangs on the value of ψ, with the complete wff $\ulcorner(\phi \wedge \psi)\urcorner$ being assigned the value 1 iff ψ also gets the value 1.

If every atom in our propositional language L is treated ambiguously in this way, then every sentence in L is assigned the value 1, while when no atoms are treated ambiguously, we get only the familiar classical valuations. To see what happens in between these extremes, we need to examine how the ambiguous atoms contribute to the values assigned to various wffs systematically. But the resulting valuations on our propositional language L turn out to be quite familiar. Consider an LP valuation assigning one of the values T, F or B ("both") to each atom. The unambiguous atoms are assigned the fixed values 1 and 0, respectively, in our ambiguous valuation, while B is reserved for the ambiguous atoms. It turns out that the initial agreement between our ambiguous valuation and the corresponding LP valuation, concerning which *atoms* receive a designated value, extends to all the sentences in the language. (See [1].)

To see why this is so, we need only reflect on how the values we assign to instances of the atoms affect the values assigned to wffs including connectives. Of course, if an atom ϕ is unambiguous, then either all instances of ϕ are assigned the value 1 or all are assigned the value 0. Now suppose the atom ϕ is ambiguous. Then both ϕ and $\ulcorner\neg\phi\urcorner$ will receive the value 1 in our valuation, because disambiguating explicitly, we treat ϕ by itself as an instance of the "true" atom ϕ^+, while the instance of ϕ appearing in $\ulcorner\neg\phi\urcorner$ is treated as an instance of the "false atom" ϕ^-. We thereby impose the value 1 on both wffs, and similarly, for all wffs of the forms $\ulcorner\neg\cdots\neg\phi\urcorner$. $\ulcorner(\phi \wedge \psi)\urcorner$ is assigned the value 1 if and only ϕ and ψ are both assigned the value 1, whether ambiguously or unambiguously. $\ulcorner(\phi \vee \psi)\urcorner$ is assigned the value 1 if and only if at least one of ϕ and ψ is assigned the value 1, whether ambiguously or unambiguously.

The only complexity we need to worry about here is how the property of being "ambiguous" in this way propagates through the language. It's obvious that LP and the corresponding ambiguity valuation match in the case of '¬', that is, on which wffs in the set of atoms closed under the operation of prefixing a '¬' are assigned a designated value. But what about '∧' and '∨'? Given an ambiguous assignment assigning one of the fixed values 1 and 0 to some atoms and treating the rest as ambiguous, we will say that an arbitrary wff in our ambiguous assignment has the *property* B (which we no longer think of as a "truth value"), when it has the property that all and only the ambiguous atoms have. That is, if the wff, its negations, the negations of those negations and so on are all assigned the *value* 1. Similarly, an arbitrary wff ψ will be said to have the *property* T iff $V(\psi) = 1$ and $V(\ulcorner\neg\psi\urcorner) = 0$, and it will have the *property* F iff $V(\psi) = 0$ and $V(\ulcorner\neg\psi\urcorner) = 1$. We can then construct *property tables* as a simple generalization of truth tables, using our three properties, denoted by T, F and B. As indicated above, the ambiguity of the wildcard atoms is exploited to assign the value 1 to the wffs including them whenever possible. We thus arrive at the following familiar tables for these *properties* of wffs:

ϕ	$\neg\phi$
T	F
B	B
F	T

$\wedge$	T	B	F
T	T	B	F
B	B	B	F
F	F	F	F

$\vee$	T	B	F
T	T	T	T
B	T	B	B
F	T	B	F

The truth functions here are just classical. Since every wff with the property T (F) gets the value 1 (0) in all instances, the lines in the tables for T and F are purely classical. The table for negation is obvious, given how ambiguity is applied here. For the rest, a wff ϕ is assigned the property B if and only if (given the stable "background" assignment of 1 or 0 to each "normal" atom) instances of the ambiguous atoms can be assigned the values 1 or 0 in ways that impose 1 on both ϕ and $\ulcorner\neg\phi\urcorner$. Other wffs receive 1 or 0 as a *fixed* value, so they all have one of the properties T and F. Tracing this through for '$\wedge$', it's easy to see (by putting in values for wffs) why $\ulcorner(B \wedge T)\urcorner$, $\ulcorner(T \wedge B)\urcorner$ and $\ulcorner(B \wedge B)\urcorner$ all have the property B: the negations of $\ulcorner(B \wedge T)\urcorner$, $\ulcorner(T \wedge B)\urcorner$ and $\ulcorner(B \wedge B)\urcorner$ all get the value 1, because we can impose 0 as the value of the instance of the sub-wff(s) with the property B, thereby, imposing the value 1 on $\ulcorner\neg(B \wedge T)\urcorner$, $\ulcorner\neg(T \wedge B)\urcorner$ and $\ulcorner\neg(B \wedge B)\urcorner$. Similarly, it's obvious why whenever either ϕ or ψ has the property F, $\ulcorner(\phi \wedge \psi)\urcorner$ also has the property F. Finally, since the connectives here are just the classical two-valued connectives, $\ulcorner(\phi \vee \psi)\urcorner$ can simply be defined as $\ulcorner\neg(\neg\phi \wedge \neg\psi)\urcorner$. The upshot is that in any ambiguous valuation the *property* B is coextensive with the LP truth value 'both' in the three-valued LP valuation satisfying all the sentences assigned the value 1.

Given the symmetries between LP and $K3$, it follows immediately that exploiting ambiguity in the opposite way, i.e., using it to allow us to assign 0 to both ϕ and $\ulcorner\neg\phi\urcorner$ and more generally to assign the value 0 to any wff including some ambiguous atoms whenever that can be done given the "background" of a fixed assignment to the rest of the atoms, gives us an ambiguity semantics for $K3$.

Finally, combining both uses of ambiguity (applying them to a pair of disjoint ambiguity sets, one used to impose the value 1 whenever possible while the other is similarly used to impose the value 0) produces an ambiguity semantics for first degree entailment (FDE). A wff ϕ is assigned the property T, B, N ("none") and F, respectively, when ϕ has the value 1 while $\ulcorner\neg\phi\urcorner$ has the value 0, when ϕ has the value 1 while $\ulcorner\neg\phi\urcorner$ also has the value 1, when ϕ has the value 0 while $\ulcorner\neg\phi\urcorner$ also has the value 0 and when ϕ has the value 0 while $\ulcorner\neg\phi\urcorner$ has the value 1.

At this point, some explanation of the motives that lurk behind this approach to LP and FDE is due. First, drawing a line between the truth values 1 and 0 and the "properties" T, B, N and F focuses attention on the preservationist suggestion that a consequence relation can preserve properties other than truth values. This point is also connected to Scott's result, noted above: broadly speaking, any property whose domain is the set of sentences of some language L can be used as the basis for a consequence relation that *preserves* that property. Of course, it's much easier to determine what pairs of subsets of L the consequence relation $\vDash_L$ holds for when we are given rules for assigning the property to some proper subset of the sentences (we can call these the "atoms") and recursive rules for extending such assignments to all of L. More practically, we often explicitly specify a class of valuations based on an assignment to some finite subset of the atoms, and the values of the sentences of L all of whose atoms are assigned a determinate value in every member of that class, i.e., the sentences whose construction requires only those atoms.

What these ambiguity semantics are meant to show here is that we can give very different accounts of what the *properties* preserved by a given consequence relation really are. In the ambiguity semantics for FDE sketched above, the properties T and F involve a *stable* relation between a wff and the truth values 1 and 0 (that is to say, the relation is fixed regardless of whether the wff appears on its own or as a sub-wff in any other wff), while the properties B and N do not.

A nearby observation goes a step further. There is not just an abstract symmetry but a rich *similarity* between the properties B and N: both reflect the ambiguity of the truth-value assigned to instances of some atom(s). Though this ambiguity is applied quite differently in generating a particular evaluation on L, the same *semantic resource* of ambiguous atoms could be used to generate a range of different evaluations, based on different choices of how to use the ambiguity of each atom. For instance, whether it is used positively, in an effort to impose the value 1 whenever possible, or it is used negatively, to impose the value 0.

This gives us a very different picture of what is going on with FDE than the picture we gather from either Dunn's four-valued semantics in [4] or his semantics where the valuation function is weakened to a relation, e.g., in [3]. Our ambiguity approach is based solely on the familiar truth values and classical truth-functional semantics for the connectives, and simply proposes a new way for atoms and sentences to be "connected" to truth values, that is, systematically but (in general) ambiguously, rather than univocally. As we've seen, on this approach the difference between the properties B and N is just a matter of how the ambiguity of individual ambiguous atoms is *exploited*.

Given a list of atoms to be treated ambiguously, we can consider separately just how the ambiguity of each is to be used when it comes to producing a 1/0 assignment for the sentences of L. This perspective makes the connection between a world w and its Routley $*$-world much closer than it looks to be in Dunn's semantics. Rather than involving two distinct valuations, in which the values B and N "trade places," ambiguity semantics produces $*$-worlds or $*$-interpretations by changing how we apply (or exploit) the ambiguity of certain atoms. More broadly speaking, from this perspective it is also natural to consider the full range of valuations that result from all the ways of combining our two different *uses* of ambiguity to each ambiguous atom, producing every 1/0 valuation that results from treating some of our ambiguous atoms as having the property B and the rest as having the property N.

3   Ternary Remarks

Here we turn to some more speculative remarks about how this preservationist perspective on FDE might contribute to the interpretation of relevance logics. Since the ternary semantics for relevance logics were first proposed, the question of how to interpret the ternary relation has been a hard nut to crack. In the usual treatment, we have a modal frame in which every world combines an FDE valuation with a modal accessibility relation to deal with the semantics of '$\rightarrow$'. However, while a subset of the worlds (the *normal worlds*) have a binary accessibility relation, the rest (the *non-normal worlds*) have a ternary accessibility relation and a modified truth condition for '$\rightarrow$'. Namely, at a non-normal world w, $\ulcorner \phi \rightarrow \psi \urcorner$ gets the value 1 *iff* every pair of worlds w', w'' such that $Rww'w''$ is such that, if $\vDash_{w'} \phi$, then $\vDash_{w''} \psi$. Finally, the theorems of these logics are the sentences that hold in all *normal* worlds in all models, not in all the worlds.

Since the underlying "truth-functional" logic is FDE there are no theorems at all in the sub-language of the atoms and wffs formed from them using '$\neg$', '$\wedge$', and '$\vee$'. On the heterodox version of the semantics for FDE above, any such wff can have the property N at some world, ensuring that it gets the *value* 0. However, there are still theorems, all involving '$\rightarrow$'.

So the starting point for us here is a "space" of worlds at which the sentences of L are assigned FDE valuations, produced according to our ambiguity semantics.

But before we begin to think about semantics for '$\rightarrow$', we need to consider the place of $*$-worlds in our proposal. The inclusion of these worlds in the set of valuations on L making up the "frames" of a modal conditional logic is standard practice in the semantics of relevance logic, and their presence in a modal frame for the '$\rightarrow$' makes a difference to the truth and falsity of conditionals, even though the star world semantics for negation actually gives us the same possible valuations as the 4-valued Dunn semantics and our ambiguity semantics. This is because the star world semantics for negation ensures the presence, in every modal frame, of a "mirror image" world w^* for each world w, at which the properties B and N "trade places."

As we've noted, however, this sort of "switch" or reversal of B and N is a natural fit from the ambiguity perspective. We capture it by allowing the roles of the two ambiguity sets to be ambiguous in their own right, that is, for every world (i.e., for every valuation in the frame) with left and right ambiguity sets L and R there is an otherwise indistinguishable world where the set of atoms L is the right ambiguity set and the set R is the left ambiguity set. (A broader approach would begin with a single set of ambiguous atoms and generate valuations for every possible allocation of those ambiguous atoms to the properties B and N; adopting this perspective would constrain the allowed frames even further.)

As in the usual semantics, the "normal" worlds have a binary accessibility relation, but there is also a set of "abnormal" worlds, which have a ternary accessibility relation. The theorems of the resulting logic are determined by the sentences true at all normal worlds. However, since a normal world can have access to an abnormal world, the accessibility relation determining values for wffs of the form $\ulcorner(\phi \rightarrow \psi)\urcorner$ is, like that for Lewis's '$\rightarrow$', binary at the first step, but unlike Lewis's '$\rightarrow$', becomes ternary thereafter.

A difficult (and persistent) interpretive question for this semantics has been the following: How should we understand the ternary accessibility relation? An answer suggested by the role of ambiguity in the proposed semantics for FDE is that it reflects an ambiguity of the *accessibility relation* at remote worlds that are greater than one step away. This provides a different gloss on the semantics, by introducing a new kind of constrained semantic ambiguity, which is ambiguity of the accessibility relation when "looking" more than one step along the relation. On such an account, every world unambiguously picks out the worlds it has access to, but the worlds those worlds have access to are "seen through a glass dimly," with the result that when we evaluate a conditional $\ulcorner\phi \rightarrow \psi\urcorner$ at a world accessible to w, the worlds at which we evaluate ϕ differ, in general, from the worlds at which we evaluate ψ. We can think of this as analogous to a quantum measurement, where in general, making a measurement changes the state of the system measured. Similarly here, when ϕ is evaluated at a world w_2 two steps away from w_0, the "path" along the frame relation that led from w to w_2 is altered at its second step, so that when ψ is evaluated, the path now leads to a different world, $w_{2'}$.

The upshot, in logical terms, is the same as the usual semantics for B. The evaluation of $\ulcorner \phi \to \psi \urcorner$ at a world w', one "step" along the frame relation from a world w, is based on evaluating ϕ and ψ at related pairs of worlds, each pair consisting of a first world w_2 accessed along that path, at which ϕ is evaluated, and a second world accessed along (what we regard as) *the same path*, $w_{2'}$, at which ψ is evaluated. This ambiguity in the accessibility relation allows failure of $\ulcorner \phi \to (\psi \to \psi) \urcorner$ at a world w for the same reason that this wff can be false in the usual semantics for B. In the latter, there can be an accessible world w' at which, due to the ambiguity of its accessibility relation as "seen" from the starting point of w, when we check for the value of ψ as antecedent we get the value 1, but when we check again for the value of ψ as consequent, we access a different world where ψ gets the value 0.

This approach fits nicely with Priest's tableau method, as presented in [6, 190f], where we only need one normal world in a counter-model, since after that all worlds introduced are assumed to be non-normal (i.e., to have ternary accessibility relations). But here, interestingly, we get the same effect by starting with any world w in the frame; any accessible world w' will have an ambiguous accessibility relation *from the point of view of w*. The advantage of this reading is subtle, because we can say that it merely imposes the same oddity in a different way, however, it does it without needing to distinguish between *normal* and *abnormal* worlds. All worlds are normal on this account, and the ambiguity of the frame relation (which, in general, picks out different valuations when testing the antecedent's value at an accessible world than when testing the consequent's value at that same accessible world) arises due to the ambiguity of what world we "encounter" along the frame relation, whenever we look two or more steps along the frame away from the starting point of our evaluation.

The upshot is just a minor variation or gloss on the standard semantics for B, and by itself, is obviously not very significant. It might well be argued that adopting such an ambiguous or unstable frame relation is a stranger and more radical change to our understanding of the semantics of conditionals than the ternary relation. However, a further step takes us closer to the familiar realm of binary modal frames and "relative possibility." We can set aside ambiguity of the frame relation, in favour of a kind of "instability" in the valuations we find at points of the frame, with the advantage of dropping the distinction between normal and non-normal worlds altogether. We've already identified a very close link between the property B and the property N, by grounding both in ambiguity. The only difference between these properties lies in how the ambiguity is exploited to generate 1/0 valuations on L. The fundamental character that underlies the properties B and N here is the ambiguity of certain atoms. On this proposal, all that changes between the pairs of worlds accessible to a given world w in the ternary case is a switch (or, more generally, an alteration or instability) of which ambiguous atoms are treated as having the property B and which are treated as having the property N. Allowing such changes in the valuations we encounter at worlds two or more steps along the accessibility relation, when testing for the truth values of the antecedent and the consequent of a conditional, would have the same effect as the standard ternary relation, under the condition that the two worlds related to our w be such that (partly or fully) exchanging the *values* B and N in the standard semantics for FDE in each of the two worlds transforms each into the other. If we restrict our consideration to exchanging the property B for the property N and vice versa for every wff, the effect on our 1/0 valuations would be equivalent to requiring any two worlds accessed via the ternary relation to be a pair of worlds that bear the $*$-world relation to each other.

In either case, the resulting conditionals pass the first test for relevance. This is easy to see, since the sentence $\ulcorner \phi \rightarrow (\psi \rightarrow \psi) \urcorner$ is invalidated at a world w such that, at some accessible ambiguous world w', ψ gets the property B (so that $\vDash_{w'} \psi$) as antecedent and the property N (so that $\nvDash_{w'} \psi$) as consequent. The upshot of this approach seems likely to be close to B, but I suspect there may be different interactions between negations and conditionals here, since this approach effectively grounds both in $*$-worlds.

4 Conclusions

Preservationist logic arose from the recognition that the preservation of properties other than consistency and satisfiability can be used to constrain the consequences of classically trivial premise sets (see [7]). The preservation of ambiguity measures (as described in [1, 2]) provided a link between the preservationist approach and the multi-valued paraconsistent logics LP and FDE. Here we have presented a new treatment of ambiguity logic that makes the link to these logics more direct by defining the *properties* preserved by the $K3$, LP and FDE consequence relations — by appeal to ambiguity measures. We have also suggested a preservationist, ambiguity-based reading of "star-worlds" and the ternary semantics for relevance logics. How far this approach can be extended remains an open, and, I hope, an interesting question.

References

[1] Bryson Brown. Yes, Virginia, there really are paraconsistent logics. *Journal of Philosophical Logic*, 28:489–500, 1999.

[2] Bryson Brown. Ambiguity games and preserving ambiguity measures. In Peter Schotch, Bryson Brown, and Raymond Jennings, editors, *On Preserving: Essays on Preservationism and Paraconsistent Logic*, pages 175–188. University of Toronto Press, 2009.

[3] J. Michael Dunn. Intuitive semantics for first-degree entailments and 'coupled trees'. *Philosophical Studies*, 29:149–168, 1976.

[4] J. Michael Dunn. A sieve for entailments. *Journal of Philosophical Logic*, 9: 41–57, 1980.

[5] Henry E. Kyburg Jr. Conjunctivitis. In M. Stone, editor, *Induction, Acceptance and Rational Belief*, pages 55–82. D. Reidel, Dordrecht, 1970.

[6] Graham Priest. *An Introduction to Non-Classical Logic*. Cambridge University Press, 2nd edition, 2008.

[7] Peter K. Schotch and Raymond E. Jennings. On detonating. In R. Sylvan, G. Priest, and J. Norman, editors, *Paraconsistent Logic: Essays on the Inconsistent*, pages 306–327. Philosophia Verlag, München, 1989.

[8] Dana Scott. On engendering an illusion of understanding. *Journal of Philosophy*, 68(21):787–807, 1971.

Received 13 November 2016

Ambiguity and the Semantics of Relevant Logic

Nicholas Ferenz*

Department of Philosophy, University of Alberta, Edmonton, AB, Canada
<ferenz@ualberta.ca>

Abstract

The American Plan semantics for relevant logic attempts to avoid using the star world operator of the Australian Plan semantics for relevant logic because of the philosophical issues surrounding star worlds. However, on one natural approach to creating an adequate four-valued semantics, the American Plan semantics requires dual worlds that behave in ways similar to star worlds [11]. Because of the differences between dual and star worlds, I claim that informal reasoning with non-classical truth values motivates the existence and nature of dual worlds. Moreover, I show that this account of informal reasoning can be formally modeled in such a way as to be compatible with a variety of metaphysical views.

Keywords: American Plan semantics, four-valued, negation, relevant logic, semantics

1 Introduction

The Australian Plan for the semantics of relevant logics is a possible worlds structure in which each world is two-valued [9, 10]. That is, every sentence receives either the truth value True or the truth value False. While the Australian Plan might be appealing for this reason, the set of worlds of every model is closed under the Routley star operator. There are philosophical difficulties with the star operator, some of which may be found in [4]. The American Plan semantics, as found in [12, 11, 7], seemingly avoids these philosophical concerns by not including a star operator in its formal semantics. In the American Plan, the possible worlds are four-valued; every sentence receives one of the truth values True, False, Both and

---

*I would like to thank Katalin Bimbó for organizing the *Third Workshop*. I would also like to thank the anonymous reviewer, whose insightful comments led to the improvement of this work.

Neither. Unfortunately, allowing the two extra, non-classical truth values into the picture did not fully solve the problem of the Routley star. As pointed out by Restall, the American plan semantics, on one natural construction, must be *closed under duality* [11, p. 148]. That is, for every world in a model, its dual world is also in the model. Here, dual is meant something like a maximally compatible world such that for every sentence A, if a world makes A true (false), its dual does not make A false (true). The requirement that the models are closed under duality raises its own interesting set of philosophical issues, but these issues are different from those raised by the star operator of the Australian Plan. I offer a solution to the philosophical problems posed by the dual worlds.

The formal properties of dual worlds will be explained by an account of informal reasoning. I shall write $w \vDash A$ to denote that A *is true at* w, and $w \vDash \neg A$ to denote that A *is false at* w (with the model M being implied). We say that a wff A receives the truth value True if and only if A is true and not false. The wff receives the truth value Both if it is both true and false. The cases for the truth values False and Neither are similar. The formal properties of dual worlds, where the dual of w is written as w^*, are defined as follows:

$$w \vDash \alpha \text{ and } w \nvDash \neg\alpha \quad \text{iff} \quad w^* \vDash \alpha \text{ and } w^* \nvDash \neg\alpha$$
$$w \nvDash \alpha \text{ and } w \vDash \neg\alpha \quad \text{iff} \quad w^* \nvDash \alpha \text{ and } w^* \vDash \neg\alpha$$
$$w \vDash \alpha \text{ and } w \vDash \neg\alpha \quad \text{iff} \quad w^* \nvDash \alpha \text{ and } w^* \nvDash \neg\alpha$$
$$w \nvDash \alpha \text{ and } w \nvDash \neg\alpha \quad \text{iff} \quad w^* \vDash \alpha \text{ and } w^* \vDash \neg\alpha$$

What is in need of explanation is (1) the formal properties as defined and (2) the requirement that the four-valued models are closed under duality. I will show that both (1) and (2) can be explained by starting with an informal account of reasoning. The explanandum (2) can be refined, as the dual worlds are *only* required in a single valuation condition in the models. I will show that the informal reasoning supports this somewhat puzzling fact.

I have extended the ambiguity-measure preservation semantics of Bryson Brown, which he defined for **FDE**, to a possible worlds semantics for the relevant logic **B** and its extensions, in [5]. In Brown's semantics for **FDE**, sentences receiving either the truth value Both or the truth value Neither are considered ambiguous, and disambiguating these sentences produces classical models. The worlds in the models I have defined are given an interpretation which I believe captures informal reasoning about non-classical truth values, or, in this case, a specific kind of ambiguity. The informal reasoning will show that our intentions concerning the truth value of a conditional sentence can change depending on whether we are trying to determine if the conditional is true or if it is false. This fact will explain the required existence of dual worlds in the models. The nature of dual worlds will be determined by how we reason given these differing intentions.

In Section 2, I will focus on the role of ambiguity in my models. My quite idiosyncratic use of the term 'ambiguity' will be explained, for I use the term to describe an attitude towards sentences. While what could be called 'disambiguations' in my account are similar to David Lewis' disambiguations in his "Logic for Equivocators" [6], I will distance myself from Lewis by highlighting the differences in motivation and in what is called ambiguous on each account. I will, however, wait until my semantic models are explicated before fully discussing the differences between my account and Lewis' "Logic of Equivocators." That being said, the primary difference between the accounts is found in what is called 'ambiguous' by each account. On Lewis' account ambiguity is more or less the typical ambiguity in natural language, and an ambiguous sentence is only true or false *after disambiguation*. In contrast, I call a sentence ambiguous when we treat it as having one of the non-classical truth values of the four-valued semantics. That is, when we have reason to assert (or deny) both a sentence and its negation. In Section 2, I will also illustrate how my use of the term differs from Brown's by demonstrating how his models for **FDE** work. In Section 3, I will make the further distinction between three kinds of ambiguity (as I use the term). My models for **B** and its extensions will be reproduced in Section 4. Finally, in Sections 5 and 6, I will offer an account of informal reasoning and demonstrate that the models in Section 4 capture the essence of the account. By showing that the models capture the informal reasoning, I will explain the nature and existence of dual worlds.

While I will use the terms 'world' and 'possible world', I believe that the account given here is compatible with a use of other terms in the literature. As the particular term used has little bearing on my arguments here, I will not discuss the potential differences between using each term.

2 Ambiguity

Sentences receiving either of the non-classical truth values (Both and Neither) will be called ambiguous. I stress that this term is meant epistemically and not metaphysically. J. Michael Dunn noted that the use of the word 'ambiguity' might suggest that the interpretation of relevant logic which I propose would eliminate the possibility that a sentence is, say, both true and false, and that calling something both true and false is a mistake.[1] However, I only wish to call sentences ambiguous when we have cause to believe that they have non-classical truth values. We might have a cause

---

[1]Dunn suggested the phrase "the logic of politeness" for this reason.

to believe a sentence is both true and false, *because it is* both truth and false. Other times our epistemic limits might warrant the assignment of non-classical truth values. In this way, I aim to remain metaphysically neutral. Thus, the term 'ambiguity' is perhaps not well suited to my aims. While I will continue to use the term as is, my intention is to use the term to describe sentences that we have cause to assign one of the non-classical truth values Both and Neither. Nevertheless, I believe my argument works whether or not we call some sentences ambiguous. The argument relies only on what it means to treat a sentence as having the truth value Both or Neither.

The models I have defined for **B** extend Brown's ambiguity-measure preservation semantics for **FDE** by creating **FDE**-like possible worlds and relating them with the usual ternary relation. (It was my original hope that the formal properties of ambiguous sentences could help explain the ternary relation, but I have so far been unable to use ambiguity in this way.)

Brown's models work as follows. A consequence relation that preserves levels of ambiguity is defined. Here ambiguity is used to describe sentences which must undergo a formal disambiguation in order to be treated classically. That is, an ambiguous sentence P, e.g., one that is treated in **FDE** as both true and false, must be disambiguated by splitting its instances into two sentences, P_t and P_f, such that the former is assigned the truth value True and the latter is assigned the value False. In this way, we are able to consistently (and in some sense classically) model sentences such as $A \wedge \neg A$. For example, if A is a propositional variable, then we may model $A \wedge \neg A$ by treating the first instance of A as true and the second as false by splitting the instances of A into instances of A_t and A_f. Doing so produces the new sentence $A_t \wedge \neg A_f$. This sentence has a classical model.

Given any set of sentences Γ, Brown considers the possible ways of producing a consistent image (or consistently deniable image) of Γ. Brown defined a three place relation:

> A set of formulae, Γ', is a consistent image of Γ based on A (which we write $\mathrm{ConIm}(\Gamma', \Gamma, A)$) iff A is a set of atoms, Γ' is consistent, and Γ' results from the substitution, for each *occurrence* of each member α of A in Γ, of one of a pair of new atoms, α_f and α_t. [3, p. 176]

For example, if Γ is already consistent, then $\mathrm{ConIm}(\Gamma, \Gamma, \emptyset)$ states that by treating no sentences as ambiguous we produce a consistent image of the already consistent Γ. However, we also have $\mathrm{ConIm}(\Gamma, \Gamma, A)$, if Γ is already consistent, for treating additional sentences as ambiguous does not produce inconsistency. We should, therefore, confine our interest to the smallest A's that produce consistent images of a set of sentences. Brown defines an *ambiguity set*:

[t]he ambiguity set of Γ, $\mathrm{Amb}(\Gamma)$ [is]

$$\{\, A \mid \exists \Gamma'\colon \mathrm{ConIm}(\Gamma', \Gamma, A) \wedge \forall A', A' \subseteq A, \neg \exists \Gamma'\colon \mathrm{ConIm}(\Gamma', \Gamma, A') \,\}$$

This is the set of smallest sets, $A_1, \ldots, A_n$, where for each A_i there is some Γ' such that $\mathrm{ConIm}(\Gamma', \Gamma, A_i)$.   [3, p. 176]

The level of ambiguity of a set of sentences, then, is the smallest number of atomic propositions which must be treated as ambiguous in order to make the set of sentences classically consistent.

Brown then defines a notion of an acceptable extension that preserves the level of ambiguity by ensuring that no acceptable extension requires treating additional atomic sentences as ambiguous in order to create a consistent image. For any set Γ, the ambiguity sets of its extensions will be subsets of its own ambiguity sets. Formally, this is symbolized as follows.

$$\mathrm{Accept}(\Delta, \Gamma) \ \Leftrightarrow\ \Gamma \subseteq \Delta \ \&\ \mathrm{Amb}(\Gamma \cup \Delta) \subseteq \mathrm{Amb}(\Gamma).   [3, p.\ 177]$$

The notion of an acceptable extension plays a key role in Brown's construction. Brown shows that we can create a consequence relation by quantifying over all acceptable extensions of a set. That is, $\Gamma \vdash_{amb} \alpha$ if and only if, for every extension Γ' of Γ, adding α to Γ' produces an acceptable extension of Γ', i.e., $\mathrm{Accept}(\Gamma' \cup \{\, \alpha \,\}, \Gamma')$. It turns out that this consequence relation is the same as the consequence relation of **LP** [3, p. 177].

From here, Brown shows how we can dualize the notion of a consistent image to that of a consistently deniable image. Treating some atomic sentences as ambiguous, by treating some instances of the atomic sentences as new, true sentences and the rest as new, false sentences, enables us to project a consistently deniable image of a set of wffs. Doing this ensures that inferences to tautologies are not trivially valid. Brown defines the dual notion as:

Let $\mathrm{Amb}^*(\Delta)$ be the set of minimal sets of sentence letters whose ambiguity is sufficient to project a consistently deniable image of Δ. We require that any sentence from which Δ follows be an acceptable extension of every acceptable extension of Δ, where acceptability is now consistent deniability.   [3, p. 181]

We write the dual notion of an acceptable extension as $\mathrm{Accept}^*(\Gamma, \Delta)$. With this notion of acceptable extension in place, we can again define a consequence relation. $\Gamma \vdash_{amb^*} \Delta$ if and only if every acceptable extension of Δ is such that adding Γ produces an acceptable extension.

The trick to producing a consequence relation equivalent to that of **FDE** using these two defined consequence relations is quite simple.

Then $\Gamma \vdash_{FDE} \Delta$ if and only if every such consistent image of Γ can be consistently extended by some member of each non-trivial image of Δ based on a disjoint set of sentences letters. [3, p. 182]

In other words, an argument is **FDE**-valid if and only if every way of disambiguating the premise set to produce a consistent image classically implies every consistently deniable image of the conclusion set.

An atomic sentence is ambiguous on Brown's account when we have to treat some of its instances as true and others as false in order to consistently assert or deny a set of sentences. With his notion of ambiguity, the logics **FDE**, **LP**, and others can be adopted by someone who denies that any sentence can have non-classical truth values. One of Brown's intentions was to show that paraconsistent logics can be adopted by non-dialetheists (cf. [2]). Brown's ambiguity-based approach is, therefore, metaphysically light, as it is compatible with many different metaphysical views. This fact was what motivated me to extend the ambiguity approach to relevant logics that extend **B**, which contains **FDE** as its non-implicational fragment.

3 Further Distinctions

In contrast to Brown, I call a sentence ambiguous if and only if we have reason to assign one of the two non-classical truth values of **FDE**. I have thereby introduced a particular technical definition of ambiguity that departs in some ways from Brown's. The non-technical definition includes some sentence which is not ambiguous in the technical sense that I require. For example, the sentence "I will turn right here" could mean that I am turning to the right direction at this location or that I am turning at this moment in time. When the sentence is uttered when I am turning right, it would be considered true no matter which way the sentence was disambiguated. Thus, this example shows that the definition of ambiguity I am using is distinct from the usual conception of ambiguity and, as will be shown, is therefore, distinct from the use of ambiguity found in Lewis' [6]. My use of the term ambiguity comes from the use of the term in Brown's models; however, I take the term to describe merely an attitude towards sentences.

We can make a further distinction between those sentences we have reason to assign the truth value Both and those we have reason to assign the truth value Neither. We shall call a sentence *ambiguously both* when we have reason to assign to it the truth value Both, and *ambiguously neither* when we have reason to assign to it the truth value Neither. Additionally, we shall call a sentence *generally ambiguous* when we have reason to assign one of the truth values Both and Neither. Ambiguity, then, is meant to be an epistemological notion rather than a metaphysical one.

Sentences are generally ambiguous for one of two reasons. The first is that the sentence is ambiguously both (neither) — i.e., we have reason to assign to it the value Both (Neither). The second is that we have reason to treat the sentence as either ambiguously both or ambiguously neither, but our reasons do not warrant a choice between the two.

I can now further explicate one of the claims of this paper using this distinction. I claim that logical implication with respect to generally ambiguous sentences (i.e., those receiving non-classical **FDE** truth values) depends on *both* (1) how the ideal agent reasons with sentences that are ambiguously both and (2) how the ideal agent reasons with sentences that are ambiguously neither. That is, when reasoning with an implicational sentence, with a generally ambiguous antecedent or consequent, one must consider both possible ways of treating the generally ambiguous sentence. In particular, when a generally ambiguous sentence occurs as a subsentence of an implication, the role of the implication in our reasoning is determined by considering the case wherein the subsentence is considered both true and false and the case wherein it is considered neither true nor false.

4 Models for B (and its Extensions)

The models I construct here are similar to the models already available in the literature, but differ on one semantically significant detail. Namely, these models have 2 levels of analysis or description. This difference enables one to interpret the non-classical **FDE** truth values in a way similar to Brown's models for **FDE**. The first level of description is a set of *atomic sentences* which behave like the atomic sentences of the American Plan semantics. The second level of description is a set of what I will call *sub-atomic sentences*, the behavior of which will account for the behavior of the atomic sentences.

While the syntax of the language contains only the atomic sentences, the relational semantics will additionally require sub-atomic sentences. The denumerable set of sub-atomic propositions, for notational simplicity, will be written as

$$\mathbb{P}_1, \mathbb{P}_{-1}, \mathbb{P}_2, \mathbb{P}_{-2}, \ldots, \mathbb{P}_n, \mathbb{P}_{-n}, \ldots.$$

The atomic propositions, also denumerable in size, will be denoted by

$$\alpha_1, \alpha_2, \ldots, \alpha_n, \ldots.$$

Definition 4.1. A *model* $\mathfrak{M}$ is a structure $\langle W, N, R, P, f \rangle$ where W is a set of worlds closed under duality, N is a set of non-normal worlds, and R is a ternary relation between worlds. The function f assigns a pair of distinct sub-atomic sentences to each atomic sentence such that (1) every sub-atomic sentence occurs in some pair and (2) no sub-atomic sentences occurs in more than one pair. A convenient choice is to use the enumeration provided above such that:

$$f(\alpha_1) = \langle \mathbb{P}_1, \mathbb{P}_{-1} \rangle$$
$$f(\alpha_2) = \langle \mathbb{P}_2, \mathbb{P}_{-2} \rangle$$
$$\vdots$$
$$f(\alpha_n) = \langle \mathbb{P}_n, \mathbb{P}_{-n} \rangle$$
$$\vdots$$

Finally, P is a function from the non-zero integers to subsets of W. That is, P assigns sets of worlds (or propositions) to the sub-atomic sentences.

We will use the notation P_n to denote the subset of $\wp(W)$ which is assigned by P when applied to n. That is, P_n will represent the set of worlds at which the sub-atomic sentence $\mathbb{P}_n$ is true.

To denote that A is true at the world w in the model M, I shall write $w \vDash_M A$. (For simplicity, I shall henceforth drop the M unless doing so would cause confusion.) A wff is valid in a model if it is true at every normal world, and valid in a class of models if it is valid in every model belonging to the class.

Definition 4.2. Let w be a world in a model.

(1) $w \vDash \mathbb{P}_n$ iff $w \in P_n$

(2) $w \vDash \alpha_n$ iff $f(\alpha_n) = \langle \mathbb{P}_n, \mathbb{P}_{-n} \rangle$ and $w \vDash \mathbb{P}_n$

(3) $w \vDash \neg \alpha_n$ iff $f(\alpha_n) = \langle \mathbb{P}_n, \mathbb{P}_{-n} \rangle$ and $w \nvDash \mathbb{P}_{-n}$

We can see that the sub-atomic sentences receive classical truth values such that every sub-atomic sentence is either just true or just false. There are four possible combinations of truth values for any pair of sub-atomic sentences. Given that the f function assigns pairs of sub-atomics to each atomic, and given how truth at a world is defined for atomic sentences, each atomic sentence receives one of the four **FDE** truth values at each world. If a pair of sub-atomic sentences agree on a truth value, then the atomic sentence corresponding to the pair also receives that truth value. If we take the sub-atomic sentences to represent possible disambiguations of the instances of the corresponding atomic sentence, as in Brown, then the non-classical truth values are explained by the fact that some instances are true and others false.

Where A and B are wffs built up from the atomic sentences in the usual way, the valuation is extended to the extensional connectives as follows:

(4) $w \vDash (A \wedge B)$ iff $w \vDash A$ and $w \vDash B$

(5) $w \vDash \neg(A \wedge B)$ iff $w \vDash \neg A$ or $w \vDash \neg B$

(6) $w \vDash (A \vee B)$ iff $w \vDash A$ or $w \vDash B$

(7) $w \vDash \neg(A \vee B)$ iff $w \vDash \neg A$ and $w \vDash \neg B$

The valuation of a conditional can be given as follows with no reference to dual worlds:

(8) $w \vDash (A \rightarrow B)$ iff   for every x, y such that $Rwxy$: if $x \vDash A$, then $y \vDash B$.

We are only *required* to invoke the dual worlds to evaluate the negations of conditionals. The valuation is as follows.

(9) $w \vDash \neg(A \rightarrow B)$ iff $w^* \nvDash (A \rightarrow B)$

However, note that this is equivalent to the following:

(9') For some x, y, such that Rw^*xy both $x \vDash A$ and $y \nvDash B$.

In Section 5, I will demonstrate that an explanation of (9') naturally follows from how one should reason with non-classical truth values.

In [5], I was able to show that these models are equivalent to the extant models for the relevant logic **B** and its extensions given in Graham Priest's *Introduction to Non-classical Logic* [8].

The semantics I have defined in this section are not too surprisingly different from other American Plan semantics. The only difference is one of explanation: the sub-atomic sentences, which are absent in the usual American Plan semantics, combined with the account of ambiguity in the previous section, explain the non-classical behavior of the atomic sentences. By removing the sub-atomic sentences and giving a new valuation condition for the atomics, we can construct the usual American Plan semantics with dual worlds.

The formal nature of the dual worlds is illuminated by these models. A sentence is treated as ambiguously both (neither) at a world when it receives the **FDE** truth value Both (Neither). A world's dual is a world which makes the same set of sentences generally ambiguous. Furthermore, anything ambiguously both at a world is ambiguously neither at its dual, and anything ambiguously neither is ambiguously both at its dual. That is, a world and its dual have the same set of generally ambiguous sentences, but they differ maximally on how such sentences are treated.

5 Reasoning with Generally Ambiguous Sentences

I will show that how we ought to reason with generally ambiguous sentences motivates and explains the truth and falsity conditions in the models of the previous section. To this end, I will first consider the case of conjunction, then the case of implicational sentences. I will illustrate, by means of informal reasoning, when an ideal epistemic agent should assert and deny sentences with generally ambiguous sub-sentences. In each case, I will show that the formal conditions capture our intuitions with respect to the informal reasoning.

Let us first consider the simple case of conjunction. A wff $A \wedge B$ is true if and only if both A is true and B is true. So suppose that the wff A is generally ambiguous. That is, suppose that we are analyzing the truth and falsity of a conjunction with a conjunct that we have reason to treat as either ambiguously both or ambiguously neither. It follows that this conjunction, $A \wedge B$, is true if and only if (1) B is true and (2) we treat A as ambiguously both. This is because the only way to treat a generally ambiguous sentence as true is to treat it as both true and false. On the other hand, the conjunction is false (i.e., $\neg(A \wedge B)$ is the case) if and only if either B is false or we treat A as ambiguously both (i.e., both true and false). This bit of informal reasoning merely relies on the definition of generally ambiguous sentences and a basic semantic definition of conjunction.

This informal analysis supports the truth and falsity conditions of the above section. That is, it supports the truth and falsity conditions that do not involve the dual worlds. However, the conditions are equivalent to ones which do involve dual worlds.

$$
\begin{aligned}
w \vDash \neg(A \wedge B) \quad &\text{iff} \quad w \vDash \neg A \text{ or } w \vDash \neg B \\
&\text{iff} \quad w^* \nvDash A \text{ or } w^* \nvDash B \\
&\text{iff} \quad w^* \nvDash (A \wedge B)
\end{aligned}
$$

The informal reasoning nonetheless does not require dual worlds, and this is reflected in the truth conditions in the previous section. It follows, in a quite straightforward manner, that the formal conditions are explained by the informal reasoning, which I hope to have shown is correct.

Now let us consider the arrow, for which the dual worlds will be shown to play an indispensable and explanatory role. The arrow of relevant logics is used to capture relevant implication, entailment, and other implication relations. In a relational semantics, the properties and strength of the arrow is determined for the most part by the ternary relation. Here, I will not assume any particular interpretation of the ternary relation, nor will I assume that any conditions have been placed on it, for I aim to explain the role of dual worlds in every relevant logic that extends **B** (captured by relational semantics) in a single stroke. I require only the basic idea that an arrow statement is true if the antecedent really implies (entails, etc.) the consequent. Unlike the case of conjunction, I will require an additional assumption in the informal reasoning that will follow. I require that, in the process of determining whether an implicational statement is true of false, we create or pick out situations or worlds which make some sentences true or false.

I believe that the informal reasoning I present will motivate one particular interpretation of the ternary relation more than it will motivate other interpretations. On this interpretation, a conditional $B \to C$ is true at a world w if and only if for every world w' such that $R_B w w'$, $w' \vDash C$, where "$R_A xy$ could, given some assumptions, be defined as y is one of the A-satisfying worlds closest to x" [1, p. 605]. It will be shown that switching between a world and its dual (potentially) changes the set of closest A-satisfying worlds, for any A. Moreover, this fact will be shown to mirror our intentions when reasoning with non-classical truth values.

Once again I will split up my argument into two parts. First, I'll argue that reasoning about conditionals (and their negations) with generally ambiguous, proper sub-sentences requires us to consider the cases in which the sub-sentences are ambiguously both *and* the cases in which they are ambiguously neither. Then, I will argue that the truth and falsity conditions in the models of the previous section and the American Plan semantics naturally captures this informal reasoning.

Suppose that A is a generally ambiguous sentence. Consider a wff of the form $A \to B$, which contains the generally ambiguous wff A as a sub-sentence. I will consider how we should determine whether or not the conditional is true and whether or not it is false. At least, that is, whether or not we should treat it as true or false.

First, how do we determine whether $A \to B$ is true at a world? Instead of picking out every (normal) situation or world in which A is true and determining whether or not B is also true (i.e., determining whether something like $A \vDash B$ holds), we pick out a restricted set of (closest) situations or worlds in which A is true and determine whether or not B is true in those worlds. To do this, we first pick out a set of worlds at which A is true. If A is generally ambiguous in our own world, then this piece of information restricts the set of A-worlds we pick. That is, the information in our own world restricts which set of worlds are the *closest* A-worlds with respect to the implication sentence. Picking out A-worlds from a world in which we treat A as generally ambiguous focuses our attention on the possible worlds in which A is ambiguously both. That is, using what we currently know (including that we treat A as generally ambiguous) *we pick out certain worlds* — we pick out the worlds at which A is ambiguously both.

We say that $A \to B$ is true if and only if, when we picked out the closest worlds at which A is true (by being both true and false), we find that B is also true at every one of these worlds. This shows that reasoning about arrow sentences with generally ambiguous antecedents requires us to treat the generally ambiguous antecedent as ambiguously both. In addition, the formal truth condition is explained by this informal reasoning, for the formal condition states formally what the informal reasoning asserts. However, we go about making the generally ambiguous antecedent true, the result is a world in which the consequent is true (if the conditional is true).

The case for generally ambiguous consequents is similar, but for that we must consider the dual, modus-tollens-like behavior of the conditional. The conditional is true if (and only if) every one of the closest worlds picked out in which the consequent is false (by making it ambiguously both) also makes the antecedent false, after relevant combination. This is captured in the formal condition by the presence of set-theoretic, or material, implication.

On the other hand, we say that $A \rightarrow B$ is false, i.e., $\neg(A \rightarrow B)$ is the case, if and only if the truth of A is in some sense compatible with the falsity of B. When A is generally ambiguous, we ought to consider the case where the accidental properties of our own world (say, making the antecedent both true and false) do not over-determine the truth value of the negated implication. We want to try to pick out a situation or world that, when combined with ours under certain restrictions, produces a world in which the antecedent is true and the consequent is false. If we currently treat A as both true and false, then we are limited as to which worlds can be combined with our own to produce the desired world. Thus, we must consider the case where we treat the antecedent as being neither true nor false.

If we currently treat A as ambiguously both, then when considering the non-negated conditional we pick out the *closest* worlds in which the antecedent is at least true, and we look for the consequent. Similarly, when considering the negated conditional we pick out a *potentially different set of closest* worlds at which the antecedent holds and, after a similar construction, look for the consequent at the resulting world. However, in each case the closest worlds might be different. This difference comes from a difference in our attitudes towards the conditional and its negation. When considering the bare conditional, we are trying to pick out worlds where the antecedent holds. In contrast, when we consider the negation we attempt to pick out worlds at which the antecedent holds and the consequent does not.

When we already consider the antecedent to be ambiguously both, looking for worlds where the antecedent holds produces worlds restricted by our belief that the antecedent is both true and false. However, in a world where we believe the antecedent is ambiguously both, we look for different kinds of worlds when considering the negated conditional. Often, it seems, the closest worlds ought to be similar when considering a conditional and its negation; however, it is not necessary that the closest worlds when considering the conditional are the exact same worlds picked out when considering the negated conditional.

So the question remains as to why we want to use the peculiarities of our current world when considering a conditional, but not when considering the negation of a conditional. I believe that this is explained by the difference in our intentions when picking out worlds. Contrast this with the case of conjunction. When considering a non-negated conjunction, we look only at a subpart of our world; we focus our attention only on the truth/non-truth of the conjuncts. When considering a negated conjunction we look at the falsity/non-falsity of the conjuncts. In each case, different information is pertinent to our aims. When considering the negated conditionals, the information about our own world is less pertinent than it is for non-negated conditionals. This is so because our focus is on the compatibility of the truth of the antecedent with the non-truth of the consequent, when we consider the negated conditionals. We care less about the closest worlds at which the antecedent holds (or the consequent fails).

Therefore, it is the case that, when we are reasoning with negated conditionals whose antecedent is ambiguously both, we must consider what follows when we treat the antecedent as ambiguously neither. It follows that, when we have conditionals with generally ambiguous sub-sentences, we have to consider both ways of treating each generally ambiguous sentence in order to fully analyze the truth and falsity of the conditional.

The falsity condition for the conditional in the models presented in the previous section captures this informal reasoning. The condition, $w \vDash \neg(A \rightarrow B)$ iff for some x, y such that Rw^*xy both $x \vDash A$ and $y \nvDash B$, states formally what is given informally, with the interpretation of the ternary relation given above. That is, the falsity condition for implicational statements formalizes what we do when we try to construct worlds with the intention of producing the compatibility between the truth of the antecedent and the non-truth of the consequent. It also captures the fact that we must look to the dual world in order to pick out a different set of *closest worlds*. The informal reasoning, therefore, offers an explanation as to why the dual worlds can be related via the ternary relation to different sets of worlds.

The formal conditions are, therefore, motivated by the informal reasoning. This shows that the purpose or existence of the dual worlds in the models can be motivated prior to the construction of the relational semantics. In Section 6, I will expand upon this and justify the further claim that the nature of dual worlds is also explained by the content of this section.

5.1 Lewis' Logic of Equivocators

With my relational semantics and its informal reasoning explicated, I can now compare my account with Lewis' "Logic of Equivocators" before concluding that the nature and existence of dual worlds are explained by my account. I will show that (1) the role of disambiguations in my account is similar to the role of disambiguations in Lewis' [6] and, on the other hand, that (2) ambiguity is used to describe entirely different phenomena in my account. It will be shown that the role of ambiguity, and not the role of disambiguations, is required to create the metaphysically light interpretations of relevant logic which are acceptable to those who deny non-classical truth values.

For Lewis, the term ambiguity is used much as it is used to describe the phenomenon in natural languages.

> Strictly speaking, an ambiguous sentence is not true and not false, still less is it both. Its various disambiguations are true or false *simpliciter*, however. So we can say that the ambiguous sentence is true or is false on one or another of its disambiguations. The closest it can come to being simply true is to be true on some disambiguation (henceforth abbreviated to "osd"). [6, p. 438]

In direct contrast, my account treats a sentence as ambiguous when we have reason to assert (deny) both a sentence and its negation. The typical natural language phenomenon of ambiguity is irrelevant to my account, and, as I have shown above, treats some sentences as ambiguous when my account would not treat them so.

The informal reasoning I have presented makes no essential use of ambiguity (as I use the term). Instead, what is important to the informal reasoning is that we are in a position to treat certain sentences as both true and false by, say, being justified in asserting both a sentence and its negation. The informal reasoning, then, supports both the usual four-valued semantics and the formal semantics I have given in Section 4. Furthermore, both non-classical truth values are motivated on my account by the same kinds of reasoning. That is, there are situations in which we are justified in asserting (denying) a sentence and its negation.

On the other hand, Lewis' use of ambiguity motivates the **FDE** truth value Both, for a sentence can be disambiguated into true or false sentences. However, Lewis runs into trouble while attempting to motivate the **FDE** truth value Neither. He considers allowing gappy sentences which are "neither true-osd nor false-osd" [6, p. 438]. Unfortunately, Lewis found no philosophically satisfying justification for rejecting ambiguous sentences that are gappy-osd as well as true-osd or false-osd. Allowing these kinds of disambiguations of ambiguous sentences gives us seven truth values instead of four [6, p. 438]. Lewis' ambiguity-based account, therefore, provides poor motivation for the four-valued semantics, while my account gives the four values a unified motivation.

For Lewis, the ambiguity-based approach is needed only by those whom he calls "pessimists," who believe that we cannot fully disambiguate anything (or most things) [6, p. 439]. Again, this ties Lewis to the natural language phenomenon of ambiguity. I, here, use the term merely to describe sentences such that we are justified in asserting (denying) the sentence and its negation.

The account I have given above shares with Lewis the act of disambiguation in order to produce classically true and false instances of "ambiguous" sentences. However, what I mean by 'ambiguous' is by no means the same as what Lewis means by this term. The benefits of my account, including the motivation for the non-classical truth values and justification of dual-worlds, come from my account of ambiguity, which is merely a description of the non-classical **FDE** truth values. The disambiguations on my account enable the "classical metaphysician" to accept the relevant logics motivated by my idiosyncratic use of the term 'ambiguity'.

6 Dual Worlds Explained

Given that the truth and falsity conditions mirror informal reasoning with generally ambiguous sentences, there is more than mere formal reasons for including dual worlds in the semantics. Furthermore, I have shown that the informal reasoning gives dual worlds a particular purpose and that this purpose is reflected in the fact that dual worlds are only required for one of the falsity conditions, i.e., the falsity condition for implicational sentences. Having shown this, I make the further claim that the nature of the dual worlds is explained by their purpose, which is itself explained by the informal reasoning.

In the informal reasoning, we constructed a function that took our current epistemic stance (ambiguously both or neither) towards a sentence and produced a world wherein our stance dualized in order to prevent the features of our current stance from over-determining our implicational judgments. That is, the dual worlds are those created by a function that eliminates certain accidental features of our own world from our considerations. A world and its dual are those which completely agree on which sentences are generally ambiguous, but disagree maximally on how such sentences are treated. All of the accidental features of how we treat generally ambiguous sentences are ignored in order to pick out a different set of closest worlds.

It is through this lens that the dual worlds are not merely formal tools, but exist to capture the informal account. They are explained because the informal account requires that we consider such a function. Reasoning within a particular conceptual frame is not reasoning about any particular world, but about a pair of worlds. We consider the pair at once, even though we might treat any generally ambiguous sentences as ambiguously both or ambiguously neither. I thus make the claim that a world and its dual are the fundamental units in the four-valued semantics, and that it is wrong to think of an individual world without its dual as fundamental. Thus, it is not a world that is fundamental in this account, but pairs of worlds.

A pair of worlds could, say, be determined by a particular conceptual framework under which some sentences are under- or over-determined. Reasoning within a particular conceptual frame, then, is reasoning about a particular pair of worlds. We cannot ignore either of the worlds when determining the value of implicational statements. If we did, as we have seen, the way any world accidentally treats a generally ambiguous sentence is too restrictive. We must be able to pick out a different set of worlds. In short, both ways of treating generally ambiguous sentences are relevant to the truth value of implicational statements.

The models I have reproduced here show how this informal reasoning supports an interpretation of the relevant logics that is metaphysically light; there is no prior metaphysical commitment required when it comes to the non-classical truth values. However, it seems that the informal reasoning supports the four-valued semantics in general, for it shows that reasoning with the American Plan's non-classical truth values leads to the four-valued models.

If the account of informal reasoning I offered in Section 5 is taken seriously, one can explain the formal properties of the four-valued relational semantics for relevant logic prior to the construction of the formal semantics.

References

[1] Jc Beall, Ross Brady, J. Michael Dunn, A. P. Hazen, Edwin Mares, Robert K. Meyer, Graham Priest, Greg Restall, David Ripley, John Slaney, and Richard Sylvan. On the ternary relation and conditionality. *Journal of Philosophical Logic*, 41:595–612, 2012.

[2] Bryson Brown. Yes, Virginia, there really are paraconsistent logics. *Journal of Philosophical Logic*, 28:489–500, 1999.

[3] Bryson Brown. Ambiguity games and preserving ambiguity measures. In Peter Schotch, Bryson Brown, and Raymond Jennings, editors, *On Preserving: Essays on Preservationism and Paraconsistent Logic*, pages 175–188. University of Toronto Press, 2009.

[4] B. Jack Copeland. On when a semantics is not a semantics: Some reasons for disliking the Routley-Meyer semantics for relevance logic. *Journal of Philosophical Logic*, 8:399–413, 1979.

[5] Nicholas Ferenz. A preservationist approach to relevant logic. Master's thesis, University of Waterloo, 2014.

[6] David K. Lewis. Logic for equivocators. *Noûs*, 16:431–441, 1982.

[7] Edwin D. Mares. "Four-valued" semantics for the relevant logic R. *Journal of Philosophical Logic*, 33:327–341, 2004.

[8] Graham Priest. *An Introduction to Non-Classical Logic*. Cambridge University Press, 2nd edition, 2008.

[9] Graham Priest and Richard Sylvan. Simplified semantics for basic relevant logics. *Journal of Philosophical Logic*, 21:217–232, 1992.

[10] Greg Restall. Simplified semantics for relevant logics (and some of their rivals). *Journal of Philosophical Logic*, 22:481–511, 1993.

[11] Greg Restall. Four-valued semantics for relevant logics (and some of their rivals). *Journal of Philosophical Logic*, 24:139–160, 1995.

[12] Richard Routley. The American plan completed: Alternative classical-style semantics, without stars, for relevant and paraconsistent logics. *Studia Logica*, 43:131–158, 1984.

 Received 31 October 2016

www.ingramcontent.com/pod-product-compliance
Lightning Source LLC
LaVergne TN
LVHW062310060326
832902LV00013B/2144